The Conservation and
Improvement of
Sloping Land

The Conservation and Improvement of Sloping Land

A manual of soil and water conservation
and soil improvement on sloping land

Volume 3 : *Practical Application*
– Soil and Water Conservation

P.J. STOREY

CRC Press
Taylor & Francis Group
Boca Raton London New York

CRC Press is an imprint of the
Taylor & Francis Group, an **informa** business

First published 2003 by Science Publishers, Inc.

Published 2019 by CRC Press
Taylor & Francis Group
6000 Broken Sound Parkway NW, Suite 300
Boca Raton, FL 33487-2742

CRC Press is an imprint of Taylor & Francis Group, an Informa business

First issued in paperback 2019

No claim to original U.S. Government works

ISBN 13: 978-0-367-44664-2 (pbk)
ISBN 13: 978-1-57808-234-6 (hbk)

This book contains information obtained from authentic and highly regarded sources. Reasonable efforts have been made to publish reliable data and information, but the author and publisher cannot assume responsibility for the validity of all materials or the consequences of their use. The authors and publishers have attempted to trace the copyright holders of all material reproduced in this publication and apologize to copyright holders if permission to publish in this form has not been obtained. If any copyright material has not been acknowledged please write and let us know so we may rectify in any future reprint.

Visit the Taylor & Francis Web site at
http://www.taylorandfrancis.com

and the CRC Press Web site at
http://www.crcpress.com

LIBRARY OF CONGRESS CATALOG CARD NUMBER : 2002283724

FOREWORD

This book has been written by Peter Storey, a man with many years of practical experience in developing countries. It is a manual on soil and water conservation and is written as a practical guide for a wide range of people working in rural development and related subjects.

For the last forty years the world has been able to produce food surpluses and farmers in Europe and the USA have been paid subsidies not to farm some of their land. Why then is it so important that we conserve our soil and water resources? There are a number of reasons. First, the amount of land available for food production is finite and this land is steadily being lost through soil erosion and other forms of land degradation. In addition, land is being lost to food production as it is diverted to other purposes such as housing, roads, airports and golf courses. At the same time, the world's population is continuing to increase - it is estimated that it will increase from its present level of approximately six billion to approximately 8.9 billion by the year 2050 - and, with this increase, the demand for food and other related agricultural products will grow and more land will need to come into production.

Secondly, the present high rates of agricultural production have been achieved largely through the use of irrigation, the heavy use of fertilisers, insecticides and herbicides, combined with the use of new, high-yielding crop varieties. Unfortunately, much of this is not sustainable: the water resources in India, China and several other major food producing countries are rapidly being depleted, while systems of monoculture and heavy use of chemicals are leading to a number of problems, including pollution of soil and water supplies, soil acidification, salinisation and the long term loss of soil fertility.

Thirdly, even though there may be sufficient food worldwide, many poor people - even in those countries that are exporting food - do not have enough money to purchase what they need, nor do they have the opportunity to find employment outside agriculture. The only way that many millions of people in developing countries can exist is by farming whatever land is available to them. For these people it is vital that the available land is conserved in a productive state.

To acknowledge that more must be done to conserve our soils and water resources is one thing, but to do something about it is another. Most of the problems of land degradation are at present to be found in poor, developing countries, where there are few trained and experienced extension workers and little money is available for projects. For those wishing to tackle land degradation problems, there are few practical text books which can provide the inexperienced field worker with a sound understanding of the problems and what to do about them.

The author of these books, Peter Storey, with his years of experience on soil and water conservation projects in the developing countries, has recognised the problem and has successfully put together very comprehensive manuals which should provide an invaluable guide to anyone working in this field. Not only does he provide the reader with practical advice on how to carry out various conservation practices, but he also provides the basic information that every field worker should have about soil, including their formation, chemistry, physical attributes and management. He also provides much useful information on related subjects, including vegetation, livestock management, surveying and approaches to extension. All this is done in easy to read English, with the use of technical terms being kept to a minimum.

The books are based very much on the author's own practical experience which was gained mainly in Taiwan and Nepal, but this is liberally supplemented with examples that he has found of other workers in Africa and other parts of the developing world. While some of the approaches and suggestions made in this book may not be conventional, they are all based on practical experience and have been proved to work under the conditions given.

Both volume 2 and volume 3 should provide invaluable manuals to any rural developments worker faced with the basic problem of farmers in the developing world—how to gain the maximum production from their land while at the same time conserving its soil and fertility.

David Sanders
President
World Association of Soil and Water Conservation.

PREFACE

Why this book?

We learn of mud slides occurring more and more often such as the one "in Venezuela in December 1999 where over 40,000 people were killed and billions of dollars of damage caused by mudslide". As excess amounts of run-off water erode soil and cause landslides, so adding to the waters bulk, houses are swept away, roads are broken, bridges and fields are buried, lives and livelihoods lost.

This author has seen where wrongly made terraces collapsed in a typhoon, carrying the houses below down into the raging river.

I have seen other faulty terraces where the topsoil (living soil) had been buried, so that the hillside was now sterile.

All this and other bad land management resulting in formerly agricultural land being degraded, causes economic migration of people, sometimes to city slums, but more often to clear other land, resulting in the loss of increasingly scarce natural habitats. **Surely it is time to reverse this trend.**

In the Philippines are terraces which are 3,000 years old and still fertile. Properly built modern terraces can produce far greater crop yields than slope land, and be durable, while the currently fashionable contour strip systems have been shown to be worse than doing nothing. A recent report from the Philippines stated that they are now finding that the SLOPE system of contour strips is actually causing landslides. I am told the World Bank last year realised that the Vetiver grass contour strips were failing and so their program to promote the system has been stopped. Studies in Honduras showed that at best, contour strips do not reduce erosion.

The problem with building good terraces is that they have been slow and expensive to build, how to get over that problem? I believe the Good News step by step system of terracing I have used is Practical, Desirable, Doable and Durable and can be done by two farmers working together in stages. The first stage stopping the loss of soil and plant foods and resulting in much better crops on the developed land, so encouraging the farmers to continue the stages until all their land is terraced in a durable way. The Good News System and other soil and water conserving methods are to be found in this very comprehensive book.

Natural soil development and conservation

Soil has not developed simply by the breakdown of rock into small enough particles to create a sandy, silty or clayey layer with some soluble salts. It has developed because micro organisms, (very tiny plants and creatures) and macro soil organisms such as earthworms and termites, have mixed and digested soil particles turning them into plant foods. Then simple plants started to grow, and as their remains further improved the soil other plants grew. The plants gradually covered the soil and protected it from the sunlight, from drying out quickly, and from harmful erosion. More and more plant remains were mixed into the soil and the soil became more and more fertile. The plant remains rotted down into humus and plant and animal remains with the humus improved the soil strucutre, binding the soil together and increasing its ability to both absorb and retain rainwater while allowing excess water to drain through to renew the aquifers (water reserves in the ground). This process takes a very long time, about 600-700 years to develop 1 inch (2.5cms), and so we should conserve, not waste the soil.

Under jungle (natural) conditions, sustainable biomass (the amount of plant material) is produced and slowly increases. There is far more biomass produced in a jungle even on poor land than is the case when the land is cropped commercially, and it is produced in a sustainable and improving way.

The trees and other plants protect the soil from UV light (the burning sunlight) which destroys the organic matter in the soil surface, from heat, and from wind and water erosion. The soil is cooler and moister and this favours the soil structure and the life within the soil. The tree roots go deep into the subsoil and enter cracks in the underlying rock and so hold the soil on the slopes. There is a natural mulch of decaying plant matter on the soil surface. The plant canopy (covering) over the area results in a cooler micro climate which attracts mists, dews and rainfall. When land is bare it is much less able to attract such moisture.

We need to produce a field environment as close as possible to natural conditions while enabling sustained or improving economic crop production. I believe that on slopelands we can do that by terracing slopes in a way which is desirable, do-able and durable; which keys the terraces to the slope, keeps the living topsoil on top, and breaks up the soil pans. A system which incorporates some tree growth, so that as the tree roots penetrate deeply they increase the stability, and bring up minerals from the subsoil, their leaves adding the minerals as well as organic matter to the surface soil. The trees should not compete with the crops.

In addition using green manuring, inter-cropping and mulching, as explained in volume 2, to keep the soil protected and increase the organic matter. All these factors encouraging the soil micro organisms which improve the soil texture and release plant foods.

From an article in Appropriate Technology. Vol 29. No 1 March 2002.
> Grain production in China is expected to fall, as water shortages begin to bite. Consequently, grain imports may have to rise.
> This scenario emerges from a survey which reveals that China's water situation is far worse than realised, writes Lester Brown of the Earth–Policy Institute. The water table under the North China Plain, which produces over half of China's wheat and a third of it's maize, is falling faster than thought---.
> The study, conducted by the Geological Monitoring Institute (GEM) in Beijing, reported that Heibei Province in the heart of the North China Plain, the average level of the deep aquifer dropped 2.9 meters (nearly 10 feet) in 2000. Around some cities in the province it fell by 6 meters.---
> As thousands of wells run dry, so the three rivers that flow eastward into the North China Plain – the Hai, the Yellow, and the Huai – are drying up during the dry season. Lakes are disappearing too.---.
> Whatever it does, China will almost certainly have to turn to the world market for grain imports. If it imports even 10 percent of it's grain supply - 40 million tons - it will become overnight the largest grain importer, putting intense pressure on exportable grain supplies and driving up world prices.

The full article and additional information from the Earth Policy Institute, 1350 Connecticut Av. NW; Washington DC 200-36, USA www/earth-policy. org

Soil and water conservation are closely linked. The value of water conservation is not only in the amount of water saved but:-

a) When water runs off land it usually takes soil and plant foods with it.
b) The more the water is retained in the field the longer the soil is covered and protected by growing crops.
c) A longer growing season results in better crops which means more organic matter is produced, which, if returned to the soil directly or as manure improves the soil structure and fertility. Better soil structure leads to better resistance to erosion and better water retention. Better fertility produces even better crops so even more organic matter and so an upward spiral. The greater crop residues can be used to mulch the soil, further reducing evaporation and blocking the harmful effects of UV light from the sun which destroys soil organic matter in the top 1-2 inches (2.5-5.0 cms) of soil.

Soil conservation reduces water run-off and maintains the depth of the soil and so the ability of the soil to retain rainwater. As topsoil is eroded whether by wind or water the crops are poorer and the growing season reduced. The underneath subsoil is usually less able to absorb the rainfall so water run-off is greater.

Unlike many books on soil and water conservation these books are written from personal and practical experience over 30 years, written by a practitioner for practitioners. It also draws on 207 other sources of information, much of which is difficult or impossible to obtain. It is a book that the author would wish to have if he was starting again in soil and water conservation, or in any aspect of rural work depending on the use and management of slopeland.

ACKNOWLEDGMENT AND DEDICATION

I would like to acknowledge all those who have helped me gain an appreciation of the importance of, and experience of - sustainable yet practical, soil and water conservation and good soil management.

Those people range from the time of my first lectures on tropical agriculture under Geoffrey Masefield at Oxford in 1963, to often illiterate, though intelligent farmers in the third world in the years since then.

I want to thank the people and institutions which have supported me throughout the last 38 years. Particularly those who gave encouragement and back up, when administrators and others who did not appreciate the importance of sustainable development, have tried to belittle, block, or divert what I learnt was the only responsible approach.

Daphne Jackson, Francis Shaxson and Alex and Peter Herring who encouraged me when I wondered if the books would ever be published.

People who prayed for vision, perseverance and ability for me to finish the book.

Andre DeBatt, Tom Cox and Bill Broekhuizen helped with improving the writing, my brother-in-law Edward Watson gave valuable help with the computer work. Bruce Creighton helped by copying half frame slides onto a floppy disk. Graham Eastham, Bob Schelmedine and David Perry helped with the computer and scanner. David Sanders, President, World Association of Soil and Water Conservation, who kindly agreed to write the foreword.

Most importantly the patient support of my wife who has had to put up with my moans, groans, and frustration throughout the time, particularly as I have been unpaid while writing these books. The whole process has been long and costly; it is our hope and expectation that the books will provide a better realisation of what can, and should be done, and enable people and organisations to conserve and improve the basis of life on planet earth.

I am grateful for what I have learnt from and with some of the multitudes of ordinary people of the two thirds world, many of whom are my friends, who struggle to exist, who deserve much more of the rest of us.

I also acknowledge my Christian faith which has inspired and enabled me through the often difficult years of work and to write the books, which has been for me a difficult and onerous task. As Edgar Stoetz wrote in *Beyond Good Intentions:*

> a serious gap in the myriads of volumes available is any serious attempt to relate theories to practice and address them to the practitioner in the field. It would seem that practitioners do not write and theoreticians remain in the abstractions of their theories.[55]

As a practitioner I know why practitioners do not write. I would rather be out for a month in the worst weather in a monsoon or typhoon observing what is happening, and at other times working under a blistering sun to prevent or reverse the harm being done, than spend a day writing.

I have tried to follow the demonstrated teaching of Jesus who "had compassion on people, and taught them many things- to love our neighbours as ourselves, and even love our enemies." He taught that we should judge people fairly and show mercy; he condemned hypocrisy and corruption and was a friend to the despised and rejected people. He believed and lived what he taught and so taught with credibility. The more people follow that teaching and example, the better place the world will be for everyone.

THIS BOOK IS DEDICATED TO ALL OF WHATEVER FAITH OR NONE WHO ARE CONCERNED TO CARE FOR OTHERS, TO TRY TO MAKE THE WORLD A BETTER PLACE FOR THOSE LIVING NOW, AND FOR THOSE OF FUTURE GENERATIONS.

INTRODUCTION

Soil and Water are The Worlds Most Important Resources.

HOW TO CONSERVE AND IMPROVE THESE VITAL YET DIMINISHING RESOURCES WHICH ARE THE BASIS OF LIFE ON EARTH?.

This and volume 2 SOIL AND WATER CONSERVATION offer a positive response to the challenge.

In the 39 years since I started working in Asia I have seen an escalating loss of soil and of soil quality, of land and of water sources, as rainfall runoff and landslides not only remove fields, but bury other fields, destroy roads, and increase the volume of the contents of rivers, so that they burst their banks, break bridges, and wash other land away. The roads and bridges are rebuilt, but little serious consideration is given to the causes of the problems. It is building a path around the problem rather than addressing the source of the problem.

At the same time, it is easy to blame governments for not acting to address the cause instead of the symptoms, but when most attempts of soil conservation fail for various reasons, governments are naturally reluctant to support what is likely to fail. There has sometimes been a little encouragement for green manure crops, and some encouragement for the making of contour strips, by for instance the World Bank, but as I show in this book contour strips are not really the answer and can in fact make the situation worse.

Subsidised chemical fertiliser and pesticides have been covering up the loss of soil fertility, digging deeper wells or piping water from elsewhere have covered up the growing shortage of water, but they do not provide a lasting solution, the situation will continue to deteriorate. Without effective land husbandry there will be increasing rural poverty; leading to the poor selling their children, or the children having to go to work to help their families. Rural people moving from worn out lands, to clear the remaining forests, or move to live in poverty in the slums of the cities, leading to slavery and exploitation. Overcrowding and poor services in the slums and shanty towns, lead to health problems which threaten the population as a whole. The increase in numbers of an underclass threatening political stability.

Is there no hope? I believe there is, but it depends on appropriate lasting solutions, not quick fix methods which are claimed to be sustainable but are not.

There is a need for more practical research into better land management, this book shows approaches that have worked. While the approaches have to be tailored to be location specific, in some cases practical farmer based conservation can be sufficient, in others outside help will also be needed, the present trends can be reversed.

The loss of both soil and soil quality

Because of the shortsighted way one third to one half of the worlds croplands are being managed, the soils on these lands have been converted from a renewable to a non renewable resource.

Assuming an average depth of remaining topsoil of seven inches (18 cm's), or 1,120 tons per acre = 2,800 tons per hectare, and a total of 3.1 billion acres = 1.24 billion hectares of cropland, there are 3.5 trillion tons of topsoil with which to produce food, feed and fibre. At the current rate of excessive erosion, this resource is being depleted at the rate of 0.7% per year = 7% each decade. In effect, the world is mining much of its cropland, treating it as a depletable resource, not unlike oil.[102 pp6-23]

While erosion is normally thought only to be caused by wind and water it is also caused directly and indirectly by soil management. I have seen areas in Australia where the farm houses are standing higher than the fields after soil built up under thousands of years of forest was continually cropped, using up the reserves of organic matter and humus. This is happening to a lesser degree on land which was malarial jungle in Nepal. I have seen that when the dark topsoil is cropped in the wrong way the organic matter is used up and what is left is not much more than white sand or very infertile subsoil.

Indirectly erosion is caused or increased by wrong land management, which uses up the organic matter and humus which helps to bind the soil, and the soil texture is spoiled, so that it is easily blown or washed away.

Not only is the rate of loss increasing, but the speed of increase of rate of loss is also increasing.

This is slowly being realised and there is much talk of the need for sustainability, unfortunately it is mostly talk.

> 'What is happening today is that nature in some parts of the world can take no more. Land has been pumped for all it is worth and is crying 'halt'.
> The degradation of agricultural land 'is the most widespread threat to habitability worldwide'
> says Jodi L Jacobson in an article in the UN Fund for Population Activities magazine Populi, 'vast areas are becoming unfit for human habitation---. Refugees from land degradation often migrate from region to region, cultivating one plot of marginal land after the other, exacerbating the problem'.[96 pp17-18]

> There is convincing evidence that '*If the soil is destroyed, then our liberty of choice and action is gone, condemning this and future generation to needless privation and dangers*'. These words from W.C.Lowdermilk (1953) highlighted the importance of the soil in the rise and fall of nations in his treatise, *Conquest of the Land through Seven Thousand Years*.[135a]

Unfortunately the above warning has not been taken seriously.

Soil fertility

In addition to the loss of soil, the over reliance on chemical fertilisers is mining and spoiling the soil that remains. When I was working in Taiwan during the late 1970's, it began to be appreciated that 'chemical agriculture' was not sustainable. To produce the same amount of rice grain, farmers were needing to apply five times the amount of fertiliser as was used 20 years earlier. On a BBC television documentary in 1992 a similar account was given about agriculture in the Punjab which is a major food producing area of India.

A report in the journal 'International Agricultural Development' of January/February 1998 page 14 stated:

> three decades after the 'green revolution' blossomed for the first time in the north western part of India, there is a growing realisation of the pitfalls and unsustainability of hightech farming. With the spectre of starvation, malnourishment and even hunger deaths looming over certain pockets of India, the wisdom of continuing the 'green revolution' is being questioned with increasing intensity by farm experts and development strategists.---.

> Though the fertiliser consumption has increased by eight times over the past 25 years, the output of cereals has not even doubled---.
> In Punjab, the first Indian state to use 'green revolution' technology, 85% of the land has no more ground water left to exploit. In the neighbouring state of Haryana, water logging and salinity affects nearly a third of all land. In the Western Uttar Pradesh, an overdose of chemicals in the form of fertilisers and pesticides have played havoc with the soil fertility. A national Academy of Agricultural Science report warns of a dangerous situation if micronutrients are not replenished and green manuring is not practised. [113 p14]

There is good news, my experience has shown sustainable conservation and soil improvement of such land can be achieved, however, when the soil has gone nothing can be done.

Water loss

Without water crops cannot grow, without water we cannot survive.

Water wars are the next threat to peace in countries such as the Middle East and the Indian sub continent.

In Delhi, the Indian capital city, water pressure is so low that water has to be carried to upper floors. The rich can afford to pay for private water, the majority cannot and the situation is rapidly getting worse.

Similarly with irrigation systems, the flow in rivers is steadily decreasing, except during the flood seasons which are getting worse, causing more damage and death. Wells and boreholes have to be made deeper, and so water is more and more expensive. More of the deeper water is saline and so will reduce crop yields and will spoil the soil.

Life and prosperity needs soil and water. Urban as well as rural populations need cheap enough food and water, without which wars will follow. Cheaper goods depend to a large extent on cheaper labour. Cheaper labour depends on cheap food and water. If people cannot afford to buy goods the producers cannot afford to produce them.

Sustainable development

In the last 15 years there has been more and more talk of sustainable development, however there is still very little truly sustainable soil use. How can there be sustainable development if the basis of rural development – soil and water – is not managed sustainably?

In the July/September 2001 issue of Intermediate Technology it was reported that:

> Earlier this year, scientists at the National Oceanic and Atmospheric Administration laboratory in Boulder, Colorado USA reported that a huge dust storm from northern China had reached the United States "blanketing areas from Canada to Arizona with a layer of dust." They reported that the mountains along the foothills of the Rockies were obscured by the dust.
>
> The dust storm did not come as a surprise. On March 10, 2001, *The Peoples Daily* reported that the season's first dust storm – one of the earliest on record – had hit Beijing. Along with those of last year, this storm was among the worst in memory, signalling a widespread deterioration of the rangeland and cropland in China's vast northwest. ---.

WATER

> In addition to the direct damage, the northern half of China is literally drying out as rainfall declines and aquifers are depleted by over pumping. Water tables are falling almost everywhere, gradually altering the regions hydrology.[190 p48]

Much more information from the Earth Policy Institute website; *www.earth-policy.org* or email *jlarsen@earth-policy.org*

> Recognizing the problem, top-level policy makers in many countries have devoted substantial portions of their budgets to finding a solution. But the sums allocated are often insufficient, costs are too high, and many of the conservation techniques attempted have been ineffective and are inappropriate for small farmers. Soil erosion continues at an accelerating and alarming rate.[1 pp111]

> It should be emphasised that sustainable agriculture does not represent a return to some form of low technology, 'backward' or 'traditional' agricultural practices. Instead it implies an incorporation into current systems of recent innovations that may originate with scientists, with farmers or both.
> It is also not only about food production and productivity, but also about increasing the capacity of rural people to be self-reliant and resilient in the face of change, and about building strong rural organisations and economies. (*'The criteria for a sustainable agriculture can be summed up in one word: Permanence; which means adopting techniques that maintain soil fertility indefinitely; that utilise as far as possible only renewable resources that do not grossly pollute the environment; and that foster...biological activity within the soil and throughout the cycles of all the involved food-chains.'* Balfour, 1977 Soil Association Newsletter.)[149 pp7-8]

Since 1963 I have been learning about the problems of soil and water loss, both of quantity and quality. I have worked in difficult situations and learnt how it is possible to stop the loss and how to improve even the worst land.

Sustainable soil use and soil conservation also results in sustainable water use and water conservation. This book together with volume 3 discusses the needs, and gives a comprehensive description of methods which will enable sustainable soil and water management. This would allow us to use the land in cultivation more effectively so that the need to destroy more species of plants, animals, and people is at least reduced. The livelihood of rural people will be improved, resulting in reduced need for large families, so contributing to balancing the worlds population with the world's resources.

Derek Balinsky of the Zimbabwe Agricultural Society has rightly said:-

The prudent farmer, in any part of the world, is the man or woman who studies first his available resources in terms of soil and water, finance and manpower, his climate, and his possible market and his distance from that market.
        Without doubt the real wealth and the greatest heritage of any nation is the soil.
It takes nature countless ages to produce productive agricultural soil. Conversely we have evidence that men can completely destroy the same productive soil in a matter of a few years, leaving it not only unsightly, but rendering it useless to both man and beast, neither of which can survive without soil. *In essence, therefore the wanton destruction of the soil is a more serious crime against humanity than the most potent weapon used during wartime.*   Soils in the tropics are very young in terms of soil formation processes and low in organic matter. The maintenance of soil structure, therefore, is one of the most important factors to be considered in it's utilization.' (Italics by this author)

A Kikuyu proverb should speak to all of us
'God will not ask your race, nor your sex, nor your birth. But he will say to you:- "What have you done with the land I gave to you"?

Volume 2 Soil improvement. First look at the soil, how it develops, what makes fertile soil, what spoils soil. Soil chemistry. Plant nutrition. Recognizing and treating mineral deficiencies. When is fertiliser use appropriate. Ways to improve soil texture. Improving the soil management. Examples of soil improvement. Soil Micro organisms, their improtant potential, and how to increase that potential.
Volume 3 Soil and Water Conservation. Takes the subject of better land husbandry further. The book first points out the sort of things which have been, and are being used, but are failing to deliver what is required. It then explains the causes of erosion, the theory and practice of soil and water conservation, practical and lasting ways to construct terraces, including such details as how to build different kinds of terrace walls. Nothing grows from the top down, as also with soil improvement. We have to help the farmers who are the key people in most soil and water conservation, so we have to use methods that are appropriate for farmers to use and to develop for their own situations. The book first points out the sort of things which have been, and are being used, supposedly for soil and water conservation, but are failing to deliver what is required, many are often definitely harmful. It goes on to provide a better understanding of soil erosion. It explains the causes of erosion, and the theory and practice of soil and water conservation. It explains in detail the way of achieving practical and sustainable soil and water conservation, practical ways in which farmers can themselves make and use practical and lasting ways to construct terraces, including such details as how to build different kinds of terrace walls.

Both books have shared chapters on matters related to both subjects. Explaining in more depth the roles of forestry, agro forestry, bamboo culture, livestock, and grasses and legumes for soil conservation and soil improvement. The practicalities of slopeland management, practical land classification, surveying, research and development, extension policy and methods.

As I wrote I considered the many people who cannot afford a library of books, so the books cover all the facets of this wide subject sufficiently to give balance. To supply enough information to enable the reader to know if he or she will need additional resource material, on for instance, Zero cultivation, suitable plants for green manures, the construction of small dams? Will a qualified surveyor be needed to mark out contours properly? Is planting trees the answer to soil and water conservation? I have avoided the common fault of incomplete explanation, such as stating that 'the run off water should be directed to a grassed waterway', without describing the sorts of plants which are suitable for such waterways.

These books will reduce the need for people to purchase or research many other books. The books have lots of useful information I have compiled from many sources often sources available no where else They also contain a large glossary of terms covering not only technical words used and explained in the books but others likely to be contained in related books, reports and articles. This will enable the readers and translators to use material which though not technically difficult is hardly comprehensible to many people who may benefit from the information.

There will be faults in the books, many will feel that I am attempting too much; or the books are sometimes too wordy because of avoiding technical jargon as much as possible. I the author, am a practitioner not an academic.

As Edgar Stoesz once wrote "a serious gap in the myriads of volumes available is any serious attempt to relate theories to practice and address them to the practitioner in the field. It would seem that practitioners do not write and theoreticians remain in the abstractions of their theories."[55] As a practitioner I find writing a very slow painful and inefficient process, my hope is that my practical outlook compensates for my lack of writing skill.

Why should I should write these books?

My first experience of soil conservation was in 1965. At that time I was involved in agricultural extension work, and in teaching and managing the training farm at the Yu Shan Mountain Peoples Agricultural Training Centre in Taiwan. Prior to that I had observed horrific soil erosion in the mountains of Taiwan caused during Typhoon Gloria in 1963.

When I took on the job of farm manager the farm was scarcely using its mountain side land, and what was used was an example of 'how it should not be farmed'. As that land was like the land most of the graduates would have to farm, it was important that it should be developed in a sustainable way. The razor grass covering the land was cut and composted, appropriate terrace lines were marked out and the larger rocks moved onto them. Terraces were then built keeping most of the top soil on top. Where the land was too rocky the soil was scraped off and put into the spaces between the rocks, bananas and Green leaf desmodium was then planted.

Trials were carried out of different conservation practices, with all details such as labour required, and crop yield for each method, recorded on sign boards. The work was not only of interest to our own students, it was also interesting other mountain people, who came to visit the farm, observe and ask questions. We heard that Government experts were puzzled as to why mountain farmers were starting to make terraces, which they would not do when the government was encouraging them. Later a high level group including FAO representatives and the Government Minister of the Interior on a tour of agricultural establishments visited and were *very* complementary.

A few years later I was directing the East Taiwan Aborigines Agricultural Service Project.

I asked if we could have a piece of degraded land for three years to see if we could find a way to improve the situation? we would then hand it back. We had serious problems but overcame them and the land was transformed.

My next assignment was in the high hills (4,000-6,000 feet ASL) which comprise about half the farmable land of Nepal. I noticed how similar in many ways were the problems for farmers in the two countries. However areas of very poor or very erodible land which in Taiwan would be used to grow Bamboo forest were rapidly deteriorating further, and landslides were resulting.

I wanted to try planting bamboo on such land, and improved grasses, and fruit and nut orchards, on steeply sloping land. Such land was used being used to grow maize and wheat; with a high degree of rainy season soil loss, up to 2½ inches (6.4cms) per year.

After two years I was given the use of land in a hospital compound which had the topsoil removed over twenty years previously. Even after that long time only a few clumps of thin grass and odd thorn bushes were growing. It looked hopeless, the land was hard red laterite subsoil, impossible to dig until broken up by pickaxe. Farmers looking over the wall said we were wasting our time, as we started to construct terraces, half terraces and contour bunds. We also planted grass and legume trial plots. On the slightly better area we dug out holes and planted fruit and nut trees.

The following year we were able to lease an area of very stony waste slopeland, again by the side of the road. Again we were told it was a complete waste of time. I replied we were trying to see if the only land landless people could afford could be made economically viable. Again we made trials on the land, digging holes between the most stony land, adding what soil there was and planting the most likely fruit and nut trees.

A range of grasses and legumes and combinations was planted. We constructed three replications each of full terraces, half terraces, contour strips and control bare slopes. On the two very different areas of land half a mile apart we planted 37 different species of fruit and nut trees, plus bamboo and leucaena.

Three or four years later when showing some visitors round, two old farmers called me over to the roadside. They informed me that they had said two years before, I was a fool, 'the land would never be worth bothering with'. Now they were amazed, they knew that we had not used chemical fertilisers, nor used more compost than ordinary farmers used. If we 'could get as good results on this land any land could be improved. How could we do it? Could we show their village? The government should be doing this kind of work, they only used good land'.

Both areas of land were transformed and fruit and nuts, vegetables, field crops and fodder grasses and legumes grown well. Later the government district agricultural officer brought his staff to see our work; and later arranged for us to do soil conservation trials in nearby areas. He pointed out that the farmers were starting to dig up the contour strips the government had been promoting, and so it was important to show successful methods.

We were then asked to have trials on farmers fields, we started and showed economical ways of transforming degrading land into improving land and were invited to teach in other areas. Unfortunately I was then told that as I did not have a degree in agriculture my work visa would not be renewed , consequently someone else should take over my work while before leaving Nepal I should write a book on 'Vegetable Growing in Nepal', and a book on 'Bamboo a valuable crop for the hills' which I did. My successors not having the same appreciation of the need to keep the work simple and appropriate using minimum manpower, the work became less appropriate and so no longer appreciated and was then terminated.

A related matter was to encourage the propagation and planting of bamboo.
I tried different methods of propagating with very pleasing results. After a few years I was given a rocky hillside below a mission hospital and eventually we planted 24 species of bamboo, mostly stem cuttings; using different methods of propagation the only such collection in Nepal; in 5 years we had a bamboo forest.

We were also invited to carry out trials on planting bamboo to stop landslides from further developing. This also gave useful encouraging results and was written up by a Swiss organisation which was working with the Nepal government.

I and my staff were invited to give talks on soil conservation and bamboo propagation in other parts of Nepal. As previously mentioned I was asked to write a book 'Bamboo a valuable crop for the hills'. It has been published in both English and Nepali. I later wrote 'BAMBOO' a valuable multi purpose and soil conserving crops which is for wider use than Nepal with more information.

One of the important developments of my work in Nepal was the 'Good News step by step terracing system', this can provide a method for two or three farmers to cooperate and stop the loss of soil and plant nutrients quickly. The process continues until all their land is developed into a fertile and sustainable system of terraces. The 'Good News multi-use level', and the 'Good News Bulldozer conversion for walking tractors' are other developments which could be of much value in many places; all are described in the two practical volumes.

While I have only worked in Taiwan and Nepal what I have seen of slope lands in India and what I have heard of other countries in Asia such as The Philippines, Indonesia, Malaysia and Sri Lanka, also parts of Africa and the steep hillsides of Latin America give me the impression that the experience I have had is relevant in many places.

Gandhi once said

'I object to violence because when it appears to do good, the good is only temporary – the evil that it does is permanent'

I would like to make a statement on the way we manage agriculture.

'I object to unsustainable agriculture because when it appears to do good, the good is only temporary-the evil that it does is permanent'.

As a Senegalese farmer has stated

'When the land is no longer fruitful, and the authorities have abandoned us peasants, we can only survive by using guns'.

'Hunger and disease are merciless killers and desperate poverty is at the root of many wars plaguing poor countries---.'

I hope and believe that the knowledge and experience I have gained, and the 265 illustrations in this book, can be used in many places, to produce durable soil and water conservation, and to improve the existing soil.

As mentioned before, practitioner's seldom write books, this book is written by a practitioner, I do not claim to be an academic so there will be faults in the book. No doubt there will be criticism, but surely it is better that practitioners are encouraged to write practical books. If there are people who feel that my 213 references are not detailed enough, I can only say that I do not have a library available to me, only my own books and records. There are quotations used which are from quotations in technical publications, I have quoted all the information given in them, I have done the best I can. I have been unpaid for the seven years it has taken me to research write and produce the illustration, rewrite and rewrite. I have done it because I am convinced the books are needed, I hope that readers will appreciate that.

CONTENTS

LIST OF PLATES

(For better clarity the below-mentioned text photographs have been reproduced on art plates at the end of the book.)

PART 1

CHAPTERS RELATING DIRECTLY TO
SOIL AND WATER CONSERVATION

CHAPTER 1

WHAT WILL THE FUTURE BE?

1.1 A question to ask a farmer in a poor rural area

What kind of future do you want?
TO BE A CONTENTED, WORKING FARMER,
OR
To be a landless, workless, beggar?

Now is the time to decide ! However in many cases while the farmer would like to change the future, he is unable for various reasons. It may be that in the past when an area of land no longer produced worthwhile crops the people cleared good fresh land, but now that is no longer possible. The people may not know how to conserve the soil, in fact few agencies which claim to know are able to produce evidence of economical and lasting soil conservation and soil improvement. Too many projects have failed and so confidence in change for the better is often non-existent. It could also be the case that the society is against change, perhaps because of attempts in the past which have only made things worse, I have seen where government soil conservation projects have caused landslides which carried houses with them. In other places where the so called conservation work has buried the top soil and resulted in useless infertile terraces. It may be that practical lasting demonstrations and explanations are needed to show that poor farmers do not need to end up as landless, workless beggars. This book can help those who want to enable rural people to have a lasting future, to show the way, involve farmers. Then to invite others to come and see, and discuss with those fellow farmers how what works with them may perhaps with local modification, be suitable in their situation.

1.2 What do we need for our health and well being?

1. FOOD. 2. DRINKING WATER. 3. WATER FOR WASHING. 4. WATER FOR CROPS AND VEGETABLES. 5. FUEL FOR COOKING AND KEEPING WARM WHEN IT IS COLD. 6. BUILDING MATERIALS FOR HOUSING. 7. FODDER FOR ANIMALS. 8. WEALTH TO PAY FOR CLOTHES, TOOLS, HEALTH CARE, SOCIAL AND RELIGIOUS NEEDS.

1.3 To get them what do we need?

For 1. Good soil, to grow the food for ourselves and our animals.

For 2. 3. & 4. Sufficient water close to us, and the rain water soaking into the soil, rather than much of it running away, the land and the water springs not having enough for the needs of our crops and other needs.

For 5. Wood or bamboo, or extra crops to pay for electricity or other fuel or pay for a biogas unit. To grow them we need good soil, and the right amount of soil moisture for as long as possible.

For 6. Again we need wood or bamboo or other crops to sell to buy the building materials. To grow them we need good soil, and the right amount of soil moisture for as long as possible.

For 7. Fodder crops, and fodder trees which may include bamboo – to grow fodder to feed the work animals and to produce milk, meat, perhaps wool for our own use and for products to sell. To grow them we need good soil, and the right amount of soil moisture for as long as possible.

For 8. Extra crops, for money or goods to pay for clothes, tools and other things we need; to grow them we need good soil, and the right amount of soil moisture for as long as possible.

LOOKING AT THIS LIST WE SEE THAT THE MOST IMPORTANT NEEDS ARE GOOD SOIL, AND WATER WHICH IS NOT FLOOD WATER, AND LASTS AS LONG AS POSSIBLE IN THE PLACE WHERE IT IS NEEDED.

IF WE DO NOT HAVE THESE WE CANNOT HAVE HEALTH AND WELL BEING UNLESS WE CAN MOVE TO SOMEWHERE WHERE THEY ARE AVAILABLE.

IN MORE AND MORE PLACES THERE IS NO LONGER ANYWHERE GOOD TO MOVE TO.

Soil conservation means the use of land so that the maximum permanently sustainable production is achieved. Too often the soil is regarded as something permanent, to be treated in any manner which will produce crops by the most convenient and economical methods. This is a dangerous mistake; the soil if not properly cared for and managed, is no more permanent than buildings or tools. Soil like tools, can become spoilt, worn out and removed from the farm. The difference is that when a tool becomes worn out or stolen it can be replaced; soil cannot normally be replaced.

The world situation is alarming.

- ❏ There is climate change, the destruction of forest lands, the spread of deserts and the creation of new deserts.
- ❏ The disappearing species of plants, animals, birds, butterflies and fish and people.
- ❏ People concerned about disappearing plants and animals may not realise that there are also disappearing tribes of people, because other people clear their habitat. Some of the clearing is done by large companies, but much is done by peasants who need land because their land is worn out or because they themselves have been displaced.
- ❏ The pollution of air and water. We learn of how water is polluted by agricultural chemicals, and becoming a scarcer commodity, of wells needing to be deepened at regular intervals, and of water sources drying up or becoming brackish.
- ❏ The spread of cities, more and wider roads, airports, industrial estates, landslides and land being buried by rocks sand and debris, the increase of degraded land.
- ❏ The problems of what to do with nuclear waste with a half life of millions of years.
- ❏ The rapidly widening gap between the rich and the poor which should concern all fairminded people.
- ❏ A question that all people need to ask is: WILL WE HAVE A FUTURE? WILL OUR CHILDREN HAVE A FUTURE? IT IS THEIR FUTURE WE ARE BUILDING OR DESTROYING NOW!

We all love our children and we want them to love us. What will they think of us when they have children of their own?

THE RURAL PEOPLE OF THE TWO-THIRDS WORLD HAVE TO ASK THEMSELVES - IF WHEN THEIR CHILDREN GROW UP:

- ❏ The soil is too poor to feed them.
- ❏ There is not enough fodder for the animals and to make manure for the fields.
- ❏ The fields are falling down the hillsides.
- ❏ There is not enough wood for firewood or for building and repairing houses.
- ❏ The rainwater runs off the land and so the springs of water and the streams have less and less water and start to dry up.

WHAT WILL LIFE BE LIKE THEN?

> There will not be enough food.
>
> It will be very hard work to grow what we can.
>
> Water will be scarce.
>
> Our children will say you wasted the land, why should we care for you?
>
> The young people will have to leave and go somewhere else where they can live and support their own children. We will be left alone.

We will be lonely, sad, tired, poor and hungry, life will be very hard.

What happens to the rural people of the two-thirds world will have effects on the one-third world.

CAN WE DO ANYTHING ABOUT IT? I AM CONVINCED WE CAN. THAT IS WHAT THIS BOOK IS ABOUT.

I BELIEVE MY EXPERIENCE CAN HELP OTHERS, BOTH RURAL PEOPLE AND THOSE WHO WANT TO WORK WITH THEM, TO A BETTER FUTURE, AND REDUCE THE PRESSURE ON THE REMAINING WILD HABITATS.

As an example of a developing country we can consider Nepal in a report in 1999:

> Over 80% of Nepalis are engaged in agriculture and related activities. Agriculture also contributes 43% of Nepal's gross domestic product (GDP). However, the country cannot produce enough food for its population of 20 million. Out of its 75 districts, 41 are classified as food-deficient---.
>
> Once a producer of three tonnes of food grains per hectare, Nepal's productivity is now less than two tonnes a hectare. This situation compels the kingdom to import hundreds of thousands of tonnes of food grains almost every year. In the last five years imports have risen by four per cent per annum.
>
> Production of enough food is one thing; the distribution system is another. A lack of regular and inexpensive transportation hinders the distribution of food from surplus areas to food deficit ones. Rural Nepal, where most of the agricultural activity takes place, is one of the most poorly served areas of the world in terms of roads and other basic infrastructures. Thus the people are fed with imported and air transported food, where transportation costs can be four times higher than the cost of the food itself. Often transport subsidies exceed the total agriculture budget in some of the food deficit areas.
>
> The most important aspect of food security is people's entitlement to food. More than a quarter of all Nepalis are unemployed, and a large number fall below the absolute poverty line. In this situation there is little hope that everyone will have access to sufficient food, even if there is enough available in the market.
>
> Some urban Nepalis may be able to afford imported and expensive foodstuffs, but rural men, women and children and the deprived classes will continue to suffer unless something is done. Nepal is currently facing a food deficit, but it may be even more serious in the years to come.[199.]

Since the above statement, Nepal has been plunged into crisis by a revolt which started in the very poor far west region but soon spread and threatened the whole country. Unless countries like Nepal with help from the developed countries face up to the urgent need to reverse the loss of soil quantity and quality there will be more poverty and hopelessness. Poor hungry people will become desperate people, and more desperate people are easily led to resort to desperate measures. The good news is that the situation can be reversed, but mind sets will have to be reversed first.

I had a normal agricultural training. When I went to Taiwan in 1963 I did not know much about soil microorganisms. I had the conventional ideas that what was needed to improve third world agriculture was improved livestock, and improved management of the livestock. Improved crop varieties and the needed fertilisers, pesticides and equipment to help the improved plants to grow their best. I knew that we should keep the land in good condition, and keep up the humus and organic content of the soil; providing we did that intelligent fertiliser use could only be beneficial. So I was basically in favour of green revolution techniques, and that by using improved seeds, fertilisers and pesticides and perhaps some organic manure we could improve both crop yield, improve the soil, and make good profit. However, I found by experience how fast humus is lost in tropical conditions, and how the rate of loss is accelerated by the use of fertilisers,

Fig. 1 The soil is bare, the rain drops are enemies instead of friends, they dig at the soil, and steal and remove soil and plant foods.

I found that fertiliser use on typical tropical hill soils can actually reduce yields to an uneconomic level. I found how important humus is, and I found by accident the importance of micro organisms. Also that micro organisms and humus go together.

While I do not feel that fertiliser use is always wrong or that we should never use pesticides, I have learned that their use should be very limited if we are to have long-term economic agriculture, sustainable agriculture. I have also learned that we can achieve amazing results while using no fertiliser and very restricted pesticide use. I found that the less we use pesticides the less we need them.

In Taiwan I saw the development of hydroelectric dams at very great expense, I later saw how rapidly they silted up following deforestation and non-conservation farming on the steep hillsides. I then went to Nepal and saw the same mistakes being made there. It seemed folly to spend millions of pounds on dams before controlling the soil erosion which would fill up the dams. I saw that the future of the country depends on caring for the soil which provides the basic needs of most of the people.

After 16 years in Taiwan, I worked in Nepal from 1980-1990, then again from 1994-1996. While there I learned more of how we can bring to life land which was practically dead, and which could scarcely support a few sparse weeds. With conservation and soil improvement; healing the earth with the help of compost, soil bacteria and microfungi (soil microorganisms), suitable rotations, and the earthworms which came as conditions for them improved.

Good soil is more than a few chemicals mixed with ground up rock, it is living soil.

In Figure 1 the soil is bare, the soil has become poor and weak.

Raindrops the soil needs, become enemies instead of friends. They dig at the soil, steal soil and plant foods and take them away.

Soil, plant foods, and water, are all needed by the hill fields; but often they are carried away to other places. At first they go slowly, we don't notice the loss.

Plant foods are like sugar and salt when they are in water we cannot always see them, as plant foods and soil are carried away the soil becomes poorer. The crops become poorer and soil loss becomes more and more.

The longer the loss continues the harder it is to stop the soil loss until the land becomes useless.

1.4 What can we do?

The following are some basic suggestions as an introduction to the subject.

1.4a Keep the soil and the plant foods, do not let them be taken away

Do not plant up and down the slope, water flows easily down hill. It gains speed and power to erode the soil and so is able to wash the soil and plant foods away. Plant across the slope.

When hoeing, push the soil into lines across the slope. This blocks or slows the flow so that erosion is prevented or reduced. Create stone or grass bunds across the slope.

More permanent methods. On very steep slopes mini terraces can be made. Long growing crops such as tea, pineapple, or banana are planted.

OR, fruit, nut, or fodder trees are planted on individual mini terraces, with interceptor ditches to bring surface water to the trees.

OR, if very steep or very erodible, plant the area to bamboo forest.

When land is less steep make normal terraces using the correct methods shown in this book.

Farmers may have to work hard now so that they may enjoy ease in the future. Saving our soil is better security than saving money. If they are farming in such a way as to cause loss of soil they are stealing the food of their children, perhaps killing their grandchildren . The soil produces the food for the future.

During the years from 1963 to the present I have witnessed pendulum swings of fashion in many aspects of rural development, someone has called them "The Holywood Syndrome". The latest swing in land management is from concentrating on supposedly soil conserving 'contour strips' and all will be well, to just 'managing the soil better' and all will be well.

With all pendulum swings the right balance is in between the extremes. Both are needed but when the topsoil has gone or is too shallow to absorb and retain moisture it is too late to think about improving it.

In the Philippines there are terraced hillsides which have been growing crops for 3,000 years. If they had not been terraced they would not have been able to. In contrast we can see similar hillsides which have been farmed as bare slopes or contour stripped for 50-100 years or less which are now eroded, degraded and infertile.

Providing we can make terraces which will not slide, keeping most of the topsoil on top, which can be made by farmers with a reasonable amount of labour for the perceived benefit, it should be clear that the only sensible way to conserve crop land is by making terraces. I believe that what is taught in chapter 6 should enable people working in a range of conditions to terrace their crop land.

1.4b Improve the soil

An area of jungle grows much more plant material and keeps on doing it, or even getting better, compared with crops grown on the same soil, we need to consider what we can learn from the ways the soil behaves differently.

We can improve the soil by increasing the organic matter (rotted plant material) in the soil. By growing a range of crops rather than just the same kind all the time. By protecting the soil from sunlight and heavy rain drops as much as possible. By increasing the microorganisms (tiny plants and animals too small to see) in the soil. By improving the structure of the soil so that air and water can penetrate into the soil, plant roots can grow deeper and wider, and the soil will not dry out so easily. See volume 2 Soil Improvement.

We can and should help farmers secure their future. A life which builds up and improves is one with meaning and value. A life which spoils or destroys has no meaning or value.

Two ways of looking at life are:

"I pass through the world just once.
If there is any good that I can do or any kindness I can give,
I should do it now and not delay it as I will not pass this way again."

The opposite view believes in reincarnation after death, that the person will be reincarnated as someone else. If that is so and we spoil the future now what will future lives be like?

If we believe in either of these views we should be doing what we can to make the world a better place now.

CHAPTER 2

SOME OF THE WRONG IDEAS AND PRACTICES CONCERNING SOIL CONSERVATION

2.1 Introduction
2.2 Poor understanding of the real situation and poor planning
2.3 Some impractical or unsustainable 'conservation' practices being promoted
   a) Planting fruit trees on the edge of terraces
   b) Lazy terracing, including contour strips on hillsides
2.4 Theoretical surveying
2.5 Impractical or inaccurate methods of surveying land
2.6 Wrong ideas

2.1 Introduction

Unfortunately much that is carried out as soil conservation work fails in the long or short term, why is this? What can we learn from our failures? This chapter considers some of the faulty approaches commonly used so that we may learn from mistakes.

> Despite decades of effort, soil and water conservation programmes have had surprisingly little success in preventing erosion. The quantitative achievements of some programmes can appear impressive. In Lesotho, all the uplands were said to be protected by buffer stripping by 1960; in Malawi (then Nyasaland), 118,000 km of bunds were constructed on 415,000 ha between 1945 and 60; and in Zambia (then Northern Rhodesia), half the native land in eastern province was said to be protected by contour strips by 1950 (Stocking, 1985). In Ethiopia, during the later 1970s and 1980s, some 200,000 km of terracing were constructed and 45 million trees planted (Mitchell, 1987).
>
> Ironically, many programmes have actually increased the amount of soil eroding from farms. This is because these impressive achievements have mostly been short lived. Because of a lack of consultation and participation, local people, whose land is being rehabilitated, find themselves participating for no other reason than to receive food or cash.
>
> Seldom are the structures maintained, so conservation works rapidly deteriorate, accelerating erosion instead of reducing it. If performance is measured over long periods, the results have been extraordinarily poor for the amount of effort and money expended (Saxson et al, 1989; Hudson, 1991; Reij, 1991).[133]

It is easy to criticise, harder to learn from mistakes and try again till we do find ways which are effective, lasting, and can be carried out by ordinary farmers. It is important to know the sort of mistakes which have been made, and to realise the importance of careful thought concerning the possible constraints which may block take up of methods, and how to remove or reduce those constraints. It is also important to find ways to encourage ordinary farmers to take up and further develop ways which have earned credibility, as appropriate, and desirable, for their conditions.

2.2 Poor understanding of the real situation and poor planning

I believe that when conservation methods fail it is not just a problem of lack of maintenance, the main problem is unsuitable design. The design or method is not first proved and demonstrated to be practical, it has not earned credibility, and so convinced the local people that the design is sustainable and good, so that farmers are pleased to participate in work or do it themselves.

A major problem is that in many cases soil conservation generally lacks credibility, so farmers are not really interested in it. The conservation work is something that outsiders are interested in, rather than the farmers being convinced that it is worthwhile for them. Distrust of outsiders also puts a question mark in local peoples' minds. Rural people, particularly, the rural poor have often learned to mistrust outsiders, so they may wonder what is the real reason for pushing conservation. It could even be that there is a hidden plan to take over the land after it has been improved, as has happened in my experience. Also, as the outside organisation gets most of the credit for what is achieved the local people are de-motivated.

Generally with so called conservation work, the main aim is to complete as many kilometres as possible, the quick fix approach. Unfortunately as we will see, quick fix methods such as Vetiver or Napier grass contour strips are still being advocated; such methods lead to fertility drift downwards and accelerate erosion.

1. An expatriate engineer who worked in the same area of Nepal as the author, was sure he understood why there was little terracing of the steep hillsides, except for paddy rice fields. He was of the opinion that the first people to settle in the area were intelligent people who built terraces. Those who came later were stupid and/or lazy and so they just farmed the slopes with some rough contour strips.

This kind of statement reveals an all too common attitude which assumes that farmers who do not do what seems obvious to the outsider, are stupid. Little credit is given to the possibility that the farmers could have been doing what they did do, for good reasons.

We should consider why farmers do things the way they do.

When farmers move into a new area they look for the most fertile places and clear those first. In rice growing areas they also look for places which have water which can be used for irrigation. They consider the growth of the trees and other plants, and what kinds of plants they are. They look to see how deep the soil is, the quality of the soil, and the slope and direction of the land. In addition farmers know from experience and local knowledge that certain trees and plants are indicative of good deep soils and others of poor soils.

As an example, in Queensland, Australia farmers, know that when Brigalow (a type of tree) country has the tree cover removed the soil has a very slow rate of water absorption, so erosion is more likely. If the rains in the area are heavy, such an area should be avoided, particularly if the ground is sloping.

Farmers would also tend to choose flat land or the lower and more gentle slopes. Such places would also be more likely to have deeper soil. So the first farmers moving into the area would clear the land which had better and deeper soil, and if available, land with potential for irrigation.

For ease of working, and to maximise the benefit of water, they would make paddy fields and terraces, particularly if the land was fertile enough to support a larger work force, which would then be available in the slack season, for such improvement work. It would be easy to terrace by simply dragging the deep soil level; where the slope was not too great there would be no need to move much soil.

More often the land would be terraced by slow terracing, which is simply laying trash and unwanted tree trunks and branches, with any rocks and stones, in lines across the slope. Then when ploughing, turning the soil towards the contour lines would cause the soil to gradually move down to the bottom of the strips, so over time rough terraces would form. With deep soil this terracing would be desirable, as the soil was deep with the underlying soil not so different from the topsoil, and as the process slow, there would not be much difference in fertility across the terrace, the rainwater would soak into the soil and the crops would be good. As only the ploughed surface would have been disturbed it is very unlikely that there would be any problems of landslip.

Following partial development of these areas, farmers coming into the area later, or descendants of the first settlers, would have to clear poorer land. They would find that when they used the same techniques on thinner steeper soils the results were much poorer; and unless they could find a source of water and make paddy fields, the terraces would sometimes turn into landslides; so it was better to leave the slopes as they were.

When soil is shallow and the subsoil hard, dragged down terraces have two problems.

(A) as the soil is dragged down, the top (fertile) soil is dragged first, then the under soil which covers the top soil, and lastly the sterile subsoil which goes on the top. The result is that the best soil is buried, so the crops will be planted or sown in the dead infertile subsoil, poor or no crops will be the result After all the work the land is much less fertile than before. On shallow soil even if this is achieved slowly by ploughing down to a contour line, the best soil is underneath.

The subsoil can only produce at best very poor crops, at worst not even weeds grow. The author has seen this method being used by government and non-government organisations in both Taiwan and Nepal. They then apply large applications of precious compost and manure, and add fertiliser. The local farmers look on either wistfully or derisively. After about three years of heavy manuring grass will start to grow, after a few more years other crops may also grow. The whole procedure is impractical and unless they are bribed or pressured, it is not taken up by farmers who have better things to do with their manure.

(B) The dragged down soil forms a wedge of loose soil on top of the hard subsoil, which may be shaly or slaty rock or bricklike soil. In the monsoon or rainy season, surface water runs down from above onto the wedge of loose soil. As the hard undisturbed subsoil does not absorb water easily the terraced soil becomes wet and heavy, and if the rain continues it may slide down the hill, taking not only itself but sometimes the soil below it also. I have seen a hamlet where all the houses were pushed into the river by a landslide caused by government ordered terraces made that way. Fortunately the people were warned, just in time they ran out and sheltered below a bamboo grove which stopped the landslide where they were, so that no one went down with the houses. They lost everything including their stored grain for the year as it was just after harvest. **So much for soil conservation!**

IT SHOULD BE CLEAR THAT WHILE SOIL CONSERVATION IS ESSENTIAL IT MUST BE CARRIED OUT WITH CARE.

The same results occur with slow terracing, though as the process is slow they do not show as quickly, and as less soil is loosened the risk of the land sliding is much less. In some areas the farmers regularly trim off the face of the terraces so that the wedge of better soil is gradually added to the terrace below.

When farmers have learned the dangers of terracing they will not have made terraces on unsuitable land. Rather than being stupid or lazy as the foreign "expert"thought, they were being practical and sensible.

Paddy fields

In a paddy field situation the problem of landsliding is much less likely to occur. The reason is that in the process of making the paddy fields the land is consolidated by treading, usually by bullocks or rice tractors, then wetted and tramped by people and/or animals/tractors. More water is added and more tramping and levelling is carried out until the soil is consolidated and watertight.

Even though the subsoil may not be good, the result is acceptable for a paddy field. The reasons are that paddy fields are more valuable than dry fields; and farmers know that in flooded conditions the soil reacts differently, more plant foods are released. Deposits of soil and nutrients are left by the irrigation water and so the soil improves. The soil is cooler, organic matter and humus from stubble and straw tends to stay longer in the soil than in dry fields exposed to tropical sunlight, so paddy field soil is more like that of fields in temperate countries.

Comparing Conservation Method

In Taiwan from 1965 to 1968 a trial was initiated which compared different methods of soil conservation on a hillside. These included full terraces, half terraces, rock and trash (roots, stems, and weeds,) bund contour strips, grass bund contour strips, and plain slope. (A bund is a marked strip of barrier land, contour means horizontal/level lines, so a grass bund is a level strip of grass covering a piece of land, normally about two feet wide (60 cms)). Over three years the production from the terraces increased by about 20%, from the half terraces by 1-2%. The production decreased from the two types of contour strips by about 20% and 60% from the plain slope, the rate of the decrease increasing each year. Subsequently, soil conservation work carried out at by the author in Taiwan was always with carefully made full terraces.

As already mentioned both government and non government organisations often advocate inappropriate methods. The result is as they prove impractical, fail or make matters worse, and so soil conservation is discouraged rather than encouraged.

The people who advocate or carry out these inappropriate methods intend to do good and should be encouraged, but encouraged to use appropriate and sustainable methods. They can only use the methods that they know, and follow what is currently thought to be the best way. Unfortunately the current best ways are too often taught by people who are very good with words but weak in practical understanding. We seldom hear from practitioners with long term experience. For instance I with 16 years experience in Taiwan and 12 in Nepal have read in books and articles, and heard a talk by a very well paid "expert???" about soil and water conservation in Nepal, all of which shows how little the experts looked at or understood the real situation. It is obvious that the articles and the talk have been written by persons based in Pokhara, the very pleasant tourist centre of Nepal, and so written about the area surrounding Pokhara, which is not typical of hill agriculture in most of Nepal. Furthermore 'experts' tend to disappear during the typhoons and monsoon season, they are not there to see and learn from what actually happens. At such times the weather is bad, sickness is common, and the roads are blocked by landslides so the experts take a holiday. Consequently the information is misleading to people who are trying to improve the situation. People who may live for two or three years in real conditions are often then treated as experts and pushed to write and speak before they have enough tested experience.

2.3 Some impractical or unsustainable conservation? practices being promoted:

a). Planting fruit trees on the edge of terraces.

While planting fruit trees in contour lines with the position of the trees on each contour line offset from those above and below is good, planting on contour bunds which will build up to semi-terraces is wrong.

The tree roots cannot develop properly, so the trees are stunted. In a dry period the trees will suffer more. In a storm the trees are likely to be blown over. Pruning, pest management and picking of fruit, on at least one third of the tree is difficult if not impossible. See Figure 1.

b). Lazy terracing including contour strips on hillsides.

That is, terracing by any means by which the terrace is built up on a sloping surface. That of course includes contour strips on steep hillsides designed to develop into terraces. In particularly heavy monsoons the terrace without foundations is liable to slide down the hill. Especially with soft soil, or with a sandy, shaly, gravelly or very hard subsoil.

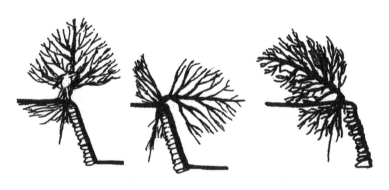

Figure 1.

Even when the soil is fairly deep and reasonable, the soil on dragged down terraces is still dead soil and takes some time to reach normal fertility.

Dragged down terracing. This is where the soil is simply dragged down or thrown down as above, until a terrace shape is made. This is commonly used by many government and NGO soil conservation projects. Figure 2.

Even where the soil is fairly deep and of reasonable quality, dragged down soil is still dead soil and takes some time to reach normal fertility. The reason is that the first soil dragged down is the best soil, the top soil. Next is the under soil and finally the dead subsoil which goes on top.

Contour strips and contour bunds. We have considered unsuitable terraces, but contour strips are encouraged more than

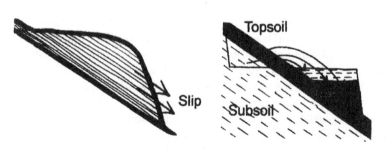

Figure 2.

terraces. Contour strips are another kind of lazy terracing. The hope is that the grass or other plants form a filter stopping the soil which is being washed down, the soil will build up and terraces form. SLOPE is one such program which is fast becoming fashionable.

Contour strips using vetiver grass have been vigorously promoted by the World Bank.

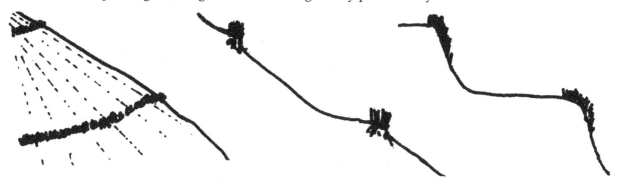

Figure 3

To quote from *Vetiver Grass,* 1990. 3rd edn. The World Bank. Washington D.C. (Ref. 1, p. 10).

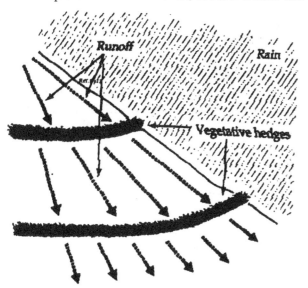

Figure 4

Figure 4 shows what happens over time in the vegetative system: the runoff drops its load of soil, the grass tillers up through this silt, and a natural terrace is created. The terrace becomes a permanent feature of the landscape, a protective barrier that will remain effective for decades, even centuries.

When the runoff reaches the vegetative hedges, it slows down, spreads out, drops its silt load, and oozes through the hedgerows, a large portion of the water soaking into the land along the way. (*This author would dispute this as the silt soon seals the pores in the soil.*)

No soil is lost, and there is no loss of water through the concentration of runoff in particular area. The system requires no engineering — the farmers can do the whole job themselves.[1, p10.]

In theory the runoff water flows evenly over the surface of the field till it meets the contour strip. There the silt is filtered out and most of the water is absorbed. In time this process turns the slope into terraces. (See Figure 4 [1 p13.])

This author would dispute the above. The soil at the contour lines will not be uniform, some will absorb more and become like porridge and slump. Others may be gravelly and slide when wet. In any case the contours will not be perfectly level or be able to remain perfectly level. In heavy monsoon rains water will accumulate along the contours and will flow to the lowest points. As the slope of the land is not perfectly uniform runoff water will move sideways as it flows and be concentrated in runlets which when they reach the contours will tend to undermine or run over the contours. At other points where the contour is slightly lower or the grass less thick the water will burst through. Finger erosion will then start and if the rain continues the water will continue to flow through those places washing away the accumulated soil and so flowing more and more rapidly. Gullies may then start to develop. See Fig. 5.

Even if this does not occur the monsoon water filtering through the grass strips will add to the water falling on the strip below. The greater film of water will cause sheet erosion which as it proceeds down the hill will concentrate into finger, then rill and ultimately gully erosion or landslides.

When we examine the drawings in the vetiver handbook we soon notice the impracticalities of the system. It shows fields which are uniformly sloping. The fields are smooth and regular, unlike most hill fields which are not smooth but have bumps and hollows.

It can be a mistake to plan for the bunds to be too solid barriers, the reason being that with solid barriers when

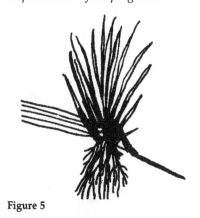

Figure 5

steady heavy rains occur the water will build up along the bunds. Even if contour bunds start level, due to variations in soil and slope they do not stay level. In heavy rain not all the water will filter through the bund, excess water will concentrate at low points and eventually burst through, this will cause more damage than if the water filtered through fairly evenly. As it bursts through, it runs down the slope being formed below the bund and gains velocity (speed and power). It flows on and joins with the water on the next strip and so gains both volume and scouring effect. It could start gullies or landslides on steep slopes, and has much more eroding and soil carrying power.

CONTOUR STRIPS AND CONTOUR BUNDS CAN BE APPROPRIATE WHERE THE SLOPE IS NOT GREAT AND THE SOIL DEPTH IS SUCH THAT AS SOIL GRADUALLY MOVES DOWN TO THE LOWER SIDE OF THE STRIP AND BECOMES LEVEL THE SOIL REMAINING ON THE UPPER PART OF THE STRIP IS STILL DEEP ENOUGH TO PRODUCE GOOD CROPS. *UNFORTUNATELY CONTOUR STRIPS AND BUNDS ARE OFTEN ENCOURAGED ON STEEPER SLOPES AND ON SHALLOWER SOILS WHERE THE END RESULT IS THE UPPER PART OF THE STRIP BEING NO LONGER ABLE TO PRODUCE ECONOMIC CROPS.* Sadly it is normally only when the soil is shallow that people feel the importance of saving what is left.

From 1981-1990 in Nepal the author paid quite a lot of attention to the use of contour strips in the area where he worked, there is more on this subject in Chapter 5.

There were few proper terraces in the area apart from paddy terraces.

The reasons being that:

- In most places the soil was not very deep and the subsoil very poor, so as explained above, the normal dragging the soil level method resulted in infertile terraces and a fertility drift from the upper part of the strips. Eventually the crops on the upper parts become too poor and then about 2/3 of the strips are not worth cultivating. At first it was disguised because the farmers would put their precious compost on the upper part of the strips knowing that much of it would be washed down by rain. In fact much of the plant foods from the manure on the sloping fields ended up manuring the fields of the wealthy farmers in the valleys below or flowing away in the rivers.

- There were traditional contour strips in the area and the Nepal government had a target for 10 miles of government style contour strips to be developed each year. It was easy to achieve targets when encouraging traditional practice along with the giving of free planting material.

In areas of land where the soil is deeper the effects were not immediately noticeable. In areas with the normal thin cover of soil (about 6" [15 cms] over an almost sterile subsoil) the effects were soon seen. Although during the monsoon there would still be very muddy water running from the hillside fields, much of the soil is filtered and held back by the contour bunds of plants, stones, or earth. Soil does build up along the lower edges of the strips. However as this soil is coming from the upper portion of the strips, those areas have less and less topsoil. On the upper parts the crops grow poorly giving the soil little shade from the strong sun or protection from the battering of the raindrops and drying by wind. Rain tends to seal the surface and run off, the shallow soil is less retentive, and what moisture is retained is soon dried by sun and wind. Harvest is difficult as the fields do not ripen evenly. The crops grown on the poorer portions leave little organic matter to improve the soil, they produce little fodder, and so less manure is available.

Eventually up to three quarters of the area of the strip is barren. Though much of the topsoil may have been saved it was concentrated in narrow bands. When the soil is shallow it should be used efficiently not concentrated in narrow bands. To correct the situation and convert to terraces will take much more work, with poorer results than if terracing had been carried out first. As there is less topsoil and fewer plant nutrients than before, the fields will be less valuable, so it is harder to justify the effort required to correct the situation.

Unfortunately few if any of the people advocating contour strips are about on hillsides during the heavy monsoon rains.

Figure 6. This is in the process of lazy terracing. In the heavy rains the soil moves down, some is stopped by the contour strips or other barriers, the rest, particularly the dissolved plant nutrients, washes on down. Total soil loss is not stopped, and the top part of the strips becomes bare and useless.

Contour strips can be appropriate in the right situations, however in hilly areas those conditions are seldom met with.

• In some places the underlying material was weak and soft such as a soft loose shale. When terraces were made, water soaked through to the shale during the monsoons. The result was that the whole terraced area became a heavy, wet, unstable mass, often developing into a landslide. This may be the reason why in some places in Nepal the hillsides are deliberately cultivated up and down the slope rather than across the slope. In such a method the surface water runs off quickly rather than soak in and perhaps cause a landslide. Light bunds slow down the speed of the flow so as to reduce the scouring effect. What seems to be contrary to normal conservation may be more appropriate in the circumstances.

There is more about contour strips and contour bunds in Chapter 5.1 and 5.4.

2.4 Theoretical surveying

An expensive theoretical survey was carried out for the land of the Yu Shan Agricultural Training Centre. It was not only very expensive, but useless. It did not take account the depth of soil when setting the height between terraces. Neither did it take into account the fact that some areas of the hillside were shallow soil, while others had extra depth of good soil. Some of that soil needed to be taken to add to the poorer areas; some areas were very stony, the stones were used for terrace walls and so in both cases the level was reduced. All this meant that the carefully planned graded terraces were neither suitable nor would they have been carefully graded on completion. Excess water would have accumulated at the lower points and then overflowed and caused erosion. Little if any would have gone to the carefully planned waterways. Another point was that there were areas of big rocks too big to break up and so not practical to terrace. These were used for other uses, they could not be used as the survey planned; they interrupted the planned water flows etc.

2.5 Impractical or inaccurate methods of surveying land

These use expensive complicated surveying levels which the farmers cannot use; or using apparently simple methods which are impractical.

There are various devices used from expensive professional surveying levels mounted on tripods, to hand levels and "A" frames. I have even seen a table with a spirit level being advocated as a means of marking out contour lines on a hillside. "A" frames are more practical but have those advocating its use tried using it on a hillside which has been ploughed or has rocks or clumps of grass, or on a windy day? If they had they would soon be disillusioned. In any case it is a slow and not very accurate method. A project near our project, taught the making of contour lines by using an "A" frame, till they discovered that the lines were coming together and then asked to borrow our Good News Level.

See Appendix 7, Surveying, The Good News Level which is a multi-purpose level, can be made and used by even illiterate farmers.

2.6 Wrong ideas

There are four points of view often heard, which I feel should be considered.

1. "Areas such as the Himalayan region are young geologically, they are gradually being pushed up. Landslides occur there mainly because of this. There is nothing that we can do about it.'

This statement distresses me because I believe that while it may be sincerely felt, it is wrong, and is discouraging efforts towards soil conservation. See Chapter 3.7.

a. Taiwan is very similar in many ways to the Himalayan region, it is also geologically young. The mountains are being pushed upwards by the Pacific tectonic plate pushing the Asian plate on which Taiwan is situated. In the same way the Himalayas are being pushed up by the Indian plate pushing up the Asian plate. The author has been to central Taiwan where there is or was still, primeval jungle. A natural mixture of huge trees, younger trees, shrubs, and undergrowth. Though the slopes are steep there is no erosion. The streams normally run clear even in the typhoons and *landslides are rare, and occur only on extremely steep slopes.*

Figure 7. A picture from Nepal

In areas near to populated areas where the land is cleared for crops and fruit trees, both soil erosion and landslides are common. In typhoon rains the rivers and streams turn brown with eroded soil.

b. The same can be seen in Nepal. After a landslide it is very common to find that land above it has been cleared, and maize is being grown. The run off from the maize field has soaked into the area which previously did not need to absorb more than the rain water falling on it. This proved too much and landslides started, being enlarged in subsequent years unless the high land was allowed to revert back to a jungle.

2. The second point of view often expressed is that: 'Local people know how to care for their land and did not abuse it before. We should leave them to their old ways, they were sustainable'.

This was the case in some communities, in other cases an equilibrium would be reached. After poor farming practices resulting in food shortages, the population would go into decline or be killed or captured by stronger neighbours, or move to a new area. As stated before, entire civilizations have declined or disappeared after the collapse of agriculture. Nepal was a closed country till the 1950's, yet it is clear that land had been systematically degraded around the area of the main trading routes. When constant cropping had degraded an area the population centres would shift to an adjoining area. In areas where there was no particular reason to stay as areas became degraded, people would move to completely new areas, clear them and soon tradesmen would follow and a new community would form. In more recent times with the population pressure this process has changed as the hill area was already developed. Hill people have moved to the jungle on the plains and cleared that. Now there is little jungle left to clear.

3. Thirdly the view about tree planting to check erosion. One extreme is to speak and act as in the film about Nepal, '*The fragile mountain*'. The film gives the impression that all that is to be done is plant trees to solve erosion problems; when of course, planting trees is just one aspect of soil conservation.

The other extreme is to say that planting trees increases erosion. While this can be the case, I think that such is probably because of improper forest management and planting. Clear felling hillsides rather than strip felling; and then planting trees where there is little or no remaining growth means the soil is open and exposed for too long. Planting pure stands of large tree species on steep slopes can result in too much weight of timber, too crowded above with no undercover, the land gives way, a landslide occurs.

If we study natural situations, there are mixed stands, at different stages of development. With the fewer very large trees able to spread their roots wide and deep into the underlying rock fissures, the large trees act to anchor and stabilise the land. Forest cover with large biomass can be seen flourishing on amazingly steep hillsides.

Figure 8. Soil and water needed for crops and for replacing the underground water essential for life is carried away down the rivers causing flooding down stream and then into the oceans. This water contains about 15% soil.

4. Fourthly the impression given that soil conservation consists only of erosion control — soil management is rarely mentioned as part of soil conservation. The author agrees with Professor Masefield who wrote:

"Soil deterioration arises in two ways, by soil exhaustion and soil erosion. The control of both is known as soil conservation."[4, p6] To reverse soil exhaustion see volume 2.

It has rightly been said:

SOIL IS THE MOST IMPORTANT RESOURCE IN THE WORLD. IT IS NON-RENEWABLE. IT IS ESSENTIAL TO LIFE ON EARTH. WHEN FOSSIL OIL RUNS OUT ALTERNATIVES CAN BE DEVELOPED. WHEN SOIL HAS GONE PEOPLE PERISH IN THE DESERT WE HAVE CREATED.

It is essential to both conserve the soil and improve it, to feed the increasing population. The improvements must be practical, appropriate to the particular situation, and locally and genuinely sustainable.

Much of what is described as being sustainable is in fact only sustainable on a short-term basis.

Another example of an inappropriate approach is when governments or big organisations plan schemes which seem feasible but cannot be trusted by farmers. One scheme in Taiwan was to bring in big bulldozers and graders and landscape an area of hillside with roads culverts, terraces etc. The farmers would pay back the cost in 10 years. The farmers did not want the project, as farmers in most parts of the world have learned not to trust governments, it was in fact done against their wishes. I talked to one of the engineers saying that it was interesting that the government was doing this work for small farmers. He replied 'of course it is not for the small farmers, in 5 years the land will no longer belong to them. The government knows that they will not be able to make the payments and they will be bought out'.

It is because of the lack of trust that I feel the Good News step-by-step terracing explained in chapter 6.7 which the farmers can do themselves at their own pace, can be an encouraging practical part of the way to a sustainable future.

EROSION

3.1 Introduction

Soil erosion is not simply a problem of landslides blocking roads and reservoirs being silted up, it is of far greater significance. Soil erosion reduces the ability of the land to support plants. We need plants for food, fibre and fuel, for paper, building materials, it also reduces the ability of poorer countries to feed themselves and pay off their debts. We also need plants to convert carbon dioxide, into oxygen and to act as carbon stores or sinks to reduce the greenhouse effect which is causing climate change, with increasing storms, flooding, and drought.

Soil erosion leads to poorer plant growth and eventually the land turning into a desert.

Plants do not have mouths and teeth, they cannot eat the soil. We could say that the roots are the mouths of the plant but the roots are really pipes with tiny holes in them. The soil cannot enter the holes but water can. The plant foods have to be dissolved in the soil water just as sugar and salt dissolve in water. In the water they can then pass through the tiny holes and move through the plants to where they are needed. Hence plants can receive nutrition only when the plant foods are dissolved.

Now consider what happens when rain falls gently onto the soil. Plant foods which are small enough dissolve into water in the soil, the roots of the plants soak up the water and so the plant foods can be received by the plant, that is good.

When heavy rains fall directly onto the soil, the soil is battered by the rain drops; they loosen it and make it soft. Plant foods dissolve into the water and if the land is fairly level and open much of the water and some of the plant food will soak down deep, deeper than the roots of the crop plants. In the dry season some of the water will rise through the soil bringing some of the plant foods back. Other plant foods will be left deeper in the soil where the roots cannot reach them unless the soil is open (looser) and the roots can go deeper.

If the land is sloping, water will start to flow away. At first it seems clear water but we must remember that just as salt water looks clear, clear water can contain plant food.

Next, as more water flows, in addition to the plant food it starts to carry away the fine lighter parts of the soil. The fine parts are the most important parts they have most plant foods and they hold the moisture in the soil. Phosphorus which is often deficient in tropical soils, is found chiefly in the finest particles of the soil, and it is these particles that are most easily carried away from the surface by running water and by wind. They also include the humus and the microorganisms and hold the soil together, so they are very important.

As these fine particles are removed erosion increases for three reasons 1. The soil doesn't stick together and so it is easily moved by the flowing water. 2. Because the soil is poorer, it has less life in it, plants do not grow well so there is less protection of the soil by the leaves and roots of the plants, the raindrops batter and break up the soil further, so that it is easily carried away by the water. 3. As the depth and water holding ability of the soil is reduced the soil dries out sooner, and so the plants do not grow as well, and easily die in dry periods. With even less plant cover, the soil is still more easily battered and eroded by water and by wind.

Next goes fine sand, then coarser sand and small stones till if the erosion continues all that is left is stones.

We can realise from what has been explained that the sooner we start soil conservation, the easier it is to stop it. The longer we leave it, the harder it is to stop it, and more of the best part of the soil will have already gone.

To improve the situation, one has to work more and harder. The longer we wait before starting soil conservation, the less the benefit from the work.

Primary erosion or geological erosion is good. It is the wearing down and breaking up of large rock masses into pieces of rock. Slowly these are further broken and worn down into gravel and sand. As erosion continues sand becomes finer and finer until it becomes like flour. When the particles become less than 0.002 mm in diameter they become clay particles.

Acids, and **small plants** like **moss and lichen** which grow on walls, start to turn the broken stone into **plant foods.** Dust from dead plants, leaves, and dead moss add to the plant foods, and in the mixture of stones, sand, rock flour, dust, plant foods and dead plants, seeds begin to grow. Their roots enter cracks in the stone looking for food and water. They grow thicker and slowly open up the cracks.

In cold weather, water in the cracks expands, and in hot weather the rock and stones expand in the day and contract in the nights and cooler weather. This all helps to crack, soften and break up rocks and smaller particles.

Very, very slowly soil develops. Small plants, then larger and larger plants grow in it. **Microorganisms** also develop in the soil. These are tiny plants and animals, too small to be seen with our naked eyes. They also help to break down and digest sand and stones, silt and clay, and dead plants. This helps to make more soil and supplies more plant foods. Then **macroorganisms** such as earthworms, ants and termites, beetles, crickets and moles, dig and burrow in the soil. This loosens it, mixes it, and helps the entry of air and water into the soil, it also helps wet soil to drain, all this helps the plants to grow better. More plants means more leaves and dead plants to mix with the soil, more plant roots entering the softer rock and cracks in the rock, opening them and allowing soil acids, and microorganisms, to enter and further break up the rock. The soil becomes deeper, <u>but</u> THIS PROCESS IS VERY SLOW. IT TAKES HUNDREDS OF YEARS, BETWEEN 600-1,000 YEARS FOR EACH 1" (2.5 cm) OF SOIL TO DEVELOP FROM THE ORIGINAL MATERIAL.

As the soil becomes deeper, more plants and still larger plants can grow. When rain falls, it's force is interrupted by the leaves and branches of the plants; so that the rainwater gently falls and soaks into the soil, instead of running quickly away down to the rivers and the sea. The deeper the soil, the more water it can hold; this means that the plants have water for a longer time and so grow better. When the plant cover of an area is great the shade effect keeps the land cooler; the moisture the plants release cools the air, and so clouds passing over are more likely to release rain, or cover the area as mist which plants can absorb through their leaves. As plants grow better and for longer periods dust is trapped by the leaves. The soil and the plants growing on it slowly become better and bigger. [Good farming also improves the soil, and this is vital to feed the growing number of people.]

Secondary erosion or accelerated erosion is bad. Instead of developing and improving the soil, it removes soil and reduces the quality of the remaining soil.

Secondary erosion occurs in three ways: 1) Erosion by water. 2) Erosion by people, animals, or vehicles, including paths roads and road construction. 3) Erosion by wind.

Soil erosion is a natural process, one that is as old as the earth itself. But today soil erosion has increased to the point where it far exceeds the natural formation of new soil. As the demand for food climbs, the world is beginning to mine its soils, converting a renewable into a non-renewable one.[148, p5.]

Some of the terminologies used when discussing erosion can be understood from the following quotation.

Morgan (1986)[200] defined soil erosion as "a two -phase process consisting of the detachment of individual particles from the soil mass and their transport by erosive agents such as running water and wind". When sufficient energy is no longer available to transport the particles, a third phase (deposition) occurs. Although many agents cause soil erosion, in Java it is caused mainly by water. Due to the tropical climate, rainfall has relatively high erosivity (the ability to erode the soil), since high volumes of precipitation (rain, this author) fall in short periods of time. Getis *et al.* (1991) also concluded that there is no more important erosional agent than water. Key factors influencing soil erosion are the erosivity of the causal agent and the erodibility of the soil (Morgan 1986).[201]

Erodibility refers to "the extent to which a soil is vulnerable (PJS able to be affected) to erosion" [202].

3.2 Erosion by Water

Let us consider a description of water erosion in the book *The Rape of the Earth* by G.V. Jacks who when he wrote it in 1939 was Deputy Director of the Imperial Bureau of Soil Science at the famous Rothamsted Research Station; and, R.O. Whyte who was at the time the Deputy Director of the Imperial Bureau of Pastures and Forage Crops at Aberystwyth. The authors drew their evidence from the USA, Australia, New Zealand, India, China, Japan, Palestine, Turkey, Russia, and Africa.

Unless the equilibrium (balance) is disturbed, a mature soil preserves a more or less constant depth and character indefinitely. The depth is sometimes only a few inches, occasionally several feet, but within it lies the whole capacity of the earth to produce life. Below the thin layer comprising the delicate organism known as soil, is a planet as lifeless as the moon.

The equilibrium between denudation (soil removal) and soil formation is easily disturbed by the activities of man. Cultivation, deforestation or the destruction of the natural vegetation by grazing or other means, unless carried out according to certain immutable conditions imposed by each region, may so accelerate (increase the speed of) denudation, that soil, which would normally be washed or blown away in a century (a hundred years), disappears within a year or even within a day. But no human ingenuity (cleverness) can accelerate the soil renewing process from lifeless rock to an extent at all comparable to the acceleration of denudation. (Can speed up the development of new soil to keep up with the speed of the loss of soil.) --this (unbalanced loss of soil) is what is now known as soil erosion. It is the almost inevitable (must happen) result of reducing below a certain limit the natural fertility of the soil - of man betraying his most sacred trust when he assumes dominion over the land.

---Until quite recently erosion was regarded as a matter of merely local concern, ruining a few fields and farmsteads here and there, and compelling the occupiers to abandon their homes and move on to new land, but it is now recognized as a contagious disease spreading destruction far and wide irrespective of private, county, state, or national boundaries. *Like other contagious diseases, erosion is most easily checked in it's early stages; when it has advanced to the stage when it threatens the entire social structure, it's control is extremely difficult. (Italics mine.)* In the main, unimportant individuals have started erosion and been crushed by it, until the cumulative losses in property and widespread suffering and want have brought governments and nations, with their immense powers for good or evil, into the fray.[7 pp 19-20]

The most significant deterioration resulting from cultivation is a reduction in the porosity and cohesion of the soil. Rainwater that was formerly absorbed by the soil, then runs off the surface, carrying soil with it and sheet erosion begins. This is usually unnoticed and there may be only a fraction of an inch removed in a season, (though there may be an inch [2.5 cm] or more PJS,) but the trouble is already well advanced, for the next layer of soil exposed at the surface is less absorbent than was the eroded layer, the amount of run-off water increases further, and the rate of erosion is steadily accelerated. In a very short time. the run-off collects in rivulets where its eroding and transporting capacity are enormously increased. The rivulets become gullies and the gullies coalesce to become chasms, penetrating through the soil into the barren subsoil. Stuart Chase in *Rich Land, Poor Land, describes how a chasm, 3,000 acres (1200 hectares) in extent and 200 feet (60 meters) deep, in*

Georgia, USA, started to grow forty years ago from water dripping unheeded off a barn roof and forming a little rill that became a rivulet that became a torrent that tore away the soil and subsoil over an ever widening area and flung whole farmsteads into a gaping wound. Other chasms and gullies, tributaries of this greatest, cover 40,000 acres in the neighbourhood. Chase likened the scenery to the Yellowstone Canyon which Nature fashioned in millions of years. Men have worked quicker; only half a century has been needed to fashion the new canyon of Georgia.[7] pp29-31.

I think it is important to note that while the drip of water from the roof was the source of the rivulet which became the chasm, it was joined by water run-off from the area. Other chasms also developed in the same area from excess run-off. Clearly the land management of the area was the problem, not the barn roof. However buildings and built up areas and roads which do not absorb rainwater, but divert it, and so increase the water flowing in other areas can be a cause of erosion. Much erosion of hillsides is caused by run-off from roads. New terracing systems which can absorb their own rainfall without problem, can be spoilt by having run-off from roads or other areas flowing onto them. If this is likely to occur, it is important to make interception ditches to divert the extra water, unless the terraces are paddyfields and so can handle the extra water.

Erosion by water can occur in two forms; A) by rainwater which cannot soak into the soil quickly enough and so flows over the surface, this is called run-off. As it flows it starts to carry plant nutrients and soil along with it. The run-off develops in several ways, these are illustrated by Figures 1-6.

B) by soil which has a loose open surface with either a weak subsoil or a hard smooth surface under shallow soil. The rainwater soaks into the soil until either the subsoil also becomes soft, or the hard under surface becomes slippery. When the weight of wet soil reaches a certain heaviness and lubricating ability the soil starts to move, and if there is nothing to stop it will become a landslide.

What happens when rain water falls on a bare sloping field.

(1)   Raindrop impact. The surface often gets sealed.

(2)   Soil particles dislodged by the raindrops block the pores in the soil, slowing down the penetration of water.

(3)   Rainwater which cannot penetrate the soil starting to flow down the slope.

(4)   Loosened soil particles are carried away by the water.

(5)   The mixture of sand, soil and water start to wear away the soil surface so increasing it's own bulk as it goes.

(6)   Plants are buried and choked by material washed down from up the slope.

(7)   Land is being eroded by swollen streams and rivers, which also cause flooding of fields and homes.

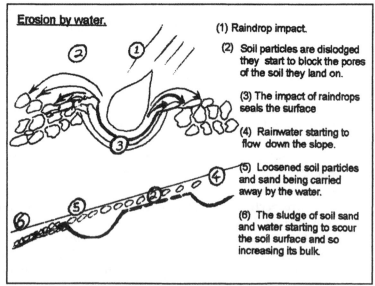

Fig. 1

As mentioned before, plant foods are like sugar and salt, they dissolve in water and cannot be seen. Apparently clear water run-off will be carrying plant foods away. As the plant foods are reduced the crops become poorer and the soil more exposed to erosion. The longer the loss continues the harder it is to stop the problem until the land becomes useless for growing crop.

In mountainous regions such as those in Japan, China, Nepal, Indonesia, the Philippines, and the Andean countries, construction of terraces historically permitted farmers to cultivate steeply sloping land that would otherwise quickly lose its topsoil. Centuries of laborious efforts are embodied in the elaborate systems of terraces in older settled countries. Now the growing competition for cropland in many of these regions is forcing farmers up the slopes, at a pace that does not permit the disciplined construction of terraces of the sort their ancestors built, when population growth was negligible by comparison. Hastily constructed terraces (*and contour strips, present author.*) on the upper slopes often begin to give way. These in turn contribute to landslides that sometimes destroy entire villages, exacting a heavy human toll. For many residents of mountainous areas in the Himalayas and the Andes, fear of these landslides has become an integral part of daily life.

Fig. 2 Plant nutrients are like salt, they disolve in water. The water running off the land carries with it plant nutrients and the finer soil particles. At first they go slowly and unnoticed, then crops become poorer and the soil coarser.

Research in Nigeria has shown how much more serious erosion can be on sloping land that is unprotected by terraces. Cassava planted on land of a 1% slope lost an average of 3 metric tons per hectare each year, comfortably below *the rate of soil loss tolerance* [the rate at which new soil is formed. PJS].

On a 5% slope, however, land planted to cassava eroded at a rate of 87 tons per hectare annually- a rate at which a topsoil layer of six inches [15 cm] would disappear entirely within a generation. Cassava planted on a 15% slope led to an annual erosion rate of 221 tons per hectare, which would remove all topsoil within a decade. Inter-cropping cassava and corn reduced soil losses somewhat, but the relationship of soil loss and slope remained the same.[147]

To quote from a description of a landslide in Nepal:

I was soaked by the monsoon downpour which had drenched the ground for days. The rain was turning soil to mud, and loosening the grip of the land beneath.---.

I ran away to the side of the stream as whole trees started tumbling down the mountain along with the mud and rocks. My father's cattle shed disappeared with the cattle as the land it stood on disintegrated and joined the landslide. Now everyone was running away, but even the land we were on was crumbling---. A great wall of mud crashed down at unimaginable speed with a deafening roar, but I was running and it missed me. Then, in horror, I saw my father's land, terraces of millet, maize, potatoes and barley, break its hold on the hillside and disappear below. Now, even if we escaped immediate danger, my family might starve to death with no land, no crops and no animals to feed us.---. In the morning we counted fifteen bodies, and from the wreckage of their houses we were able to pull out alive some of the injured. Alas, with no medical help some of these later died of their injuries--- if the landslide had come at night instead of in the afternoon we might all have been killed.[164] pp7-9.

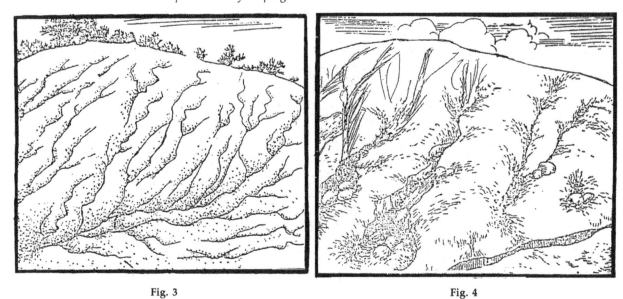

Fig. 3   Fig. 4

We have seen the effect of raindrops on bare soil, the result is that they quickly turn into muddy water. We should also know that observations in the USA showed that muddy water soaked through soil at only about 1/10 of the speed of clear water; the reduction being almost entirely because of the sealing of the surface layer of soil. They also showed that when clear water was applied, the rate at which the water soaked through the soil did not improve, unless the surface crust had first been disturbed.

Because of this surface sealing on unprotected soils, water run-off soon starts. Large number of soil particles, disturbed by the raindrop impact, are rapidly carried away by the water.

3.3 Sheet Erosion

As the depth of this run-off water increases, so does its speed of flow, causing it to loosen and carry more soil particles. If the speed of flow is increased from one foot a second to two feet a second, its soil loosening power is increased four times. In addition it's ability to carry soil increases 32 times. [6 p26.]

When more particles are being moved along in the water, the wearing effect of the run-off increases further and sheet erosion starts.

Because this type of erosion only moves thin sheets of the surface soil, its development may not be noticed for several years. Sometimes farmers talk about the soil being worn out, when the real situation is that it is being washed out. Then they may realise that the soil is more stony and the crops are poorer, ploughing is more difficult and rocks are showing on the surface of the field.

Fig. 5

As it is at first unnoticed, sheet erosion can be the most dangerous because it may not be stopped in time.

We should pay attention to the hollows in the fields, or the bottom of slope land a day or two after the rains; can we stir the soil gently and remove a layer which was laid on top of what had been the surface of the soil? Are plants at the bottom of the field buried?

It is often thought that sheet erosion is caused as a result of very heavy rain but as pointed out by T.F. Shaxson at the Soil and Water Conservation workshop of the Tropical Agricultural Association in September 2001, that is not necessarily the case. Very often the reason or a contributing reason is that due to a hard dense subsoil, or a plough pan formed by the bottom of the ploughs sliding over and compressing the soil underneath, or due to subsurface soil compaction due to the weight of animals or agricultural machinery rainwater has difficulty penetrating. As it cannot penetrate deeper the surface soil is quickly flooded and so water flows over the surface. Nowadays this can often be seen in the UK where monsoon rain does not occur but the weight of heavy tractors and machinery compact the soil. Water carries soil particles from higher to lower areas and bare patches develop on the higher areas.[191]

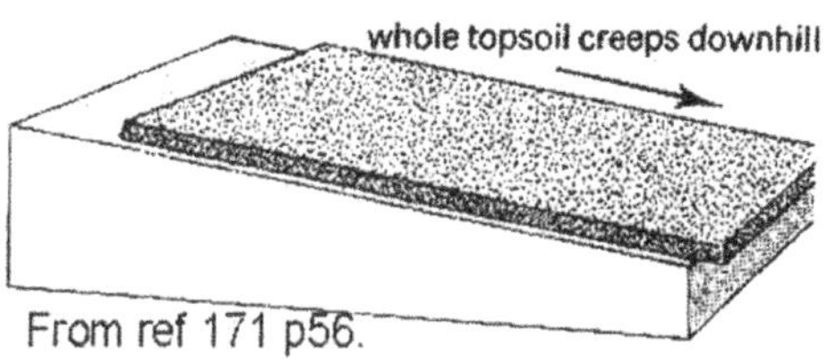

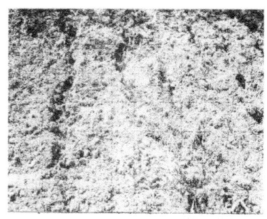

Fig. 6

The common form of sheet erosion described above only removes soil from the surface of the land, however if it is not stopped it develops into rill or finger erosion.

Another form of sheet erosion is where the whole topsoil moves downhill. It is usually a gradual process but occasionally can occur quite quickly like a landslide where only the topsoil moves. See Figure 6

3.4 Rill or Finger Erosion

What starts as sheet erosion usually develops into rill or finger erosion, the reason being that as the slope and the surface of the land is usually uneven, as run-off increases, rather than an even sheet flowing down the slope the flow will tend to move to the hollows and furrows. As more run-off moves down over those areas the speed of erosion increases and so the furrows become deeper.

While rill erosion removes some topsoil, once the rills are formed most of the cutting action is downward. This kind of erosion may start to erode the subsoil as well as the surface soil. Rill erosion increases the rate at which the soil dries out after rain, so may shorten the growing season. See Figures 5-10.

Fig. 7 Erosion starts with sheet erosion then becomes rill erosion.

3.5 Tunnel Erosion

It occurs where due to a deterioration of the soil surface an impermeable (waterproof) surface develops. When water can penetrate such a surface it can carry clay particles under the top crust down the slope and eventually tunnelling develops. The tunnels gradually deepens and widens until they eventually collapse and form a gully.

Common sense is needed when considering what methods will be suitable in a given situation. Digging a pit to understand the soil structure and under surface material will help.

3.6 Gully Erosion

When rill or finger erosion becomes too deep to be corrected by normal cultivating work (ploughing and harrowing) it develops into gully erosion. Gully erosion also occurs when small ditches or streams which have carried water for many years without eroding and growing larger, have increased flow of muddy water which cuts away at the sides, changing a stable condition into an unstable situation.

Gullies may start from tracks made by the movement of animals, people or vehicles, or from where water accumulates in hollows which now have more run-off than before. Perhaps because forest land has been cleared, exposing land to the

Fig. 8 The rills become deeper and then if the subsoil is hard the rills widen.

more direct effect of rain and wind, resulting in less of the rain soaking into the ground. Perhaps because of bad crop management, perhaps because of overgrazing leaving the land bare, or when animals are allowed to graze on slopeland when the soil is wet, or a combination. This loosens some of the soil and at the same time compresses other soil so that the rain water cannot soak in easily.

Gullies can also start from landslides, even very small ones, which often create a new hollow on a hillside so concentrating the flow of surface water. In addition the landslide may expose a springline. Water from the springline added to the rainwater increases the problem and often results in the cutting of a deep gully. The leakage of the springline affects the water supply in the dry season.

Gullies lower the water table (the depth at which the underlying material is wet and from which water can rise to the surface soil) in the surrounding land, so the growing season is shortened.

3.7 Landslides

Some landslides are caused where the land is being pushed upward by volcanic forces, or by the underlying tectonic plates lifting, as is the case with areas such as the mountains in the Andes, Taiwan, and the Himalayas. Slowly the slopes become so steep that the soil simply falls down. While this does occur, it does not occur as often as is sometimes suggested.

Most landslides occur after natural vegetation has been weakened or removed. It is common to find that above a landslide is an area which has been cleared and planted to a crop such as maize. At first there is a good soil structure with a fairly high level of decayed plant material and humus. The cultivated soil tends to hold together in clumps (it does not collapse and break up) and this blocks the movement of the water. The crops grow well and give fairly good protection from the rain drops. As time goes on the organic material in the soil disappears and the soil structure becomes poor and weak, it does not hold together and the rain soon dissolves and erodes it. The crops are poorer, they give the soil less protection to the effect of the rain, wind and sun, their roots do not help bind the soil as well, and the plants have less effect in blocking the movement of surface water. Run-off from the cultivated area flows together and becomes heavy and as it soaks into the soil the added weight and the lubricating effect of the water may cause some slippage, opening up the ground so that much more water soaks in under the surface; or, the water may flow till it meets a loose area of soil or cracks in the ground and soaks in there. The result is part of the soil in an area of steep land becomes heavy and soft with the water, and becomes more like porridge or rice gruel, until it collapses and slides down the slope. Water keeps on flowing into the area of loose soil, and, if the rain continues, a gully is formed; or the greater area of loose soil becomes full of water and a further stage of the landslide occurs.

Fig. 9 As the season progresses more soil goes the grooves made by the plough on the the subsoil are clearly shown.

Fig. 10 Later even the subsoil has gone, leaving only the underlying stone.

Fig. 11 From the 'Journal of Soil and Water Conservation' 1/1981

As more and more land is cleared and cultivated, more and more erosion occurs, and in the rainy seasons the clear streams and rivers become like liquid chocolate. I took a typical sample of such river water in Nepal and allowed it to settle in a glass tumbler overnight; when the water cleared 1/7 of the volume was soil. It is the soluble soil and the soluble plant nutrients which are most easily dislodged, the hard soil and stones move less easily. So it is generally the soil with the most potential fertility which comprises a high proportion of what is removed. Eventually we are left with rocky, stony, infertile fields. Government figures in Nepal put the soil loss as 240 million tons of soil leaving the country each year.

3.8 Factors affecting erosion by water

Erosion by water varies according to local conditions. Below are listed factors which are involved in order to decide the likely effect on a particular area.

Meaning of abbreviations:- [vb] = very bad; [b] = bad; [nsb] = not so bad; [g] = good; [vg] = very good.

1. Type of weather

Does the rain come down as:
Big drops? [b] Little drops? [nsb] Fine misty rain? [g].
Heavy rain? [vb] Light rain? [nsb].
Suddenly heavy? [vb] Slowly increasing from a gentle start for at least the first day? [nsb]
For a longer period? [vb] For a shorter period? [nsb].
In areas of monsoon climate: The main rain coming suddenly? [vb] First some rainy days about a month before the main rain so that the grass and weeds have grown to protect the earth before the big rains come? [nsb]. Gentle rain occurring fairly regularly? [g].
Does the rain come after hot windy weather? [b].
 "   "   "   "   " cool misty weather? [nsb].
 "   "   "   "   " warm weather with light rain? [g].
When the rain comes is there a strong wind blowing it? [vb]. Some wind? [b]. No wind? [nsb].

2. Field conditions

Is the soil hard dry and dusty when the rain comes? [vb or b].
 "   "   " already damp on the surface? [nsb].
 "   "   "   "   " for at least 6"(15 cm)? [g] (unless it is a very weak loose soil with no firm material underneath)? [nsb].
 Is the ground bare? [b] covered with some mulch or thin or small growing crops? [nsb].
 Or covered with thick mulch or thick growing crops? [g or vg].
 Are the crops planted up and down the slope? [vb] mixed? [nsb] or across the slope? [g].
 Are there ridges for the crops up and down the slope? [vb], across the slope? [g]. Broken ridges (ridges just round the plants)? [b].
 What kind of crops are grown? Erect plants widely spaced like maize? [b]. More bushy plants like peanuts or sweet potatoes? [nsb] Agroforestry? [nsb]. Permanent cover like grass which is not overgrazed particularly in the rainy season? [g]. Taller plants under-planted with ground covering plants established before the heavy rains come? [g]. Are the last two situations combined with agroforestry? [vg].
Is the ground cultivated and so loose at the period of heavy rain?[vb].
What kind of soil is it? See appendix 8.
What is the texture of the soil?

3. Soil texture

It is mainly concerned with the size and shape of the mineral particles of the soil. See also chapter 4, figure 3.

Particles are sand, silt and clay and they are described in this way:
 Sand 0.05-2 mm (particles visible)

Silt 0.002-0.05 mm (particles hardly visible)
Clay less that 0.002 mm (particles not visible)
Clayey soils have more than ½ (50%) clay particles.
Silty soils have   "   "   "   "   silt   "
Sandy soils have   "   "   "   "   sand   "

Loamy soils are mixtures of sand, silt and clay, they can also be described as clayey loam, silty loam, or sandy loam according to which is the most noticeable part of the loam.

Generally clayey soils stick together more and so erode less easily. Silty sandy soils erode more easily depending on the amount of organic matter and humus in the soil. Humus acts as a sticking agent holding soil particles together.

4. Soil erosion depends much on the infiltration rate of the soil, that is how fast water will soak into it. Also on how quickly it passes through it. This last point depends partly on the soil itself but it can also depend on the underneath material.

A sandy soil will let water pass through more easily so there is less run-off. A silty soil less easily. A dry clayey soil which is cracked or broken up will absorb water quickly at first but then it swells, the cracks seal up and water soaks into it only slowly.

5. Another factor which influences erosion is the soil condition

If the soil is in an undisturbed jungle it will erode less than soil which has been in cultivation.
If there is more humus (the brown organic substance) in the soil the soil will erode less.
If the soil is covered with weeds or crop plants the soil will erode less.
If the soil is deeper, the soil will erode less than if it is shallow; however, if the soil is weak and loose as well as deep and much water soaks in it could turn soft and slide down the slope.
If the soil is rough, it could block or slow the flow of water so that erosion is less.
If the soil is bare, erosion will be more likely.
If chemical fertilisers have been used for some years the soil will usually erode more easily.

6. Situation of the field

Does the piece of land only need to absorb the water *falling* on it?(g).
Does the field receive water from other areas such as drainage from a road or a storm ditch passing near it? OR from higher ground? [vb].
Can this be diverted either to get rid of it [g] or to take it to an area of storage for water conservation. Or for soaking down to replenish the ground aquifers, or underlying soil. e.g. Gorkakhot interception ditch.[vg]. This was a ditch we dug across the top of a very stony piece of land with the intention of preventing surface water from a bare hillside above the land running onto the land. We found that the nearby crops were growing better, and realised that though the land could not absorb sudden excess of water, it needed it. If allowed it to be absorbed in another way the land and the crops benefitted. Better crops normally result in less erosion and make the effort of soil conserving work more worthwhile. However as we will discuss later, this would not be good in situations where to introduce water to the underlying soil could make the soil unstable.

Is the ground almost flat? [g] or slightly sloping? [nsb]. More steeply sloping? [b]. Steeply sloping? [VERY Bad].

7. Slope

Clearly the steeper the slope the more erosion will occur.

There are different ways of measuring the amount of slope. I have found it best to use the one which I think is easiest to understand. In practice I have found farmers who have not been to school can learn to use it quite quickly. Even farmers who cannot read numbers can use it if we use symbols, as we will see. The method is explained in chapter 14.

3.9 Wind Erosion

This is affected by several factors:

Note the possibilities given for each point raised noted alphabetically are in order from the most desirable to the least desirable.

1. How exposed the site is.

What is the situation of the field in relation to the direction of the wind which blows at the susceptible period when the soil is dry. a) A sheltered field will be effected much less than b) a bare hill top.

2. The wind direction and speed.

a) A steady wind will loosen soil less than b) a gusty turbulent wind which has an average speed equal to the speed of the steady wind.

3. How exposed the soil surface is.

a) Alley cropping with a full cover of crops, grass, weeds, or a heavy mulch. b) It has a full cover of grass, weeds, or a heavy mulch. c) Partial cover of grass, weeds or crop stubble, or alley cropping is practised, with the tree or shrub leaves present at the susceptible season. d) The soil surface is bare and unprotected from the wind.

4. The soil texture / condition.

a) A good texture containing humus and other organic matter, holds well together and hence does not erode easily. b) A heavy (gummy) soil with few loose fine particles. c) A gritty soil. d) A sandy soil. e) A silty soil with many fine particles.

> NORMALLY ONLY FINE PARTICLES LESS THAN 0.1 mm ARE EASILY CARRIED. THE LARGER AND HEAVIER THE SOIL PARTICLES, THE GREATER IS THE WIND SPEED BEFORE THEY ARE MOVED.[8 p93]

Fig. 12 The results of Wind Erosion in the United States in the 1930's.

5. The soil surface condition.

a) A fairly smooth firm surface. b) A rougher lighter surface tilth following cultivation.

6. Soil moisture.

a) Moist, less easily dislodged and being heavier less easily carried away. b) Dry, easily dislodged and carried away.

7. Temperature.

a) Cool conditions. b) Hot drying conditions.

8. The length of the area in relation to the direction of the winds occurring when the soil is dry and bare.

a) Short better. b) Long worse.

Because the wind removes the finer and lighter soil particles first, wind erosion will seriously reduce the soil fertility, this reducing effect will be greater at first and lessen as there is less fine and light material to remove. Added to the fertility loss is the fact that as the finer and lighter materials are removed so the moisture retaining ability of the soil decreases. This further reduces the growth of plants, and so the drying effect of the wind will be greater than before. It is a vicious circle.

In the July/September 2000 issue of *Intermediate Technology* it was reported that:

Earlier this year, scientists at the National Oceanic and Atmospheric Administration Laboratory in Boulder, Colorado, USA reported that a huge dust storm from northern China had reached the United States "blanketing areas from Canada to Arizona with a layer of dust." They reported that the mountains along the foothills of the Rockies were obscured by the dust.

The dust storm did not come as a surprise. On March 10, 2001, *The Peoples Daily* reported that the season's first dust storm - one of the earliest on record - had hit Beijing. Along with those of last year, this storm was among the worst in memory, signalling a widespread deterioration of the rangeland and cropland in China's vast northwest. —.

In addition to the direct damage, the northern half of China is literally drying out as rainfall declines and aquifers are depleted by overpumping. Water tables are falling almost everywhere, gradually altering the regions hydrology.[190 p48] More information can be had from the Earth Policy Institute website; www.earth-policy.org or email jlarsen@earth-policy.org

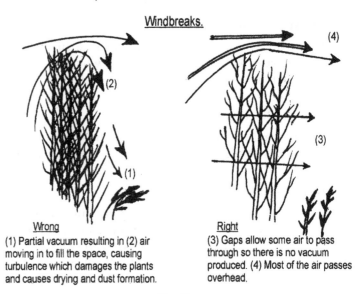

Windbreaks.

Wrong
(1) Partial vacuum resulting in (2) air moving in to fill the space, causing turbulence which damages the plants and causes drying and dust formation.

Right
(3) Gaps allow some air to pass through so there is no vacuum produced. (4) Most of the air passes overhead.

Fig. 13

3.10 The control of wind erosion

To reduce wind erosion:

1. Wind breaks should be planted across the path of the winds which cause the erosion. These are the winds which blow when the soil is dry and there are poor crops or no plants growing and protecting the soil.

    These wind breaks should not be solid wind breaks or there will be turbulence as shown in the illustration. This occurs because of the wind blowing with more concentrated force over the area behind the wind break. This causes a suction which leaves a space where there is low air pressure (a partial vacuum). Wind rushes in here to equalise the pressure and the result is worse than if there was no wind break. There should be slight gaps in the wind break so that some air gets through, then this problem does not occur.

    Soil only blows when it is dry and loose, so:
2. Try to conserve soil moisture. In some situations a mulch of straw, leaves, plastic or anything to cover the soil, will be appropriate, but if the wind is very strong the mulch may blow away.
3. Avoid cultivating the soil when it is dry, as when it is dry and loose it is most easily eroded by the wind.
4. If wind erosion is likely to occur, leave stubble and / or straw on the field if it is practical to do so.
5. During the time when there are no crops in the field, it may be better to allow weeds to grow. They can later be *ploughed in* as green manure. Light cultivation to kill the weeds will also increase erosion.
6. Better still, if it is possible, grow a drought-resistant crop such as chick peas or buckwheat; or, a green manure crop such as Sunnhemp *(Crotelaria juncea)*, so that it is still on the land during the dry windy time of the year. Leave it till it is time to plough, then cut it off. The leaves will have fallen in most cases enriching the soil, and the roots will have added nitrogen to the soil. It can then be composted or used as a mulch for the next crop. Alternatively it can be ploughed in when still soft, so that it is a good manure, adding organic matter and humus to the soil to bind it and so reduce erosion.
7. Rotate the crops so that the soil is not exhausted by having the same crop growing each year. Plant pasture crops sometimes in the rotation, when the pasture is ploughed in it helps to hold the soil. Grow cover crops between crops liable to wind erosion.
8. Plough, sow, and plant crops across the direction of the wind, if it is possible without encouraging water erosion.
9. Use nurse crops (the normal crop is sown or planted within the nurse crop) whenever possible

so that the ground is covered when wind occurs which may blow the soil.

10. In some situations it is helpful to leave the soil surface rough; cultivate so that the surface is knobbly rather than fine and smooth. This has three effects, a) it interrupts capillary action (water moving up through the soil) so that less moisture rises to the top and is evaporated. b) It interrupts and slows down the wind blowing over the surface so that it has less drying and soil moving power. The wind tends to keep to the level of the top of the lumps. c) Soil particles being carried over the surface will tend to drop into the hollows between the clumps, there they are less affected by the wind.

3.11 Factors Relating to Soil Degradation and Erosion

As mentioned earlier it is often said that landslides in the Himalayan region are the result of the land being pushed upwards by the Indian tectonic plate pushing under the Asian plate, that as the land reaches a critical angle the soil slips.

The same lifting also occurs in Taiwan yet in the remaining primeval jungle in the central mountains I have seen forested areas on amazingly steep slopes with even large trees growing. The tree roots penetrate deeply into, and wrap around the underlying rock and anchor the soil. While there are landslides, these are few and in many cases only where the angle is almost vertical, the situation is similar in Nepal.

Most landslides follow human activity, see Figure 14.

We have to be aware that the factors explained in this chapter concerning soil degradation and erosion are inter-related; the cumulative effect has a compounding result. The rate of increase of loss increases with time.

To summarise, as erosion takes place, the

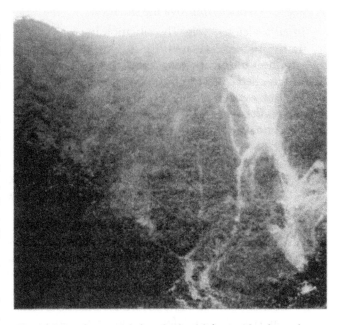

Fig. 14 It is often stated that the landslides in Nepal are due to the Himalayan uplift. Scenes uch as this show that when the land is covered with vegetation the slope is often very sleep before landslides occur. Most landslides are the results of human activ-

most fertile elements and the most water-retaining constituents, go first. This results in the protecting effect of plants being less, as the plants are poorer and thinner, and have a shorter growing period. It also means that there is more moisture loss, which results in even poorer plant growth, and so less organic matter available to reduce soil fertility and texture degradation.

With the soil bare for longer periods the soil structure is spoilt, making it much more susceptible to erosion.

Fertiliser may be used to cover/lessen the effect of fertility loss but this only works in the short term. It increases acidity, which is a problem with most tropical soils, this in turn results in a fixation of plant nutrients in a form unavailable for the plants. At the same time, it discourages microorganisms which would release the plant nutrients. The spoiled soil structure makes the soil harder to cultivate, less friendly to plant roots and more easily eroded.

The result is deterioration of all aspects needed for sustainable agriculture.

❑ Reduced quantity of fertile soil.
❑ Reduced quality of the remaining soil.

Fig. 15 A village of 37 houses lies buried under the sand and rocks carried down 3 years after clear felling of trees on a hill-side upstream from the village.

❑ Reduced water holding capacity, both due to the reduced volume of soil and the reduced ability of the remaining soil to retain water.

❑ Reduced protective effect given by growing plants from sun, rain, and wind.

❑ Poorer soil structure and so less resistance to erosion.

❑ Malnourished people with reduced energy, time, skill and intelligence of the people to work to improve the situation.

In all a vicious cycle. The situation in so many places is rushing towards catastrophe. Indeed in some areas it has already reached catastrophic proportions.

As this writer wrote in one of his annual reports as leader of the Horticulture and Agronomy Support Program of the United Mission to Nepal:

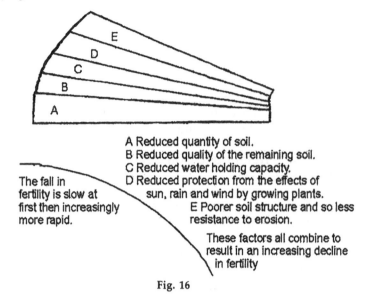

A Reduced quantity of soil.
B Reduced quality of the remaining soil.
C Reduced water holding capacity.
D Reduced protection from the effects of sun, rain and wind by growing plants.
E Poorer soil structure and so less resistance to erosion.

These factors all combine to result in an increasing decline in fertility

The fall in fertility is slow at first then increasingly more rapid.

Fig. 16

> The key to the future in areas like Palpa where the majority of the agricultural land is hillside (mostly steep hillside) land can only lie in soil conservation. The land is degrading, some at an alarming rate, some losing an average of two inches of soil in a year. As the soil degrades further the rate of erosion increases further.
>
> The aim I have felt to be paramount for the work engaged upon, is to try many possibilities with the aim of finding a range of options which farmers can fit into their own situations. Options which enable integrated, sustainable or improving systems, and so offer a future for the communities in the hills.
>
> I believe that our work indicates that the deteriorating situation in our area which has been described as an 'ecological disaster area' can be reversed. In very badly eroding or steep areas, or areas of poor soil bamboo forest should be established for soil conservation, fodder, fuel and building materials. In less infertile land and even stony land fruit, nut, fodder or timber trees should be planted. They should be planted across the slope with interception ditches to run moisture to the trees and aid absorption and water conservation. Fodder crops to be planted or sown between the trees to protect the soil as well as providing fodder till fodder from the trees and other fodder belts was available. Fodder crops could be grown on the steep north facing slopes on land unsuitable for crops but having a longer growing period due to less drying out by the sun.
>
> On less steep areas terraces must be made, contour strips are not a viable long-term solution and as an interim measure narrow contour terraces are more appropriate with intermediate narrow terraces to be made later and subsequently all the terraces enlarged until they form full terraces.
>
> The question is how to do this in a practical sustainable manner. I believe the answer is in the Good News Terracing step by step system (explained in this book).

> There is a close link between soil erosion and third world debt. Soil erosion can undermine not only a country's production capacity but its debt servicing capacity as well, for it leads to widening food deficits, mounting debt, and eventually food shortages. A nation whose people face starvation can hardly be blamed for failing to make debt repayments.
>
> A country that loses an excessive amount of topsoil needs to import more food and thereby raises the pressure on soils elsewhere.[101 p271].

Poor desperate people become rebellious people, and so revolution and war becomes likely.
It has been rightly said :

> Any ecosystem, no matter how resilient, can be pushed to a point of no return; to a threshold beyond which limiting factors become so severely operable, that recovery, in periods meaningful in the human time scale, becomes impossible. If this threshold is not to be crossed in very many areas of the world, immediate and drastic action is required.

That is the reason for this book.

CHAPTER 4

WHAT DO WE MEAN BY SOIL CONSERVATION? HOW DO WE PLAN SOIL CONSERVATION? HOW TO COMPARE FIELDS AND DECIDE WHAT EACH FIELD NEEDS?

4.1 Soil conservation means the use of land so that the optimum permanently sustainable production is achieved

4.2 Why terracing is well developed in some hilly areas but not in others where it would seem desirable

4.3 Land classification in theory

4.4 Classification systems should be flexible enough to be varied easily to match local situations

4.5 The Good News Assessment Charts

4.6 Land Classification Guide for third world slope land

4.7 The approaches to use

4.1 Soil conservation means the use of land so that the maximum permanently sustainable production is achieved.

Too often soil is regarded as something permanent, to be treated in any manner which will produce crops by the most convenient and economical methods.

This is a dangerous mistake; soil, if not properly cared for and managed, is no more permanent than buildings or machines. Just as they can deteriorate, become worn out and useless, or even removed; so soil can become degraded, worn out, and removed from the farm. The difference being that when machinery becomes worn out or obsolete it can be replaced. Soil cannot normally be replaced.

We should "IMPROVE THE LAND TO INCREASE THE LONG-TERM ABILITY OF THE LAND TO PRODUCE"

As someone has said: "PUT FERTILITY CREATION BEFORE FERTILITY EXTRACTION".

Too many aid programs and too much 'development teaching' teaches how to extract, not how to sustain the extraction. It does not give long-term security.

Soil conservation, then, includes all farming techniques that aim to maintain fertility, physical structure of soil, and the ability to absorb and retain rain water.

Where erosion damage (or soil exhaustion PJS) is extensive, the cost of reclamation is usually high, but if control methods are adopted in the early stages of erosion,(or exhaustion PJS) or better still before it commences, the cost is comparatively low, and most of the topsoil can be retained in its productive state.

The first approach to conservation farming entails the efficient use of every acre in accordance with it's capabilities. To do this, the farmer needs - an inventory of his land and its capability.--
Seldom will the adoption of a single practice do the whole job, no matter how well it is carried out.[3 p55.]

Soil conservation has many aspects—it is not only making barriers to rainfall run off, it includes soil management as explained in volume 2. Very often soil has been spoilt and so unable to absorb rainwater

water quickly enough. If we can understand that and work to improve the soil structure the need for other methods of soil conservation will be reduced and may even disappear.

To understand the sort of situation which may be faced I will quote from an article in the magazine 'Permaculture .

> Despite the best rains in 10 years this past summer, the growth of grasses and pioneers on an unplanted and ungrazed field across the fence from us was barely above a persons ankles. Our campus in, contrast, ran riot with growth due to extensive storm water harvesting earthworks, multiple perennial water supply systems, and agro forestry. Bethel Business and Community Development Centre (BBCDC) is a rural development and education institution located in the Senqu Valley in the district of Mohales Hoek in Lesotho (Southern Africa). Since being founded in 1993, BBCDC has taken permaculture as its inspiration for the design of landscape, farming systems, infrastructure and educational programs. In this article, I emphasize the role of water harvesting earthworks in conjunction with the restoration of soils. The poor growth mentioned in the introduction is due to the fact that soils have become sealed, cemented and anaerobic. This is obvious to anyone who begins to dig swales in this region or while excavating a building foundation. Compaction by hoofed animals and episodic fire has taken its toll. Shallow ploughing combined with extractive agriculture has compounded the problem.---.

> Presently much of Lesotho suffers from chronic land degradation and soil collapse. The result means loss of biodiversity, shrinking agricultural productivity, depletions of nutrient balances, falling water tables, exacerbation of temperature extremes, flash flooding, dependency and rural poverty. For this reason Lesotho was one of the first countries to sign the UN Convention to Combat Desertification.

> When we began constructing swales on the campus in 1994, a pick just bounced. For each swing, a lump of clay like a rock would dislodge, making the going slow and tough. Once the rains came, the task accelerated and allowed deepening and extension with less effort. Maturation of orchard inter-crop over several years has changed soil structure, and increased infiltration rates. Vigorous pioneer plants that grew more than 2 M (6 ft) high did not always impress visitors, who failed to appreciate the massive task their taproots were achieving. Besides breaking through plough pans and opening the soil, they provided essential organic matter for mulching and composting.

> Very careful landscape design is necessary to begin restoration because rainfall intensities up to 100mm (4 in) an hour occur during summer in Southern Africa, while infiltration rates are perhaps 1-4mm per hour on the unimproved expanses. In the uppermost reaches of a watershed or microshed,

Fig. 1 Left: A typical landscape only 1km from the BBCDC campus. Energy flows through the site are wild and produce virtually no yields. Heavy rainfall runs off and light rains evaporate. Fog capture or dew formation is inoperative. A brave attempt at soil conservation decades ago, by means of a stone retaining wall, has failed with the water cutting a new gully around it.

the 'stacking' of water and subsequent run-off during a high intensity rainfall event lasts only a few minutes. Capturing this precious water requires a Biblical sense of vigilance, urgency, expectation and preparation.[207 pp41-42]

4.2 Why terracing is well developed in some hilly areas but not in others where it would seem desirable

Very often tourists and others who have visited Nepal and have been to the principle tourist area Pokhara comment on the wonderful way in which the hillsides are terraced, giving the impression that such is the situation generally in Nepal. In fact I have even read articles stating that there is nothing to teach Nepal about soil conservation, I wish that was indeed the case. While there are parts of Nepal which have dry terraces most of the sloping cropped land is not terraced but still sloping though some have contour strips and still eroding. The reason is common to hillside cropping anywhere. The normal way to make a terrace is to drag the soil down till the terrace is formed; unfortunately there are two problems with the method.

Pokhara has unusually high rainfall and it is more widely spread throughout the year. The result it that there is plenty of water for paddy field construction, also, due to earlier light rains the cropland not terraced is covered in crops when the heavy monsoon comes, rather than a having a hard dry bare and dusty surface which quickly seals, as is most common.

With the normal way of making terraces. 1. The fertile topsoil is first moved and then buried by the much less fertile or sterile subsoil. Unless the good soil is unusually deep, much time and plenty of compost and fertiliser is needed before the terraces can grow reasonable crops.

2. Because the dragged down soil is resting on a slope it is unstable. When the heavy rains come the loose soil becomes soft and heavy and slides

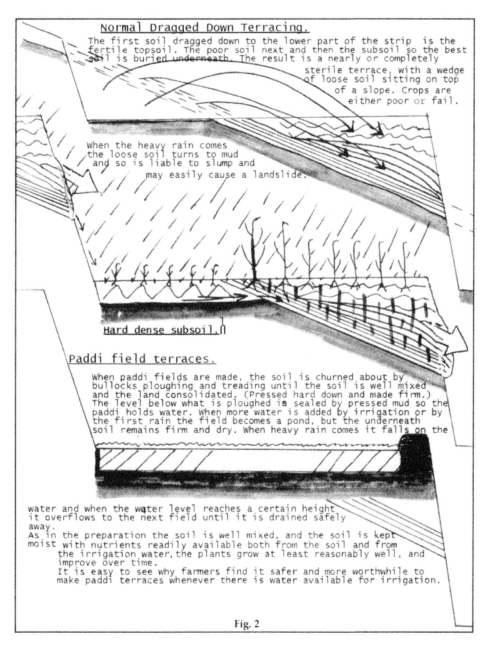

Normal Dragged Down Terracing.

The first soil dragged down to the lower part of the strip is the fertile topsoil. The poor soil next and then the subsoil so the best soil is buried underneath. The result is a nearly or completely sterile terrace, with a wedge of loose soil sitting on top of a slope. Crops are either poor or fail.

When the heavy rain comes the loose soil turns to mud and so is liable to slump and may easily cause a landslide.

Hard dense subsoil.

Paddi field terraces.

When paddi fields are made, the soil is churned about by bullocks ploughing and treading until the soil is well mixed and the land consolidated. (Pressed hard down and made firm.) The level below what is ploughed is sealed by pressed mud so the paddi holds water. When more water is added by irrigation or by the first rain the field becomes a pond, but the underneath soil remains firm and dry. When heavy rain comes it falls on the water and when the water level reaches a certain height it overflows to the next field until it is drained safely away.
As in the preparation the soil is well mixed, and the soil is kept moist with nutrients readily available both from the soil and from the irrigation water, the plants grow at least reasonably well, and improve over time.
It is easy to see why farmers find it safer and more worthwhile to make paddi terraces whenever there is water available for irrigation.

Fig. 2

down the slope; this is not the case with paddy fields. The Pokhara area is a high rainfall area and so the terraces are made for paddy fields. Other areas of terraces are usually where water is available for paddy fields. See Figure 2.

In Chapter 6 we see how terraces can be made where at least most of the topsoil is kept on the top, and the terraces are keyed to the hillsides so do not slip down. We also see how terracing can be done in a way which while stopping the loss of soil and plant foods in a very few years, produces better crops, and is economic and feasible for ordinary farmers to carry out.

4.3 Land classification in theory

Theoreticians say that the land above a certain percentage of slope should not be cultivated. However, if there is insufficient suitable land to feed the people and no alternative income to enable the people to buy food is possible, how can the people live? they have to cultivate the slopes. In such circumstances the land to be cultivated must be terraced; as we have seen in Chapter 2 contour strips are not a long-term solution on such slopeland, and in fact may increase erosion. Terracing may not seem feasible; as it normally takes too much work, the best soil is often buried, if not done properly it can result in landslides, it is slow and all the time the rest of the slopeland is being eroded. That is why we developed the Good News System of stage by stage terracing described in Chapter 6.7.

However, it is still important to have a practical way of deciding what is the appropriate way of cultivating and using land in a given situation.

Academics and engineers like to have tidy formulas for classifying land. These are only appropriate when there are constant predictable conditions; such conditions are rare on a range of slopeland situations.

4.4 Classification systems should be flexible enough to be varied easily to match local situations.

HOW TO ASSESS THE LIKELY DEGREE OF LAND DEGRADATION OF A PARTICULAR AREA AND CLASSIFYING THE LAND AS TO HOW IT SHOULD BE USED.

It is vitally important that classification systems should be flexible enough to be varied easily to match local situations; and simple enough to allow the assessment to be made and to be understood from the policy making level to the farm level. If classification cannot be used by average farmers its value is at best very limited, at worst completely useless; the systems I have seen are far too complex, and use special vocabulary.

This author considers that what is needed is a system like the one described below, which is not too technical for field workers to use, and can easily be adapted following experience in local situations.

4.5 The Good News Assessment Charts

For the system proposed we use a set of assessments each of which has seven grades.
    The assessments are based on:
    a. Soil texture.   e. Duration of periods of rainfall.
    b. Erosion resistance factors.   f. Length of slope of area.
    c. Type of rainfall.   g. Steepness of slope.
    d. Amount of rainfall.

We consider which is the most suitable grade for each factor and note down the number of that grade. As the amount of slope is always a crucial factor we include the grade for slope twice.

We then add up the grades and divide by 8 to find the average grade. That is taken as the classification for that land.

    For example we assess a piece of hillside as:

a. Soil texture	4
b. Erosion resistance factor	5
c. The type of rain	2
d. The amount of rain	3
e. The length of time of the heaviest rain falls	2
f. Length of slope.	3
g. Slope classification	_1
	21 divide by the number of grades which is

in this case 8 = 2.62 which is approximately between land class 2 and class 3 = 2-3

We then look at 2 and 3 in the classification guide which comes later in this chapter. We see that the land is suitable for cultivation with some conservation works and good soil management practices. These will include adding organic matter to the soil, and the use of suitable crop rotations; these may include a one year green manure fallow every seven years.

The conservation will include contour ploughing (ploughing as level as possible across the slope), strip cropping (level strips of crop land in between strips of grass or other crops running across the slope, where the ground is covered and not cultivated at the time) and diversion of excess surface water, using diversion ditches across the top of the fields to prevent water from other areas flowing onto the land.

If the fertile soil is deep enough, contour bunds or double check half terraces (see Chapters 5 and 6) should be made. If the fertile soil is not deep enough, double check half terraces should be made on the contour, as a first stage to the step by step good news terracing system. This should if possible be completed within 6 years.

Note To save calculating here is a table.

TABLE FOR DIVIDING THE POSSIBLE TOTALS BY 8.
7 divided by 8 = 0.87 count as class 1

	Divide		Count as Class		Divide		Count as Class
8	by 8 =	1	1	33	by 8 =	4.12	4
9	by 8 =	1.12	1	34	by 8 =	4.25	4
10	by 8 =	1.25	1	35	by 8 =	4.37	4-5
11	by 8 =	1.37	1-2	36	by 8 =	4.5	4-5
12	by 8 =	1.5	1-2	37	by 8 =	4.62	4-5
13	by 8 =	1.62	1-2	38	by 8 =	4.75	4-5
14	by 8 =	1.75	1-2	39	by 8 =	4.87	5
15	by 8 =	1.87	2	40	by 8 =	5.0	5
16	by 8 =	2.0	2	41	by 8 =	5.12	5
17	by 8 =	2.12	2	42	by 8 =	5.25	5
18	by 8 =	2.25	2	43	by 8 =	5.37	5-6
19	by 8 =	3.37	2-3	44	by 8 =	5.5	5-6
20	by 8 =	2.5	2-3	45	by 8 =	5.62	5-6
21	by 8 =	2.62	2-3	46	by 8 =	5.75	5-6
22	by 8 =	2.75	2-3	47	by 8 =	5.87	6
23	by 8 =	2.87	3	48	by 8 =	6.0	6
24	by 8 =	3.0	3	49	by 8 =	6.12	6
25	by 8 =	3.12	3	50	by 8 =	6.25	6
26	by 8 =	3.25	3	51	by 8 =	6.37	6-7
27	by 8 =	3.37	3-4	52	by 8 =	6.50	6-7
28	by 8 =	3.5	3-4	53	by 8 =	6.62	6-7
29	by 8 =	3.62	3-4	54	by 8 =	6.75	6-7
30	by 8 =	3.75	3-4	55	by 8 =	6.87	7
31	by 8 =	3.87	4	56	by 8 =	7.0	7
32	by 8 =	4,0	4				

Land which does not even fit class 7 should be left as jungle. If it has been cleared, it should be planted with wild plants and left undisturbed.

This guide is a practical basis to work from, it cannot be perfect for each situation, so can be modified to fit the local situation. For instance soils which let water through more easily will have less runoff, causing surface erosion, however what run-off there is will have more effect on such soils. Furthermore if the subsoil is weak the excess water building up in the subsoil could result in a slump or a landslide. There are variable factors and so local experience needs to be taken into account.

If, after following the suggested guidelines, we find that during the heaviest rain, run-off water is not clear, look at what is happening, then determine which of the assessments should be adjusted so that the method will be suitable.

In some conditions it may be that the 'Erosion resistance factor' may be as important as the 'Steepness of slope', and so should also be included (counted) twice when calculating the land classification.

In soil texture, clay soil is less erodible and so has a higher grade. However, as it does not soak up rainwater as easily and seals easily, there will be more run-off to lower down the slope; the chance of erosion there is increased if the rainfall is very heavy, or the uninterrupted slope is very long. Again, if this is the case, alteration can easily be made to suit local conditions.

It may appear that following observation in the heaviest rainfall, at the worst time for erosion to occur, the classification is wrong; if so it can be changed by one class up or down the scale.

It is important to realise that there will be variance between and even within larger fields; so having established locality guidelines, the farmers own natural wisdom and observation is used to make variations for his own land. To help in this sediment traps may be used, such as jars sunk into the bottom of the ditches; these can be inspected after rainstorms, and management or classification altered according to whether or not sediment is left in the jars.

A. How to identify the texture of a soil.
To decide if the soil is a clay soil, a silty or a sandy soil.

Take a handful of the soil to be tested. Slowly add water and mix it well until the ball of soil starts to stick to your hand

SOIL ASSESSMENT.

The following is a practical way of assessing the kind of soil, and a way of understanding what is meant by the terms used.

7.) When the soil is shaped by hand into a cake shape but then falls into a cone shape it is very sandy soil.
6.) If it slumps a bit but basically retains the cake shape it is silty loam.
5.) If the soil can be made into a roll but it cracks it is a sandy loam.
4.) If the roll does not crack it is a loam soil.
3.) If the roll can be bent into a curve but cracks it is a clay loam.
2.) If the roll can be bent into a curve without cracking it is a loamy clay.
1.) If the roll can be made into a ring without cracking it is a clay.

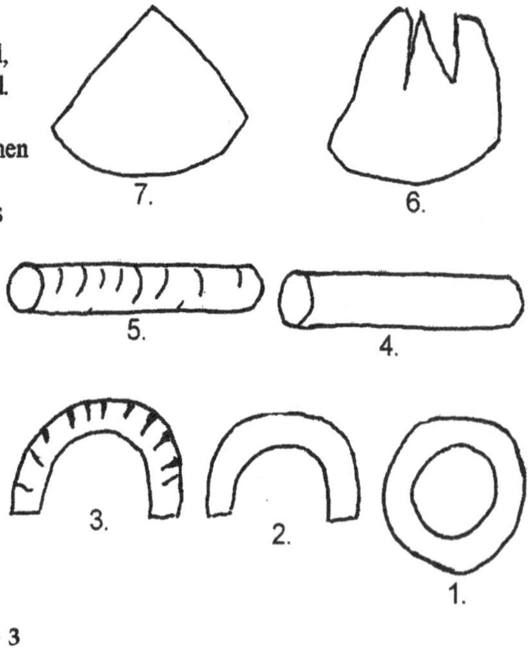

Figure 3

Soils 4,5, 6 & 7 normally let water soak through easily.
Soils 1,2 & 3 do not normally let water soak through easily.

B. Erosion resistance factor.

Note:- Trees do not always provide protection to the soil on their own, rather they may increase erosion. Anyone who has grown vegetables under large leaved trees in the monsoon season will know how the concentration of medium raindrops become many big heavy drops. These as they fall from the leaves strip and shred the vegetables under the trees. The same effect as the drops from the tree canopy drip-line hitting the soil, can cause damage which may not have occurred in the absence of the trees. The soil surface below the needles under a pine forest can become very hard and poorly absorbent. The amount of water run-off from a pine forest can be considerable.

1. Organic manure or other organic material mixed with the soil, full cover of the soil surface by crops or weeds and/or mulch, including stone mulch.

2. Organic manure or other organic material mixed with the soil, less than full cover of the soil surface by crops or weeds, and/or mulch, including stone mulch. With direct raindrop impact area over no more than one fourth of the soil surface.

3. Organic manure or other organic material mixed with the soil. Some cover of weeds, and/or newly planted crop, on or between ridges across the slope.

4. Rough absorbant surface with firm stable subsoil, newly planted crop planted on or between ridges across the slope, **or** - No cover, firm surface.

5. Newly planted crop sown up and down the slope.

6. No cover, loose surface but poorly absorbent subsoil.

7. No cover, loose surface and weak absorbent unstable subsoil.

C. The type of rain which occurs during the periods when the soil is most easily eroded by water. (The period when there is least soil cover, and the soil is loose.)

1. Mist.
2. Light rain without wind.
3. Light rain with wind.
4. Medium rain without wind.
5. Medium rain with wind.
6. Heavy rain without wind.
7. Heavy rain with wind.

D. The amount of rain falling in the wet season.

1. Less than 10 "(25 cm).
2. 10 - 20 "(25 - 51 cm).
3. 20 - 40 "(51 - 102 cm).
4. 40 - 60 "(102 cm - 152 cm).
5. 60 - 80 "(152 cm - 203 cm).
6. 80 - 100 "(203 cm -254 cm).
7. Over 100 " (254 cm).

E. The average length of time of the heaviest rainfall, during the periods when the soil is most easily eroded by water.

1. Less than 30 minutes.
2. 30 minutes - 1 hour and 30 minutes.
3. 1 hour and 30 minutes - 4 hours.
4. 4 - 8 hours
5. 8 - 12 hours.
6. 12 - 24 hours.
7. 24 - 72 hours.

F. Length of slope

1. 0-2 metres.
2. 2-4 "
3. 4-6 "
4. 6-8 "
5/ 8-10 metres.
6/ 10-12 "
7/ over 12 "

G. Slope classification.

Grade	Percentage slope	Classification
1	1-3	Slight slope
2	4-9	Gentle slope
3	10-15	Moderate slope
4	16-20	Medium slope
5	21-25	Strong slope
6	26-30	Steep slope
7	31-35	Very steep slope.

For measurement of slope angle see chapter B7.

Having used the above system to assess the class of the land, use the guide below to determine how best to manage it.

4.6 Land Classification Guide for Third World Slope Land

Class 1

Land suitable for cultivation without special practices. It must be workable or capable of becoming workable, level or nearly level, and not likely to have more than slight erosion. To keep and improve the soil structure and fertility, it usually needs suitable crop rotations and good cultivation practices.

Care should be taken to ensure that run-off water from nearby land or roads cannot cross the area; diversion ditches may be needed. This is so that the land has only to absorb its own rainfall, unless it is in an arid region, when extra water is encouraged. In that case, swales[1] should be constructed to spread the water evenly; or ditches should be made across the path of the water which may be filled with stones; the purpose of these being to run the water into the ground rather than across it's surface.

Class 2

Land suitable for cultivation with simple special practices and good management, including suitable rotations, and adding or returning organic matter to the soil.

The simple special practices will include contour ploughing and planting; contour bunds, or double check half terraces; strip cropping,[2] and simple water diversion ditches[3]; unless in an arid area where extra water should be harvested. In that case swales[1] should be constructed to spread the water evenly or ditches should be made across the path of the water, these may be filled with stones, the purpose being to run the water into the ground rather than across its surface.

Class 3

Land suitable for permanent cultivation only with the use of erosion control and soil management practices.

The management practices will include adding organic matter to the soil and suitable crop rotations, and may include a one year green manure fallow (see 7.9), every seven years.

Erosion control is very important.

Conservation will include contour ploughing, strip cropping, and diversion of excess surface water. Use diversion ditches[3] at the top of the fields to prevent water from other areas flowing onto the land. These ditches and water diversions should run into waterways which are planted or otherwise protected from erosion.

If the depth of soil which can grow crops is good, bunds or double check half terraces can be made on the contour. It is important though rarely realised, that the spacing of these barriers which result in the slow formation of terraces (lazy terracing), must be set according to the depth of the fairly good soil. This is so that when the terraces have developed and become level, there will be at least 4" (10 cm) of reasonably good soil on the upper portion of the strip, so that all the land can still grow crops. (The important consideration of correct spacing of contour works is explained in chapter 7.2 and 7.3).

If contour bunds or double check half terraces are not suitable, terraces must be made, (see Chapter 6).

When the above practices are not used, the land should not be cultivated but should be planted with permanent fodder crops, fodder, fruit or nut trees, with a ground cover of grasses and legumes.

Class 4

Land which is not really suitable for cultivation unless proper erosion control methods are used. It is normally unlikely that contour bunds or double check half terraces are suitable for this class, in which case full terraces must be made.

---

[1]Swale is a barrier of stone, soil or plant matter made across the slope with a slow slope to a ditch to carry excess water away. The idea is to divert and spread the water flowing down a slope so that it can soak into the soil. It is very useful where rainfall is not regular, when it does come is heavy, and normally quickly runs away with little penetration. It also reduces the erosion of the water pathways. For more on water use and conservation see Chapter 7.
[2]See chapter 7.2 and 7.3.
[3]See description at the end of Class 1.

If the good soil is very deep, bunds or double check half terraces can be made on the contour. It is important, though rarely realised, that the spacing of the these barriers which result in the slow formation of terraces (lazy terracing), must be set according to the depth of the fairly good soil. This is so that when the terraces have developed and become level, there will be at least 4" (10 cm) of reasonably good soil on the upper portion of the strip. So ensuring that all the land can still grow crops. (This is explained in Chapter 7.2 and 7.3).

In addition there should be diversion of excess surface water. Use diversion ditches at the top of the fields to prevent water from other areas flowing onto the land. These ditches and water diversions should run into waterways which are planted or otherwise protected from erosion, (see chapter 7 and appendix 5).

The management practices will include adding organic matter to the soil and the use of suitable crop rotations. These may include a one year green manure fallow every seven years. If this is carried out, increasingly good crop yields are possible on a sustainable basis.

If the above work and management is not carried out, the land should be planted with permanent fodder crops, or fodder, fruit, or nut trees, with a ground cover of grasses and legumes.

Class 5. a) see also 5. b) below.

This is land which is not suitable for cultivation because it is too steep or too rocky or stony or too easily eroded. It is best used for grazing, or for fruit tree, fodder tree including bamboo, or nut tree plantations. It is good if half moon shaped mini terraces are dug for each tree, to stop and absorb run-off. Interception ditches should also be cut to run water from the area between the trees towards the tree.

If due to shortage of food such land must be cultivated, terraces should be made. Conventional terracing will be impractical, but the Good News step by step terracing explained in Chapter 6 is practical for most situations. If the land is terraced, and these terraces are cared for, the soil will not erode. However as land is obviously very scarce or such land would not be farmed, it should have the fertility maintained and improved. This is done by adding organic matter to the soil and using suitable crop rotations, which may include a one year green manure fallow every seven years. If this is carried out increasingly good crop yields are possible on a sustainable basis.

In addition to the above, there should be diversion of excess surface water. Use diversion ditches at the top of the fields to prevent water from other areas flowing onto the land. These ditches and water diversions should run into waterways which are planted or otherwise protected from erosion (see appendix 5).

In very rocky situations it is best to collect the soil from on top of rocks and from small hollows in the rocks. This is thrown into the spaces between the rocks and fruit trees or bananas planted into these areas. Greenleaf Desmodium (Desmodium intortum) or similar creeping legumes should also be planted as a permanent cover crop. It will spread over the rocks so keeping the area cooler and producing valuable animal fodder or it can be composted and the compost added to the fruit trees.

Class 5. b) see also 5. a) above.

This is land which is not suitable for cultivation because it is both steep and unstable. The underneath soil is weak and loose such as when it is shaly, sandy or silty, instead of firm or rocky. The difference between it and 5/a) is that if we make terraces in the normal way, when heavy rains fall, the water accumulating in low or soft patches will soak into the soil underneath. The soil will then become heavy, wet and soft. If the rain continues, it will slide down.

We can use this land in two ways.

One is that if we can have a source of water we can make paddy terraces. These are tramped and puddled to consolidate them and make them watertight, so that water will not soak into the undersoil. **Of course, there is always the danger that if the sealing is broken water will get underneath, and so each season the puddling must be done thoroughly.** On this class of land paddy terraces would not be suitable where there are earthquakes or tremors or if there are giant worms or freshwater crabs which make holes allowing water to soak into the undersoil.

The other way is by making mini terraces. These are no more than one metre wide. Due to their narrow width and shallow penetration of the slope there is not the same danger of excessive water penetration deep into the hillside. Crops such as asparagus, pineapple, banana, papaya, peanut, sweet potato, mulberry bushes for fruit and juice, fodder for animals or for cultivation of silkworms. Passion fruit or grapes may also be suitable. As the soil is loose the terrace walls should be covered by plants such as Greenleaf Desmodium *(Desmodium intortum)*. It is a valuable leguminous livestock feed, a soil improver, and it can be cut and ploughed in as green manure. It also roots into the terrace walls, binding them and protecting them from rain, wind, and sun.

If these terraces are cared for, the soil will not erode but as the land is obviously very scarce or such land are not farmed, it should have the fertility maintained and improved. This is done by adding organic matter to the soil from crop residues, from plant material grown on the terrace walls, or from the manure produced after the crop residues and fodder has been fed to livestock. Suitable crop rotations and intercropping should also be used. The rotation may include a one year green manure fallow every seven years or less, if this is carried out increasingly good crop yields are possible on a sustainable basis. Appropriate agro-forestry planting may be carried out with legume trees, most kept to shrub size, and grown on some of the terraces or, as in the Good News terracing on the terrace walls; their leaves for shade and manure, and their roots to help stabilise the soil. See chapter B2.

In addition to the above, there should be diversion of excess surface water. Use diversion ditches at the top of the fields to prevent water from other areas flowing onto the land. These ditches and water diversions should run into waterways which are planted or otherwise protected from erosion.

If mini terraces are not made, the land should have permanent crops, for instance grass and legume cover, though it should not be grazed in the wet season. Alternatively it should be planted with fruit, nut, fodder trees with grasses and legumes planted as a cover crop; these could be cut and fed to livestock in exchange for manure, or cut and used to make compost for manuring the trees. The trees should be kept from growing too big and heavy on such land.

The other alternative is bamboo forest. As we shall see in chapter B2 bamboo leaves can also be used as animal fodder.

Bamboo has amazing ability to hold land from erosion by water and to hold it from sliding.

This class of land should not be planted with trees which grow large and heavy, or landslides could result when the ground is wet and soft. On the other hand bamboo forest will hold the soil.

If conventional forest is planted the trees should be of mixed age and size and they should be thinned and replanted, not left to grow to be a great weight on an unstable hillside. Some people have pointed to forests grown on such land as an example of how planting forests can cause landslides. It is really an example of how the foresters planted and managed the forest in the wrong way. There are unstable slopes covered in jungle which do not have serious landslide problems. The trees are mixed, including some large trees which have very deep roots, and others which bind and protect the surface.

Class 6a)

This class of land is basically the same as class 5/a) but will normally be steeper or in other ways more susceptible to erosion. Terracing is even less likely to be economical and will require more skill than for class 5/. Special note should be taken of the way the walls should be made, (see chapter 6 for detailed description of how terrace walls should be made).

This is land which is not suitable for cultivation because it is too steep or too rocky or stony or too easily eroded. It is best used for grazing, fruit tree, fodder tree including bamboo, or nut tree plantations. It is good if half moon shaped mini terraces are dug for each tree, to stop and absorb run-off. Interception ditches should also be cut between the trees to check surface run-off and run the water towards the trees.

If due to desperate shortage of food such land must be cultivated, terraces should be made. Conventional terracing will be impractical, but the Good News step by step terracing explained in chapter 6 is practical for most situations. If the land is terraced carefully, and the terraces are cared for, the soil will not erode. However as land is obviously very scarce or such land would not be farmed, it should have the fertility maintained and improved. This is done by adding organic matter to the soil from crop residues, from plant material grown on the terrace walls, cover crops, or from the manure produced after the crop residues and fodder has been fed to livestock. Suitable crop rotations and intercropping should also be used, which may include a one year green manure fallow (the land not cultivated after sowing the green manure), every seven or less years. If this is carried out increasingly good crop yields are possible on a sustainable basis. For more on soil management and improvement see Volume 2.

In addition to the above, there should be diversion of excess surface water. Use diversion ditches at the top of the fields to prevent water from other areas flowing onto the land. These ditches and water diversions should run into waterways which are planted or otherwise protected from erosion (see appendix 5).

Where the soil is very stony, fruit and nut trees such as mulberry and guava, custard apple, avocado and walnut and chestnut may be grown.

In very rocky situations it is best to collect the soil from on top of rocks and from small hollows in the rocks. This is thrown into the spaces between the rocks and fruit trees or bananas planted into these areas. Greenleaf Desmodium (*Desmodium intortum*) or similar creeping legumes should also be planted as a cover crop. It will spread over the rocks, so keeping the area cooler and producing valuable animal fodder and so manure, or it can be composted, the compost or manure used as manure for the fruit trees.

Alternatively it may be better to plant bamboo which seems to like rocky areas, as we shall see in chapter 8. Bamboo leaves can also be used for good animal fodder.

Bamboo has amazing ability to hold land from sliding and protect it from erosion by water, and/or from wind erosion

If the land is used for grazing stocking rates should be kept low, and grazing should not be allowed in the rainy season as the animals hooves compact the soil, cut up the surface, and they slowly push the soil down the slope.

Burning off should not be allowed.

This land is often stony; stony land is often thought of as useless land, yet it can be quite productive. If the need for food is so great that the land is terraced, the terraces may grow surprisingly good crops if manure or compost (which can be made in the area) is worked into the soil. This is because if half the soil is stones all the manure is concentrated in the half which is not stones, so this part is twice as rich in manure as normal soil would be. The plant roots go between the stones feeding on the rich plant foods, also stony soil is normally looser soil and contains air, so the roots spread easier and further. The stones on the surface protect the soil and act as a mulch keeping the soil moist, they also shelter and encourage the soil micro organisms.

These same factors apply when the land is planted to trees. Nut trees such as walnut and chestnut thrive in stony soil; guava and mulberry also do surprisingly well. Bamboos do very well in rocky areas; we have found that the best growth has been when the bamboo roots were climbing over and round rocks.

This land classification may of course also be used for general forestry; however if it is, clear felling should not be allowed, but only thinning or strip felling on the contour. Coppicing species should be preferred. Bamboo forest is usually preferable for many reasons as described in chapter B2.

Class 6b)

Class 6b) land is basically the same as class 5b) steep and unstable, but will normally be steeper or in other ways more susceptible to erosion. Terracing is even less likely to be economical and will require more skill than for class 6a). It is very unlikely that there will be water for paddy terraces, and, as it will be more difficult to seal them from leaking and then collapsing; if the land must be cultivated mini terraces would be the most appropriate (see below). Special note should be taken of the way the walls should be made, see 6.14.

This is land which is not suitable for cultivation because it is both steep and unstable. The underneath soil is weak and loose, such as with shale or sand or silt, instead of firm or rocky. The difference between it and 6a) is that if we make terraces in the normal way, when heavy rains fall the water accumulating in low patches will soak into the soil underneath and it will become heavy, wet and soft. If the rain continues it will all slide down.

So though cropping is not advised, if the land is to be cropped, mini terraces must be made. These are no more than one metre wide. Due to their narrow width and shallow penetration of the slope there is not the same danger of excessive water penetration deep into the hillside. Crops such as asparagus, pineapple, bananas, peanuts, sweet potatoes, mulberry bushes for fodder for animals or for silk worms are grown. Passion fruit and grape vines may also be suitable. As the soil is loose the terrace walls should be covered by plants such as Greenleaf Desmodium (*Desmodium intortum*). It is a valuable leguminous livestock feed, a soil improver, and it can be cut and ploughed in. It also roots into the terrace walls, binding them and protecting them from rain, wind, and sun.

If terraced and the terraces are cared for, the soil will not erode but as the cropland is obviously very scarce or such land would not be farmed, it should have the fertility maintained and improved. This is done by adding organic matter to the soil from crop residues, from plant material grown on the terrace walls, or from the manure produced after the crop residues and fodder has been fed to livestock. Suitable crop rotations and intercropping should also be used. The rotation should include a one year green manure fallow every seven or less years. If this is carried out, increasingly good crop yields are possible on a sustainable basis.

In addition to the above, there should be diversion of excess surface water. Use diversion ditches at the top of the fields to prevent water from other areas flowing onto the land. These ditches and water diversions should run into waterways which are planted or otherwise protected from erosion.

If it is used for grazing stocking rates should be kept low, and grazing should not be allowed in the rainy season as the animal hooves cut up the surface, and push the soil down the slope. Burning off should not be allowed.

This land is often stony, this is often thought of as useless land yet it can be productive. If the need for food is so great that the land is mini terraced the terraces may grow surprisingly good crops if manure is worked into the soil. This is because if half the soil is stones all the manure is concentrated in the half which is not stones, so this part is twice as rich in manure as normal soil would be. The plant roots go between the stones feeding on the rich plant foods, also, stony soil is normally looser soil and so the roots spread easily and deep down. The stones on the surface protect the soil and act as a mulch keeping the soil moist, they also shelter and encourage the soil microorganisms.

These same factors apply when the land is planted to trees. Nut trees such as walnut and chestnut thrive in stony soil, Guava and mulberry also do surprisingly well. Bamboos do very well in rocky areas, we have found that the best growth has been when the bamboo roots were climbing over and round rocks.

If planted with fruit, nut, or fodder trees, grasses and legumes should be planted as a cover crop. This should be fed to livestock in exchange for manure for the trees, or cut and used to make compost for manuring the trees. The trees should be kept from growing too big and heavy on such land.

It may also be used for general forestry, however if it is, clear felling should not be allowed, but only thinning or strip felling on the contour. Coppicing species should be preferred. Bamboo forest is usually preferable for many reasons as described in chapter B2.

Class 7

This land should definitely NOT be cultivated. It is too steep for terracing.

If not too steep it may still be used for fruit fodder or nut trees. In Taiwan slopes up to 40% are planted with plum and peach trees, with thick ground cover of wild plants and grasses under and between the trees. In winter this is cut and allowed to rot down where it grows, and there does not seem to be erosion. It seems a sustainable practice.

Citrus can also be grown very successfully but then we dig out the ground around the trees to make terraces; to give the trees space to bush out all round and to conserve the water and nutrients. While the trees are developing the terraces are cropped with adzuki beans, peanuts and soybeans. It is good to dig water interception ditches to stop run-off from building up, and to bring surface water to the trees. Setaria grass and Greenleaf desmodium, stylo or glycine can be planted for ground cover. This is cut and placed in heaps to rot. In spring it is dug in round the dripline of the trees. Half sacks of fruit are transported down the slope on hooks attached to small pulley wheels running on a fencing wire cable.

In addition to the above, there should be diversion of excess surface water. Use diversion ditches at the top of the fields to prevent water from other areas flowing onto the land, these ditches and water diversions should run into waterways which are planted or otherwise protected from erosion. See appendix 5.

If the above management is not used the land should be forest; Bamboo forest protects holds and improves the soil. Stem or culm cuttings can start to be harvested in 4 years from planting, many species can grow on extremely steep slopes. See chapter B2, and this author's book *Bamboo: A Valuable Multi purpose Crop*.

Very often land which is steep, very erodible, or very poor would be used for forest planting, but due to it not being sufficiently accessible by tree harvesting equipment and timber trucks is not so used. Such land can be often be used for bamboo forest. Bamboo species cover a very wide range of climatic and soil conditions. Bamboo is very versatile in its uses. It does not require expensive equipment; it can be cut with a heavy sickle, khukri, jungle knife or axe. It can be harvested and carried, slid, dragged or thrown to places where small hand bogeys can transport it to jeep trails. Then jeeps with transport frames can transport it to mass transit points for trucking, or to rivers for rafting to the market. Paper or bamboo plywood or chipboard factories could be sited at the side of rivers to convert the bamboo into valuable commodities.

I have been generally unimpressed by coniferous forests. In Nepal the run-off from pine forests 17 years old was surprising in both quantity and in the amount of soil still carried in the water. If forests are planted on steep slopes species which can be coppiced would seem preferable.

In very rocky situations it is best to collect the soil from on top of rocks and from small hollows in the Rocks. This is thrown into the spaces between the rocks and fruit trees or papaya or bananas planted into these areas. As the rainfall on the rocks will also flow onto these plots melons or pumpkins may also be grown and can spread over the rocks. Greenleaf Desmodium *(Desmodium intortum)* or similar creeping legumes should also be planted as a cover crop. It will spread over the rocks so keeping the area cooler and producing valuable animal fodder or it can be composted and the compost added to the fruit trees.

It is important that class 7/ land is used as suggested, rather than the common opinion that it is worth little, and consequently is burnt off then sown with maize or similar crops. These yield a crop for one or two seasons after which the land is worthless for many years, eventually it will be completely useless. Much of the ash from burning off is washed away in the rains, what is not, tends to seal the surface of the soil, so causing more run-off, run off from such land can cause problems to other areas. Such treatment is unsustainable whereas the suggestions recommended are sustainable.

4.7 The approaches to use

In the past and even in some places still today, the approach was/is that soil conservation was to be achieved by various soil conservation works. The land was surveyed, bulldozers and graders were used, or gangs of labourers were taken to hillsides and told to build contour lines or contour ditches. The programs were often large and required much outside investment. The problem was regarded as basically a physical one.

When working on smaller scale, methods once designed and perfected in a research area were then taken by extension workers and applied with the assistance of (usually outside) funds. The land users were expected to be grateful in the future if not at that time.

Unfortunately in most cases such programs have given very disappointing results.

Soil degradation like other facets of rural development should not be carried out without involving at least some of the other aspects. To be successful and therefore sustainable rural development programs need to be multi-disciplinary. That does not mean that a soil conservation program has to contain health and agricultural work, but if it does not it should work where there are other organisations involved in the other aspects. Perhaps it needs to encourage such organisations.

Meeting one or two aspects of rural development such as soil and water conservation while vitally important is seldom sufficient for real lasting development.

> It has since been shown that this approach usually does not work for two reasons. Firstly, land degradation is usually only a symptom of other problems – the physical manifestation of land being badly managed and put to the wrong use because of economic, social, political, legal or other pressures. Unless these real causes of degradation are tackled, there is little hope of success. Secondly, the solutions offered were often unattractive to the landusers as they usually involved them in additional work but did little to solve their immediate problems of improving yields, increasing their incomes or lessening the risks of farming.
>
> In the 1980s approaches to soil conservation began to change in many countries. It became widely acknowledged that, ultimately, the way a country's land is managed and used depends upon the perceptions and actions of its many thousands of individual landusers. These people have the ability [in many but not all cases PJS] to bring about fundamental changes in land use for the better. It was seen that the challenge was to create the conditions that would provide motivation for this to happen.[177.]

From this came the late realisation, not only in soil conservation but in all rural development work, that it was important to involve the rural people and it became fashionable to develop many approaches which were to do this. Participatory Rural Appraisal and Planning, Participatory Technology Development etc

New buzz words sprouted so that glossaries were needed to help understand the jargons. This author is not against the idea of involving and consulting the landusers, that should have been obvious for all rural development work. In some of the more down to earth agricultural projects participation of the people concerned had alway been regarded as basic. I am also wary of the use of fashionable approaches in order to receive funding where in fact only lip service is paid to real participation, sadly such approaches occur too often.

> Promising as this participatory approach may be, participation by itself does not appear to be enough to overcome all the problems of land degradation. Given the opportunity, land users may well be able to identify the underlying problems and work out possible solutions. However, the solutions may not be within their reach without financial and other forms of help. Also, the solutions which are developed may not be sufficiently attractive for them to adopt for strong social, institutional or economic reasons. For one reason or another, it seems that some form of incentive is still generally required, besides just participation, before landusers are prepared to make the necessary changes in land use and management.[173.]

While I agree with the above, the incentives may not always be from outside. If methods such as the Good News Step by Step Method of terracing can be used as taught in this book, which stops erosion fairly quickly and results in much better crops on the terraces, the farmers have incentives of their own. They are then more likely to continue the work each year until all suitable land is terraced and their livelihoods are much improved. Other farmers observing such work will have incentives to take up the work themselves.

However there are often other factors involved such as the landusers having no rights to the land. They may fear that after they have improved the land it will be taken from them, or the rents increased to a level which removes too much of the benefit resulting from their work. An important incentive can be for tenants to be given greater security, so that they are assured that they will really benefit from improvement they make.

Another fashion at this time is that:

> all that is needed is for governments or development agencies is to provide the facilities for farmers to work together and they will then be able to develop their own solutions and implement the required measures largely by themselves. I think it is partly because of this approach that so little research is currently being done, in spite of the fact that land degradation is now recognized as a growing problem in many parts of the world.

> While farmers do much innovative work on their own and are constantly coming up with new solutions, there are limits to what they can achieve without outside help. I believe that there is now a pressing need to support them with more and better scientific [this author would say 'practical' PJS] research into the processes of land degradation and its control.

> Also, if extension workers are to win the respect of the farming communities in which they live and work, they must be given the necessary technical training and be equipped with the skills and equipment that field work needs, to be able to operate effectively. Unfortunately, some now see the role of soil conservation extension workers only as "facilitators" whose job it is to create the conditions for the landusers to be able to do everything themselves.[178.]

Natural soil development and conservation

Soil has not developed simply by the breakdown of rock into small enough particles to create a sandy, silty or clayey layer with some soluble salts. It has developed because micro organisms, (very tiny plants and creatures) and macro soil organisms such as earthworms and termites, have mixed and digested soil particles turning them into plant foods. Then simple plants started to grow, and as their remains further improved the soil other plants grew. The plants gradually covered the soil and protected it from the sunlight, from drying out quickly, and from harmful erosion. More and more plant remains were mixed into the soil and the soil became more and more fertile. The plant remains rotted down into humus, and plant and animal remains, with the humus improved the soil structure, binding the soil together and increasing its ability to both absorb and retain rainwater, while allowing excess water to drain through to renew the aquifers (water reserves in the ground). This process takes a very long time, about 600-700 years to develop 1 inch (2.5cms), and so we should conserve, not waste the soil.

Under jungle (natural) conditions, sustainable biomass (the amount of plant material) is produced and slowly increases. There is far more biomass produced in a jungle even on poor land than is the case when the land is cropped commercially, and it is produced in a sustainable and improving way.

The trees and other plants protect the soil from UV light (the burning sunlight) which destroys the organic matter in the soil surface, from heat, and from wind and water erosion. The soil is cooler and moister and this favours the soil structure and the life within the soil. The tree roots go deep into the subsoil and enter cracks in the underlying rock and so hold the soil on the slopes. There is a natural mulch of decaying plant matter on the soil surface. The plant canopy (covering) over the area results in a cooler micro climate (the area local to the soil surface) which attracts mists, dews and rainfall. When land is bare it is much less able to attract such moisture.

We need to produce a field environment as close as possible to natural conditions while enabling sustained or improving economic crop production. I believe that on slopelands we can do that by terracing slopes in a way which is desirable, do-able and durable; which keys the terraces to the slope, keeps the living topsoil on top, and breaks up the soil pans. A system which incorporates some tree growth, so that as the tree roots penetrate deeply they increase the stability, and bring up minerals from the subsoil, their leaves adding the minerals as well as organic matter to the surface soil. The trees should not compete with the crops.

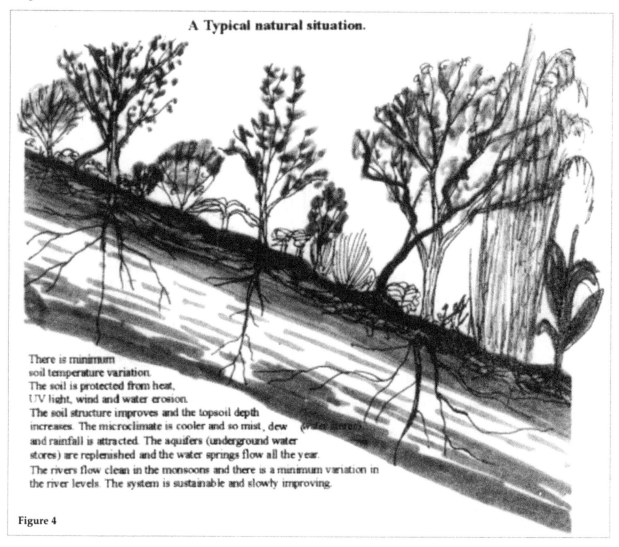

A Typical natural situation.

There is minimum
soil temperature variation.
The soil is protected from heat,
UV light, wind and water erosion.
The soil structure improves and the topsoil depth
increases. The microclimate is cooler and so mist, dew
and rainfall is attracted. The aquifers (underground water
stores) are replenished and the water springs flow all the year.
The rivers flow clean in the monsoons and there is a minimum variation in
the river levels. The system is sustainable and slowly improving.

Figure 4

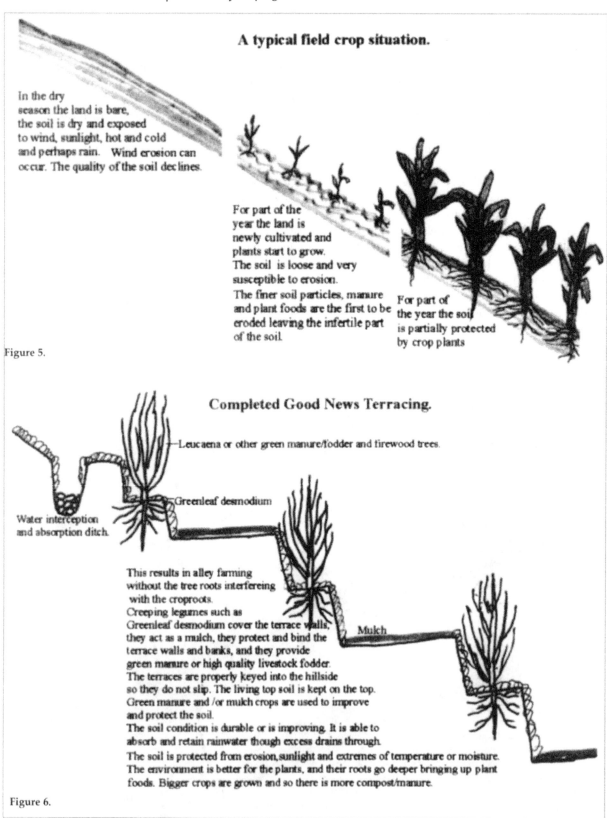

A typical field crop situation.

In the dry season the land is bare, the soil is dry and exposed to wind, sunlight, hot and cold and perhaps rain. Wind erosion can occur. The quality of the soil declines.

For part of the year the land is newly cultivated and plants start to grow. The soil is loose and very susceptible to erosion.
The finer soil particles, manure and plant foods are the first to be eroded leaving the infertile part of the soil.

For part of the year the soil is partially protected by crop plants

Figure 5.

Completed Good News Terracing.

Leucaena or other green manure/fodder and firewood trees.

Greenleaf desmodium

Water interception and absorption ditch.

This results in alley farming without the tree roots interfereing with the croproots.
Creeping legumes such as Greenleaf desmodium cover the terrace walls, they act as a mulch, they protect and bind the terrace walls and banks, and they provide green manure or high quality livestock fodder.
The terraces are properly keyed into the hillside so they do not slip. The living top soil is kept on the top.
Green manure and /or mulch crops are used to improve and protect the soil.
The soil condition is durable or is improving. It is able to absorb and retain rainwater though excess drains through.
The soil is protected from erosion, sunlight and extremes of temperature or moisture.
The environment is better for the plants, and their roots go deeper bringing up plant foods. Bigger crops are grown and so there is more compost/manure.

Mulch

Figure 6.

In addition using green manuring, inter-cropping and mulching, as explained in volume 2, to keep the soil protected and increase the organic matter. All these factors encouraging the soil micro organisms which improve the soil texture and release plant foods.

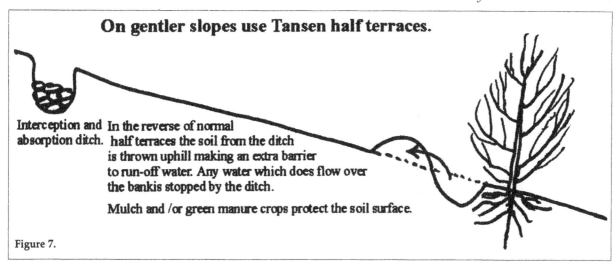

On gentler slopes use Tansen half terraces.

Interception and absorption ditch.

In the reverse of normal half terraces the soil from the ditch is thrown uphill making an extra barrier to run-off water. Any water which does flow over the bank is stopped by the ditch.

Mulch and /or green manure crops protect the soil surface.

Figure 7.

We can develop ways to conserve land and reduce water runoff using a range of methods explained in the following chapters. We should also use the soil improvement methods explained in volume 2. If we do we will be well on the way to a more hopeful future.

CHAPTER 5

SOIL AND FERTILITY CONSERVATION: SECTION ONE:- NON-TERRACING CONSERVATION

5.0 Introduction

My first experience of soil erosion in warm countries was in 1963. I arrived in Taiwan three days before Typhoon Gloria, one of the worst typhoons to hit North Taiwan. Several weeks later a friend took me with him on a trip to the tribal area he was working in. He had not been able to get to the area earlier as the roads were blocked by landslides or by the road falling down into the gorge. As we crested a hill and looked down he was horrified at what he saw. He said, "I expected to see a mess but this is horrific, I will show you a photograph of what was the view from here before. There was a valley with rice terraces sloping down to the small river near the other side. In the middle was a village with two churches and a school. Now it and the fields are completely buried." In that case it was due to the clear felling of trees by a timber company 3 years before. As the tree roots had died and so were no longer holding the soil and the soil was bare, soil, roots, stones and rocks had been washed down, increasing the volume of storm water. As it swept through it was slowed by the narrowing of the valley so dumping roots, rocks, stones and sand; most of the soil was carried away in the water. However, although in this case bad forestry was the cause, most of the erosion I have seen since then has been due to the inappropriate use of land for agriculture.

When I started work teaching and extension work at and from

Fig. 1 Most of this large village in Taiwan is buried beneath debris washed down during Typhoon Gloria. The result of the clear felling of upstream forest three years previously.

the Yu Shan Aborigines Agricultural Training Centre on the East coast region of Taiwan I saw more erosion. Sadly some of it on the hillside farm land of the training centre. After heavy rain it was impossible to enter the fields as the gate would be blocked by mud from the small area of cultivated land. The rest of the land was covered in rocks and the giant stoloniferous (creeping rooted) grass *Imperata cylindrica*. A year later I was asked to take on the job of both farm manager and in charge of students practical training in addition to the teaching I was doing. I started to make the farm a demonstration of what it should be.

We cleared the land and terraced some of it while carrying out trials of different land management practices. We had a board showing the different methods, the number of hours labour used on each, and the crop yields, the results showed clearly that it paid to terrace the land. Successive years of students terraced more land both as part of their training and as part payment for their tuition board and lodging.

Apart from our own students, on our open days other people came to see what we were doing, and we heard that both graduate students and other farmers were starting to make terraces on their own land, to the surprise of government extension workers who had been trying to get farmers to build terraces . A high level group of government officials and members of the FAO visited our centre and were very impressed. When the farm was fully developed we experienced a very serious typhoon which wrecked most of our buildings and caused serious flooding so that the other land we had by a lakeside was completely covered by the lake. However, amazing to me and all who saw it, there was very little run off from the terraced land. Only two small terraces collapsed and as they were built by two students who were really fishermen from a small island off Taiwan no one was surprised. The terraces merely fell onto the terraces below them and were later rebuilt correctly.

I later worked in the mountains, on the west of Taiwan, and witnessed more serious erosion. There among other work I encouraged the growing of fruit tree orchards, instead of cereal crops on eroding hillsides, with grass and fodder legumes covering the soil. In a few years' time, as the trees began to bear fruit the income and livelihood of the farmers who took up the advice and teaching changed dramatically.

Moving back to East Taiwan to another area, I and my staff developed a piece of worn out hillside as a trial and demonstration area. (For full details of the way the soil was transformed see Vol 2 4.23.) By that time labour in Taiwan had become much more expensive and so I had to think of a practical way of terracing. I found that there was a walking tractor in almost every village, so I designed a bulldozer conversion to fit on any of the walking tractors. At first we had problems, but eventually it proved very effective. My idea was that farmers or villages could hire the conversion kit to fit to their own tractors and use them as bulldozers to make terraces, paddy fields, roads to their villages, etc. The bulldozer is shown in the following chapter. I include drawings of the design as Appendix 4.

In (1980) we moved to Nepal, and I noticed that while physically Nepal is amazingly similar to Taiwan there were differences. A good point for Nepal was that very few pesticides were needed at that time compared with the very heavy usage in Taiwan. However, on the negative side there were too many landslides and gullies, caused by farming land which was too poor and too steep, land which in Taiwan would be bamboo forest.

Bamboo was grown in Nepal, but normally only as clumps for local use, not used as in Taiwan as a useful crop on otherwise useless land, and as a land stabiliser. I did not know how to propagate and grow bamboo, but as a sideline to my other work I set out to learn how by carrying out simple trials.

I was also concerned to see the rate of erosion on farm fields up to an average of 2" (5cm) per year.

Views often are expressed, that Nepal's agriculture is sustainable, with wonderfully built terraces covering the slopes. While this appears to be the case in the main tourist area, it is not as common in the dry hills which comprise most of the hill farming parts of Nepal. The terraces are normally only built where there is water for growing rice in paddy fields. Most of the terraces in Nepal are for rice terraces, the reason is explained in 4.2.

In the drier hill areas, while certain ethnic groups have realized the importance of terracing, it is more common for the land to be farmed simply as sloping fields or irregular contour strips. The irregular contour strips result in a staggered effect, with land

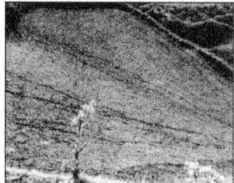

Fig. 2 This author has seen an average of 2 inches (5cm) of soil lost in one bad monsoon season. Should this continue?

sloping sideways as well as forward. This may in fact be better than continuous level strips, which can result in concentrating the runoff water which breaks through and flows strongly down to the next contour breaking through it also, and so gullies can develop.

As soon as I had some land to use for various trials I started comparing different ways of managing sloping land.

Farmers took great interest in our work and some started saying that we were transforming the worst possible land and asking me if I would show them on their land. Unfortunately because of the agreement with the Nepal government, for most of the ten years I was not allowed to work outside the trial and demonstration land. Other people did come to see and learn from what we were finding, and I did go to other areas to give some teaching.

During this time I did much observing of what had happened and what was happening to the soil.

I discovered that in the past people would move to an area, clear the best land and crop it. When the fertility began to decline they would clear more land, crop it and rest the first cropped land. This went on until the fertility of the stage two land began to decline, when other land would be cleared, and the land which had been rested again cropped. Eventually when the general fertility of the land in the area had been reduced to an unacceptable level, the people would look for a new area to develop, and eventually the community would move there.

Now there is nowhere to move to, people in the area where I worked considered that the future would be very grim unless they could find work in the towns and cities. I felt it important to look at how the situation could at least be stabilised.

Over the next ten years in addition to my main work I first observed the effect on agricultural land of traditional land management and the government soil conservation projects. I then started trials to compare the slope farming common in the area, with contour strips, half terraces and full terraces. As in Taiwan the trials clearly showed that terraces were by far the best method, but for them to be properly made took too much labour and time, so that it would take the average farmer about 20 years to terrace his land. During that time the land that had not yet been terraced would lose a lot of its soil and its fertility. As a result of this problem I developed the Good News step by step terracing system.

As well as my own observations and experience other observations have shown that terracing is the only really sustainable way of managing soil on land prone to erosion. To repeat an earlier quotation:

> Research in Nigeria has shown how much more serious erosion can be on sloping land that is unprotected by terraces. Cassava planted on land of a 1% slope lost an average of 3 metric tons per hectare each year, comfortably below *the rate of soil loss tolerance* [the rate at which new soil is formed. PJS].
>
> On a 5% slope, however, land planted to cassava eroded at a rate of 87 tons per hectare annually- a rate at which a topsoil layer of six inches [15 cm] would disappear entirely within a generation. Cassava planted on a 15% slope led to an annual erosion rate of 221 tons per hectare, which would remove all topsoil within a decade.

Fig. 3 When the monsoon rains come the water rushes over this land instead of soaking in for growing crops and replenishing the underground water. It should have bamboo cuttings planted across the gullies and Good News step by step terracing started on the other land. This could be a productive area!

Inter-cropping cassava and corn reduced soil losses somewhat, but the relationship of soil loss and slope remained the same.[147]

THIS CHAPTER LOOKS AT THE VARIOUS NON-TERRACING METHODS OF CONSERVING SOIL.

5.1 Practical soil conservation

As pointed out in chapter 2, there are many wrong ideas and hence much wrong practice in soil and water conservation work.

Despite decades of effort, soil and water conservation programmes have had surprisingly little success in preventing erosion. The quantitative achievements of some programmes can appear impressive. In Lesotho, all the uplands were said to be protected by buffer stripping

Fig. 4. This is the process of lazy terracing. In the heavy rains the soil moves down, some is stopped by the contour strips or other barriers, the rest particularly the dissolved plant nutrients washes on down. Soil loss is not stopped, and the top part of the strips becomes bare and useless.

by 1960; in Malawi (former Nyasaland), 118,000 km of bunds were constructed on 415,000 ha between 1945 and 1960; and in Zambia (former Northern Rhodesia), half the native land in eastern province was said to be protected by contour strips by 1950 (Stocking, 1985). In Ethiopia, during the later 1970s and 1980s, some 200,000 km of terracing were constructed an 45 million trees planted (Mitchell, 1987).

Ironically, though, many programmes have actually increased the amount of soil eroding from farms. This is because these impressive achievements have mostly been shortlived. Because of a lack of consultation and participation, local people, whose land is being rehabilitated, find themselves participating for no other reason than to receive food or cash.

Seldom are the structures maintained, so conservation works rapidly deteriorate, accelerating erosion instead of reducing it. If performance is measured over long periods, the results have been extraordinarily poor for the amount of effort and money expended (Saxson et al., 1989; Hudson, 1991; Reij, 1991). [123.]

I believe that it is not just a problem of lack of maintenance, the main problem is unsuitable design; and due to not first proving that the design is sustainable, good for the land and for the area, so that the farmers are pleased to participate in work or do it themselves.

A major problem is that in many cases the soil conservation lacks credibility, so farmers are not really interested in it. The conservation work is something that outsiders are interested in, rather than the farmers being convinced that it is worthwhile for them. Distrust of outsiders also puts a question mark in local peoples' minds, and the outside organisation gets most of the credit for what is achieved. The aim is to complete as many kilometres of conservation work (normally bunds or strips) as possible; *the quick fix approach*, and the program managers are praised for the amount of area covered.

Unfortunately as we will see, quick fix methods such as Vetiver or Napier grass contour strips are still being advocated; such methods lead to fertility drift downwards and can in fact accelerate erosion.

5.1a Contour strips and contour bunds are commonly advocated for soil conservation.

As pointed out in chapter 2.3 my experience has caused me to strongly disagree with their use on hillsides. They may be appropriate on relatively gentle slopes where the soil is deep or, very close together for making mini terraces. I shall explain why later in this chapter. People who advocate them feel that it is a simpler way of terracing, I call it lazy terracing. The idea is that in the heavy rains the water which is not absorbed into the soil moves down the slope carrying soil with it. The when it reaches the contour strips the soil is filtered out and the water moves on. In time as the soil is deposited the contour builds up and eventually terraces are formed. **There is ample evidence that these lazy terracing systems are worse than having nothing at**

all, unless the strips are very narrow, or the topsoil is very deep, the rainfall and/or slope never excessive, and the ground is stable.

Studies in Honduras showed that over a three year trial period vetiver grass contour strips did not reduce overall erosion.

> There were no significant differences between the sediment yield records collected between test plots defended by vetiver live barriers and those with no soil conservation treatment. Certainly, substantial amounts of sediment became trapped on the upslope side of the live barriers but much of this was compensated by accelerated soil loss on the downslope side. This compensation was suficient to ensure that, if the live barriers made any difference to overall soil loss from the plots, it was not detectable within the time frame of this study.---.
>
> Over a three year period the average yield per harvest was 1895 kg ha^{-1} on the control plots and 1876 kg ha^{-1} on the live barrier plots *(Hellin, 1999a)*. Quoted in Land Degradation and Development. **13:** 233-250 (2002) Published on line 26 April 2002 in Wiley Interscience (www.interscience.wiley.co). DOI: 10.1002/Idr.501. Ref. 209.

'Jon Hellin' is also the author of Soil and Water Conservation in Hondurasl: A Land Husbandry Approach. Jon Hellin, hellin@onetel.net.uk PhD Thesis (1999), Oxford Brooke University, Oxford, UK. A publication reviewed in the Newsletter of the World Association of Soil & Water Conservation. Volume 18 No 3 July – September 2002 an excerpt of which follows:

> Field trials also show that a widely promoted technology - live barriers of vetiver grass (Vetiveria zizanioides) - is not effective at reducing soil loss, enhancing productivity on slopes greater than 35%, or protecting hillsides from landslide damage associated with prolonged and intense rainfall. For many farmers, the decision not to adopt SWC technologies is rational. This is especially true where land and labor shortages increase the opportunity cost of establishing and maintaining the technologies.

In the autumn of 2003 a Philippino, Romo Tionco took a copy of Volume one of this book to the Philippines. There he met people who had been practising another fashionble type of contour strip for soil conservation, SLOPE.

They were keen to have the book as they said that they had found that SLOPE was actually causing landslides. They felt that the Good News step by step Terracing looked to be the answer. See also 2.3b and 5.4.

5.2 Diversion ditches

These prevent surface water from above flowing onto and eroding the land, or adding to excess water on the land, so that the total water may then be more than the land can absorb and erosion and perhaps landslides will result.

There are two kinds of diversion ditches both dug across the slope:

a) One is dug above the land to prevent water from above flowing onto a piece of land. It may be made a U shape and filled with stones, so that moderate water flow running into it may be absorbed. What cannot be absorbed will run laterally (sideways) to waterways.

b) The other kind is made at intervals down unterraced slopes which may or may not have contour strips; their purpose is to check and divert excess water which could build up and cause problems.

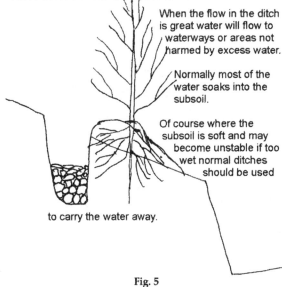

Water diversion/water absorption ditch.

When the flow in the ditch is great water will flow to waterways or areas not harmed by excess water.

Normally most of the water soaks into the subsoil.

Of course where the subsoil is soft and may become unstable if too wet normal ditches should be used

to carry the water away.

Fig. 5

Hillside ditches reduce the length of hillside slopes, and so reduce sheet and rill erosion, they also provide paths for small farm machines. The width of the ditch bottom ranges from 2-2.5 metres. Ditch spacing is affected by land slope, soil type, rainfall intensity and cropping pattern---.

The best shape of the ditch is a broad v shaped cross-section. ---. Use of the ditch resulted in increased production. Furthermore, the cross-section of the hillside canal is stable, maintenance is low, and the soil moisture content of the ditch is high.

The improved hillside ditch can easily be built by bulldozers---. The ditch bottom of the improved hillside ditch also can be used for crop production or as a farm road---.

Cover crop and mulching.

Research shows that Bahia grass *(Paspalum notatum)* and mulching, combined with hillside ditches, reduces soil erosion and run-off more than other practices. [124 p13.]

The above is obviously talking of large-scale land areas, smaller areas will require smaller ditches dug by hand or small machines.

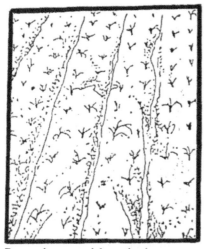

Do not plant up and down the slope as water flows easily downhill. It gains speed and power to erode the soil and so is able to wash the soil and plant foods away.

Plant across the slope. When hoeing push the soil into lines across the slope. This blocks or slows the flow, so that erosion is prevented or reduced.

Fig. 6

Contour tillage in the humid regions of the United States. Each ridge is a dam, each furrow a reservoir for tapping water and storing it in the soil.
[From Soils and Men, The 1938 United States Department of Agriculture Yearbook of Agriculture] Ref. 192 p605.

Fig. 6a

5.3 Contour farming

All sloping land in areas where heavy rainfall occurs should be managed by contour farming unless it is planted to permanent crops with complete ground cover.

This term covers all farming practices carried out on the contour. ('On the contour,' means on level [horizontal] lines across the slope.) **However, for contour strips and contour bunds on hillsides, see 5.4.**

Practices include strip cropping, contour planting, including the planting of fruit, nut and fodder trees, or other permanent crops. Contour farming also includes terracing which is covered in the next chapter, and diversion dit-ches (see 5.2) which should also be on the contour with a slight slope to drainage areas.

Strip cropping is where the slope-land is divided into strips running ac-ross the slope. Alternate strips are cultivated and crops grown, while the other strips are rested from culti-vation with fodder crops, (grasses and legumes) and small trees or shrubs grown. The soil is protected from the sunlight and weather and it stops and absorbs (soaks up) water and plant foods which flow from the cropped strips.

Strip cropping is usually carried out on slopes that are

not steep enough to warrant terracing, that is a slope of 15-20%. Strips with a poor cover crop are alternated with strips which do protect the soil.

With a slope of 0-2% the strips should be 40-50 metres apart.

2-4%   30-40
>4%   15-30

As slopes are often irregular the strips would need to be irregular widths to compensate.

This makes arable cultivation difficult. So the strips are generally

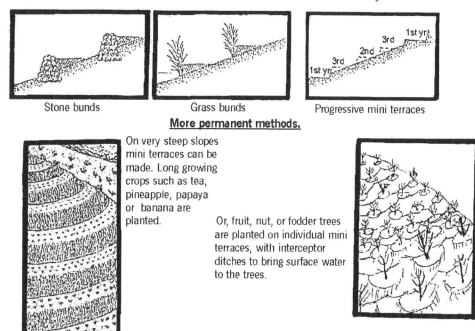

Stone bunds   Grass bunds   Progressive mini terraces

More permanent methods.

On very steep slopes mini terraces can be made. Long growing crops such as tea, pineapple, papaya or banana are planted.

Or, fruit, nut, or fodder trees are planted on individual mini terraces, with interceptor ditches to bring surface water to the trees.

Or, if very steep or very erodible plant the area to bamboo forest

Fig. 7

made of the same width (number of furrows) and the contour buffer strips are varied in width to compensate for the irregularities in the crop strips. [152]

In fact this means that the strips cannot be rotated exactly, there will be some overlap when the buffer strips are ploughed up and the intermediate strips become the buffer strips.

It would seem best if the bottom edge of the cultivated strips should be made level, and the irregularities in width adjusted at the top of the strips, so as to minimise run-off at the bottom of the strips being concentrated in hollows rather than being spread widely.

The first purpose of contour farming is to conserve plant foods, soil, and water. So contour farming is also conservation farming. However there is more to conservation farming than contour farming alone, though it is an important component. Contour farming also increases yield, and with other good management the yields will be maintained over the years. In most cases it results in fields which are easier to cultivate, also the yields are greater so less land is needed per family.

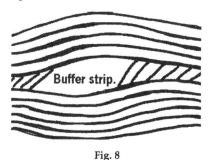

Fig. 8

When fruit or nut trees are planted they should be placed on contour lines, with the position of the trees on each contour line offset from those above and below. This is good for the trees allowing the maximum spread of their roots, it also slows down any run-off water, the run-off from one row being diverted to the trees on the row below. See Fig. 7 and Fig. 9.

Planting fruit or nut trees to make a contoured orchard. The contour lines do not have to be as accurate as for terraces, they can be made by eye judgement only.

5.4 Contour strips and contour bunds on hillsides

See first 2.3b and 5.1a Contour strips should not be confused with strip cropping which is broad strips of land along the contour (level) which are planted alternately with strips of crops which protect the soil and hold it together and strips which are cropped. Contour strips are strips of land across the slope separated by level lines often planted with grasses, or contour bunding, (bunds are strips placed along the contour to block the downwards flow of water down the slope.) The bunds may be of earth, stones, plants, or even dead plant material or suitable scrap material such as tree trunks, banana stems, or old boards. Sometimes

PLANTING FRUIT OR NUT TREES TO MAKE A CONTOURED ORCHARD.

1. Move the topsoil to the sides of the planting area.

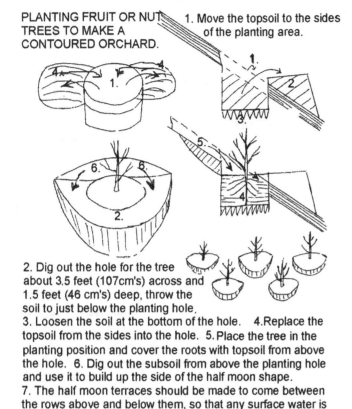

2. Dig out the hole for the tree about 3.5 feet (107cm's) across and 1.5 feet (46 cm's) deep, throw the soil to just below the planting hole.
3. Loosen the soil at the bottom of the hole.   4.Replace the topsoil from the sides into the hole. 5. Place the tree in the planting position and cover the roots with topsoil from above the hole. 6. Dig out the subsoil from above the planting hole and use it to build up the side of the half moon shape.
7. The half moon terraces should be made to come between the rows above and below them, so that any surface water is stopped by a half moon. 8. If there is no time to do all this at planting time, plant in level lines and do the terracing later.
8. The unused area between the trees should be planted with fodder legumes to cover the ground and be cut and composted or fed to livestock and returned as manure.

Fig. 9

the strips between the bunds or contour strips have barriers to stop possible sideways flow. This localises surface water which collects above the bunds and would otherwise flow to and accumulate/collect in slightly lower areas, where it may burst through the bund and cause serious erosion or even result in a gully or the start of a landslide.

Contour strips/bunds can be made for one of two reasons. 1) Where the rainfall is lighter and is spread over most of the year, the strips are to improve the ability of the land to absorb rainfall, so that the growing season of crops is extended.

2) As a means of stopping or reducing erosion of soil and soil fertility, and an indirect means of producing terraces.

As mentioned in chapter 2, the government in Nepal was and perhaps still is promoting contour strips. The World Bank has been promoting them as also other organisations. The intention of the government was good, and it is has been fairly easy to achieve annual targets, so I carefully considered their use. Farmers are accustomed to contour strips and they are easy to make. In addition the farmers are having increasing difficulty in obtaining fodder for their animals, they are also interested in diversifying their income. The government provided planting materials free or at very cheap rates providing that it was used for planting contour bunds. The materials were Napier grass *(Pennisetum purpureum)*, pineapple, banana planting stock and citrus seedlings; people were quite keen to obtain the package of plant materials.

A trial was run for 6 years, comparing four soil conservation methods, slopeland (conventional to the area), contour strips (as promoted by the government though ours had less distance between bunds), Tansen double check half terraces and full terraces, with three replications of each trial block (See volume 2.4.13e).

The result of the trial was that over the six years - against expectations - the Tansen half terraces (admittedly quite close together, the same distance as the contour strips and the terraces) had almost become full terraces. The terraced areas which happened to consist of the worst land, mostly stones with little soil, had improved considerably. The contour strips were only worth keeping as a demonstration of how unsuitable they were. However the Napier grass harboured and encouraged rats which became a problem for the whole trial area, therefore their replications, and the bare slope replications were converted into terraces.

The experience taught us that even with such short intervals between contour bunds (the normal interval was at least twice the distance we had used) soil was moving down the strip and would eventually result in the top part of each strip becoming infertile.

It also taught us that the use of Napier grass bunds was not suitable. Normally we would have expected that the crops at the lower section of the contour strip would be much better than those in the rest of the strip. This would be because of the extra soil, plant foods and moisture moving from above to add and improve the fertility of the lower part of the strips. However, on either side of the Napier grass the crops were severely stunted, while the Napier grass grew well. The Napier grass roots very effectively drew the moisture and plant foods from up to four feet away, so narrow contour strips with Napier grass bunds are not viable for crop growing. For the same reason the citrus seedlings and pineapple and banana plants planted in the centres of the bunds, had hardly developed in the six years.

In separate grass trials we found that Napier grass *(pennisetum purpureum)* with its powerful roots, when growing on poor soil exhausts the soil, and dies back within three years; whereas grasses such as Setaria *(Setaria sphacelate)*, Molasses grass *(Melinis minutiflora)*, and Signal grass *(Brachiaria decumbens)*, or the creeping legume Greenleaf Desmodium *(Desmodium intortum)*, do not. Napier grass planted in gullies or on contour bunds receives plant foods which are washed down during the rains, it then continues to thrive so it is appropriate for gullies and similar places provided that it is not near crops due to the problems mentioned here.

A further problem with Napier grass is that rats love it. They find shelter among and under its roots. Its roots and stems are sweet so there is always pleasant food available, the root zone is warmer in winter and cooler in summer. The rats are not content with that, but feed on the crops nearby. They thrive, their burrowing undermines the bund and weakens it so that when heavy rains come water pours through the breach and can lead to gully formation or even the start of a landslide. We had bunds collapsing due to rats undermining them. The government workers told us that farmers were destroying the contour bunds which they had encouraged; the rat problem could have been the reason, in the end it was best to get rid of almost all of the Napier grass.

Some large organisations such as the World Bank have been vigorously promoting the use of Vetiver *(Vetiveria zizanioides)* grass contour strips. Strips of grasses such as Vetiver grass, Napier grass and / or other plants are planted along the contour lines. The idea is to check run-off and to hold back the eroded soil so that it builds up and eventually forms terraces (Lazy terracing.)

In theory, though not in practice, the run-off water flows evenly over the surface of the field till it meets the contour strip. There the silt is filtered out and the water absorbed. In time this process turns the slope into terraces.

This author has been told by an Indian journalist that in India where the idea of using Vetiver grass originated; experienced people say that it is not suitable for contour strips. The reason is that the Vetiver grass roots yield a valuable oil. Because of this the Vetiver grass is dug up and so the contours are broken. This author heard in March 2001 that the World Bank had quietly dropped its contour strip program which has cost millions of dollars as they discovered that it is not working.

The reasons against contour strips are many, however they can be grouped into two – one that contour strips on hill slopes are not sustainable, two that as the quotation in 5.1a shows they do not improve crop yields.

A study of contour strips in Nepal showed that in the few areas of land where the soil is deeper the effects are not immediately noticeable. In areas with the normal thin cover of soil (about 6" [15cm] over an almost sterile subsoil) found on typical tropical/subtropical hillsides, the effects were easily seen.

I heard of contour strips in Lesotho where the slope was not too steep and the soil was reasonably deep where over a period of about thirty years terraces have built up. However for thirty years there was run-off and water, soil, and plant nutrients were being lost. If that had not been the case because the rainfall was gentle the soil would not have been moved to form the terraces.

However I have recently seen a report printed in 2000 that states:

> Presently, much of Lesotho suffers from chronic land degradation and soil collapse. The result means loss of biodiversity, shrinking agricultural productivity, depletion of nutrient balances, falling water tables, exacerbation of temperature extremes, flash flooding, dependency and rural poverty. For this reason Lesotho was one of the first countries to sign the UN Convention to Combat Desertification.[207] [p41.]

Fig. 10 Fertility drift on a contour strip. Eventually only about one third of the strip is fertile, the land abandoned.

POINT 1).Soil conservation is seldom carried out where the soil is deep and good; it is normally done where people are concerned at the depth of soil becoming shallow and crop yields suffering.

POINT 2.) The upper part of the strips become barren.

None of the contour strip systems I have heard of have considered the most important factor - depth of topsoil, when calculating the vertical interval between contour strips, the result is that the intervals are wider than they should be.

Although during the monsoons there would still be muddy water (carrying soil nutrients and fine soil particles) running from the effected hillside fields, soil does build up along the lower edges of the strips. However this soil is coming from the upper portion of the strips, which have less and less topsoil. On those areas the crops grow poorly giving the progressively poorer soil little shade from the strong sun, or protection from the battering of the raindrops; so the soil erodes and moves down onto the fertile soil.

The diminishing crops grown on the poorer portions leave little organic matter to improve the soil, they produce little fodder, and so less manure is available, eventually up to three quarters of the area of the strip is barren. Though much of the topsoil has been saved, what is saved tends to be the larger heavier soil particles, sand and grit, it is concentrated in narrow bands and may eventually be buried under poor soil eroded from the upper parts. To correct the situation and convert to terraces will be much more difficult, with much less worthwhile results than if correct terracing had been carried out in the first place. As can be seen in the following chapter Good News Terracing is practical, it keeps most of the topsoil on the top. While it is a stage by stage development, it comparatively quickly stops the loss of soil and nutrients, and the better crops grown on the terraces encourage farmers to continue the work.

POINT 3). Harvest is difficult as the fields do not ripen evenly.

POINT 4). The plants used for the contours are often inappropriate as already mentioned.

POINT 5). Contours do not stay level, water concentrates at lower points and eventually breaks through or flows over. When we examine the drawings showing the theory of contour strips we soon notice the impracticalities of the system. It shows fields which are uniformly sloping,, the fields are smooth and regular, unlike most hill fields which are not smooth but have bumps and hollows. The soil of hill fields is normally variable, some patches will be rocky or gravelly, others sandy, and others more sticky soil. Some quite fertile with deeper soil, others with thin infertile soil. Even if the contours are level when made, due to soil and slope variations they will not stay perfectly level.

As the slope of the land is not perfectly uniform, run-off water will move sideways as it flows and be concentrated in runlets which when they reach the contours will tend to undermine or run over the contours. At other points where the contour is slightly lower or the grass less thick the water will burst through. Finger erosion will then start and if the rain continues the water will continue to flow through those places washing away the accumulated soil and so flowing more and more rapidly. Gullies may then start to develop.

POINT 6). *Water from the strips above increases the volume of run-off on the strips below thus causing erosion.* Due to the need to discourage the sideways accumulation of water along the bunds, the bunds should not be solid barriers. Unless the soil is very open or the rain not heavy the contour bunds do not hold back all the surface water, as some water does pass through from one strip to the next, the strips have in descending order more and more water passing over them, so the likelihood of erosion increases.

The difference!

Fig. 11 Here is the contrast between terraces lower left and contour strips lower right sown with maize at the same time. The rain soaked into the terraces and the plants grow better. The terrace soil is also more fertile.

CONTOUR STRIPS - THE ANSWER?

Soil moves down, leaving a widening strip at the top which becomes barren. As some of the eroded soil and plant foods are retained by the strip a terrace bank develops. Water running down the bank accelerates and starts to scour the surface of the soil below.

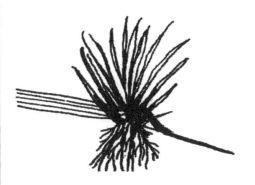

It has to be remembered that hillsides do not conform to a computerised model. The soil varies, some stony and open, some denser and quickly sealing. There are hollows and humps which divide the flow of run-off water. Contour lines are constructed on such uneven material on uneven surfaces, even if dead level at first they will change as conditions change. Where more soil is washed down the level will rise, water does not simply flow evenly down to the contour strips and gently filter through. In fact the accumulating water finds the lowest points and eventually burst through and rill erosion starts, which can quickly develop into gullies.

IS THIS SUSTAINABLE DEVELOPMENT?

Fig. 12

The contour banks will be steeper than the original slope, so runoff will start to occur earlier on the banks, this run-off water carries and moves soil particles which will tend to block the surface of the soil so that run-off further increases. The greater slope on the contour banks will accelerate the flow until it flows over the next contour, and the result as I have seen, is greater run-off where there are contour strips than without them.

POINT 7). An alternative result could be that during heavy or prolonged rain the build up the contour which has developed over a slope, becomes soft and heavy like porridge. As the water lubricates the underlying slope the contour mound starts to slide over the slope. If monsoon rains continue the field will become liable to landslide.

POINT 8). The final result can be that as the soil on top of the contour strip would now be subsoil the barrier plants would grow less well and would no longer check the run-off. The situation would be much worse than if the field had been left without the contour strips.

The rainwater running over the bare subsoil, flooding over the contours, gaining velocity down the contour banks adds volume and eroding power to the water on the strip beneath. The cumulative volume and velocity spoiling land below.

As the strip has less topsoil and fewer plant nutrients than before, the fields will be less valuable, so it is harder to justify the effort required to correct the situation.

With topsoil and reasonable subsoil together only 9-12 " (23-30 cm) which is common on hill slopes; the difference in height between the contour bunds should not be more than half the depth of the soil. This is so that after the resulting terraces have formed on the strips the whole area is still fertile, with a minimum depth of 4.5" (11.5 cm) of reasonable crop sustainable soil.

On steep slopes the contour strips would need to be so close that little cropland is left. This method may still be useful in order to have mini terraces develop naturally. When developed, if grass barriers have been used they are taken away and such crops as tea, pineapple, banana, papaya, sweet potatoes, peanuts and ginger are grown.

Perhaps grass contour strips (not using Vetiver grass) may have value where the slope is very gentle, up to maximum 10%, depending on the type and amount of rainfall, the openness of the soil, and the use the land is put to, and where the contour bunds are close together, or the soil is very deep. Places where erosion is much less of a problem or in areas of comparatively gentle rainfall. I would suggest Molasses grass (*Melinis minutiflora)* or Setaria *(Setaria sphasalata)* rather than Vetiver grass, the Molasses grass forms a thick mat. Setaria is a very good fodder grass, superior to Napier grass without the problems of Napier grass. *Making terraces on gently sloping land would not normally be feasible unless to make paddy fields. Contour strips can be suitable in some situations.*

Where both slope and rainfall are gentle, contour strips can be used in the following way.

The contour bunds are alternated every few years according to the situation. Before too much soil has been moved to the contour strips, the bunds are moved to the point on the strips where the good soil is becoming too shallow in the upper part of the strips. Soil then builds up above that position and so good crops may continue to be grown over the whole area of the field. Plants such as pineapple could be grown perhaps between two rows of Setaria grass. Then, after the ratoon crop of pineapples is harvested at between 3 and 4 years when the pineapples were due to be replanted, the contour bunds are moved, the Setaria grass clumps divided and replanted, the spare splits sold to another farmer or used somewhere else, perhaps for gully protection as well as valuable livestock fodder. In Australia:-

Setaria grass, otherwise known as South African Pigeon Grass is said to have perhaps the greatest range of adaptability of all tropical grasses. It has thick stems and an abundance of leaf.'[62 p68.]

In summary:-

Except when the slope is very gentle or the good soil is very deep contour bunds need to be closer than is normally acceptable if sustainable land management is to be achieved. If not:

(a) There will be fertility and soil drift downwards, with crops at the lower part of the strip receiving most of the nutrients and growing too lush and soft; perhaps liable to lodging and to pest and disease attack. Crops at the higher part become spindly and weak and also susceptible to pest and disease attack, but for the opposite reason. Due to crops grown on the strips maturing at different times at different areas of each strip, harvest is difficult.

In time the top part of the strips will have no topsoil, while the bottom part will have more topsoil than is needed.

(b) On hillsides, contour strips can and do actually increase runoff and loss of soil and plant foods. The sloping land does not absorb water as easily as a terrace does, so the rainwater flows, seals the surface and so even less is absorbed, thus more surface flow. This water is checked by the bund, and tends to then move sideways to a lower point where it accumulates and eventually bursts the bund. This will cause rill erosion and can lead to gully erosion further down the slope. If not the excess water soaks into the wedge of soil which has been built up on top of a sloping surface. The heavy wet soil is then liable to slip, if the rain continues this may turn into a landslide. There will be no terrace below to check it and the field may be spoilt.

(c) Contour strips on all but gentle slopes are like most lazy or quick fix methods not a long-term solution, it is better to either make double check half terraces or full terraces at the beginning, rather than after contour strips have been shown to fail.

The opinion of this writer is that normally contour strips should only be used in the form of tree trunks and/or stones and rocks placed on the contours for one to three years only; or double check half terraces for up to five years; until the first stage of good news terracing can be carried out. The stones and rocks can then be used for the terrace walls.

Normally the practice of including weeds and trash in these simple contour strips is not good. It may increase pests, diseases, and weeds.

See also chapter 2.3b and 5.1.

For full information on terracing, including how to construct different kinds of terrace walls, see chapter 6.

5.5 Deciding on the spacing between contour ditches, Tansen half terraces or terrace ditches.

NOTE:- THE DISTANCE BETWEEN TERRACES IS BASED ON A DIFFERENT CALCULATION.

If terracing cannot be done for some reason contour ditches or Tansen half terraces should be made. If rainfall is extremely heavy or the soil water absorption ability is poor, terraces may not be able to absorb all the water falling on them at certain times. If this is the case some or all of the terraces, or Tansen half terraces, should have ditches to carry excessive water away.

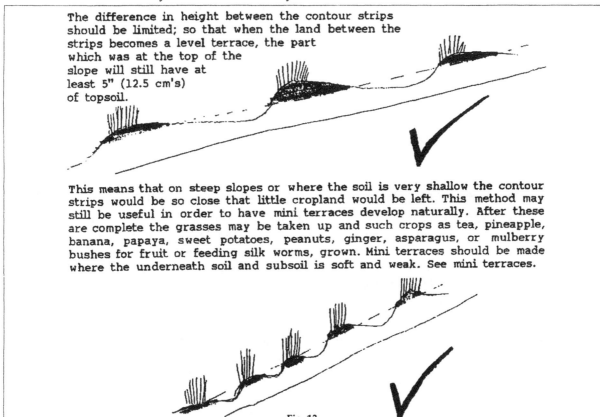

The difference in height between the contour strips should be limited; so that when the land between the strips becomes a level terrace, the part which was at the top of the slope will still have at least 5" (12.5 cm's) of topsoil.

This means that on steep slopes or where the soil is very shallow the contour strips would be so close that little cropland would be left. This method may still be useful in order to have mini terraces develop naturally. After these are complete the grasses may be taken up and such crops as tea, pineapple, banana, papaya, sweet potatoes, peanuts, ginger, asparagus, or mulberry bushes for fruit or feeding silk worms, grown. Mini terraces should be made where the underneath soil and subsoil is soft and weak. See mini terraces.

Fig. 13

On steeper slopes the risk of erosion is greater and so the ditches need to be placed closer vertically (in an up and down direction).

To help in deciding the spacing the Department of Conservation and Extension of the government of what was then Rhodesia devised a formula which is the vertical interval (the upright distance between the ditches) should be $\frac{S+F}{6}$, where S is the percentage slope, this is the number of metres(or centimetres) rise or fall for each 100 metres (or centimetres) horizontal (level) distance. See A/ below.

F is a factor which is decided according to the soil on the slope. It varies between 3 for light sandy easily eroded soil, and 6 for erosion resistant clays or clay loam.

So if the % slope is 100 and the soil is easily eroded so should be classed as 3 the formula would give

$$100 + 3 = 103 = \frac{103}{6} = 17.1$$

With the calculated distance vertically (up and down) the horizontal (level) spacing, (as a man looking over the field would notice) is closer from one ditch to the next for the steeper slope than for the less steep slope.

The chart for F was not practical for non-technical people without special charts so computed from a chart of the effect of velocity of flow I made the following table for F *values*.

0.75 Very light silty sand.
1.5 Mixed silty sand
2.0 Light sand
3.0 Coarse sand
3.0 Light sandy soil
4.5 Firm clay loam
5.25 Medium clay
6.0 Erosion resistant clay /stiff gravelly soil
7.0 Shale, hardpan, soft rock.

According to NW Hudson in Field Engineering for Agricultural Development if a channel terrace is too long the volume of run-off in the channel will become too much and the channel will start to scour . From experience it is possible to specify maximum lengths (as below).Note these are the maximum lengths in the direction of flow---.

	Sandy soils		Clay soils	
	M	Feet	M	Feet
Normal maximum	250	900	400	1200
Absolute maximum	400	1200	450	1500

This can only be a guide depending on how heavy is the rainfall expected.
THE SLOPE OF THE DITCH
For clean ditches 1 in 250 feet or metres is best. [10 p198.]

5.6 Zero-tillage, cover crops, inter-cropping, relay cropping, and mulching

These are discussed in Volume 2 in the chapter on improving the soil management, chapter 4. Obviously anything which improves the soil management results in a better soil texture which is better able to resist erosion. However these three factors also have a direct effect on reducing erosion. The more a soil is cultivated the more easily it is eroded, so zero tillage reduces erosion. Cover crops, inter-cropping, relay cropping and mulching protect the soil from wind, from the battering of the raindrops and from UV light which burns up the humus and reduces the helpful soil micro organisms.

To add to what was said about Zero Tillage in Volume 2.4 two quotations from[181].

Derpsh points out that the largest areas under no-tillage (zero tillage) are in the USA (19.3 million ha), Brazil (11.2 million ha), Argentina (7.3 million ha), Canada (4.1 million ha), Australia (1 million ha), and Paraguay (790,000 ha). No tillage, however, is only 16.3% of the total cultivated area in the USA versus 21% in Brazil, 32% in Argentina, and 52% in Paraguay.---

The following may be the main factors that induced such a rapid change in Latin America: 1) efficient and economic erosion control where erosion and soil degradation are potentially high; 2) appropriate knowledge was available through research and development, as well as farmer experiences; 3) widespread use of cover crops for weed suppression, organic matter build up, biological pest control etc; 4)consistent positive message is voiced by all sectors involved (private and public); 5) no-tillage has been the only conservation tillage technology recommended to farmers; 6) there is an aggressive farmer to farmer extension through farmer associations (alliances); 7) publications with adequate practical and useful information were available to farmers and extensionists; 8) economic returns are immediate and substantial; 9) there have been no major forces against the system; and 10) Latin American farmers must be competitive in the global market since there are no subsidies.

Derpsh offers the following outlook:

- Knowledge and information are the main constraints to no tillage adoption. Information must be relevant, locally appropriate, and useful if it is to impress farmers.
- A first step is for farmers, researchers, technicians, and extensionists to learn the latest about all aspects of the system. A mental change of farmers, technicians, extensionists and researchers' thinking away from soil degrading tillage operations towards sustainable production systems like no-till is essential.
- The superiority of no-tillage as a system over conventional has been proven worldwide. Now the system must be developed and adapted locally.
- We need to learn which soils have limitations and how we can overcome these limitations.
- We need to discover the limitations in machinery, herbicides, crop rotations, green manure cover crops, and socioeconomic constraints and find ways to overcome these.
- The attitude "it doesn't work" is not helpful to solve problems in no-tillage!!! If we are aware of the fact that no-tillage is the only truly sustainable system *(the present author does not agree)* in extensive agriculture in the tropics and subtropics, then we will have to find ways to overcome the problems and limitations.
- Lower yields may be tolerable as long as profit margins are higher, which is often related more to lower costs than to higher outputs per hectare.
- The benefits of no-tillage, such as rainwater conservation, erosion control, improvement of physical , chemical and biological conditions, reduced machinery costs, higher economic returns, lowered risks, and other benefits of the system should guarantee a steady growth to permanent no-tillage in most regions of the world. [181]

Though except on gently sloping land terracing is the best long term option it is important to use simple techniques where land is not terraced such as cultivating across the slope as the following chart illustrates.

A SEVERE RAINSTORM TEST OF NO-TILL CORN
(A 1:100 year storm, which fell in 7 hrs.)
Coshocton, USA, 1970.

Tillage type	Slope %	Rain mms.	Runoff mms.	Sediment yield kg/ha
Plowed, clean tilled, Sloping rows	6.6	14.0	11.1	50,781
Plowed, clean tilled, Contour rows	5.8	14.0	5.8	7208
No-till, contour rows	20.7	12.9	6.4	71

"No-till was helped by very good sod residue".

(metricated from a note about: Harrold L.L. and Edwards, W.M. 1971. 'A severe rainstorm test of no-till corn' in JSWC (USA), mid-1971, vol.7 p.30)

If using no till methods it is important to be able to smother weeds which can become a serious problem, not only because they compete with crops, but because they may harbour diseases and encourage pests such as cutworms.

5.7 Permanent crops

For steep areas where neither contour farming nor terracing is feasible the land should be growing permanent crops whether left as jungle or planted. If planted the plants should be planted across the slope to reduce erosion and also produce better tree or other permanent crops on the area.

A compromise may be appropriate where there are strips of crop land across the slope between strips of permanent crops or jungle.

5.8 Water courses

These should be protected from erosion by methods such as by planting suitable plants in gullies, and by making check barriers to block the flow and allow heavier materials stones, gravel and much of the sand,

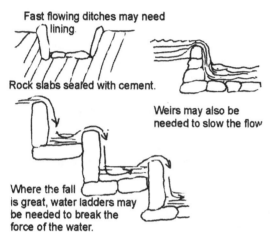

Fast flowing ditches may need lining.

Rock slabs sealed with cement.

Weirs may also be needed to slow the flow

Where the fall is great, water ladders may be needed to break the force of the water.

Fig. 14

to fall out. Along the sides of streams and rivers plants such as water tolerant bamboos should be planted to hold the soil and to confine the water to its normal course. See Appendix 5 and Appendix 6 waterways for suitable grasses and legumes. See also chapter 7.

For waterways which fall steeply, weirs or even water ladders, may be needed to slow down and break the force of the flow.

5.9 Gullies

Should be stopped at their heads after first investigating why they started. Is it the result of land being cleared above the gully? perhaps the change in land use from grass to crops such as maize? Perhaps it is the result of wrong contour farming, perhaps caused by run-off from graded terraces ditches or bunds? Or, is it from general drainage from the area due to the plant cover being reduced, this may be caused because of poorer crops? Is it the result of bad drainage from other areas, from roads, yards, building roofs, from land made less pervious (loose and open to water penetration), perhaps by vehicle, people, or animal traffic, or by bad land management, causing the land to lose its texture and crust on the top, or have a hard pan under the surface. This could be a plough pan due to ploughing at the same depth year after year so that the bottom of the plough has pressed and skidded over the under soil till a hard layer is formed which does not easily allow water to pass through.

Gully formation could in rare cases be due to new water springs appearing, perhaps because a landslide or earthquake has disrupted other spring lines, or moved or tilted the water holding or retaining rocks from which the water comes. It could be due to heavier than normal rainfall causing the underground water to find a new route.

Having established why the gully is there, can the cause be corrected? If not we need to look at controlling the gully or we may need to do both.

Perhaps it should be made into a watercourse, that means that it will be altered so that it does not grow, but is controlled, the difference being that it will be sloping much more than a normal watercourse. If it is a narrow gully like a falling ditch, steps should be made so that the force of the water is broken. The sides should be protected by either stones or stones and concrete or plants such as bamboo. At appropriate intervals according to the conditions check dams should be made, or check pools or tanks, these are to collect stones and gravel and some sand; periodically the tanks or pools are cleaned out. It may be that by the use of barriers the gully is converted into a series of soil retaining areas which may eventually become useful growing plots.

With bigger gullies check dams are made. These must be made carefully or they will make the problem greater by diverting water to the sides of the gullies and so the gullies become wider and the check dams undermined and broken. See chapter 7.

5.10 Landslides

Again these should be investigated to see, if possible, what was the cause. Consider the possible causes of gullies in the previous section. Perhaps the cause is a factor which will continue, such as drainage from a road or a new water spring which has broken out, perhaps due to land movements. If so, it may be possible to divert the water to a more suitable area such as an existing gully, or a more stable area such as an area of thick grass and trees or bamboo, or, to make a new waterway away from the unstable landslide area. If not, a protected waterway should be made in the middle of the landslide,and the landslide stopped from further developing, such as by planting bamboo.

Landslide control

First trim off loose edges and surrounding unstable areas which are likely to expand the landslide in the next year or two, then mini terraces should be made about 1.5 - 2 ft (45-60 cm) wide. These should run across the slope, however if it has been necessary to make a waterway they should have a slight slope down towards the waterway. These mini terraces are then planted with bamboo or fast growing trees with grasses and legumes planted between them. The legumes should be such as Greenleaf Desmodium, sometimes called Intortum clover, (*Desmodium intortum*); or Silverleaf Desmodium, (*Desmodium uncinatum*) though in the authors experience the latter is not as vigorous and suffers a little from insect attack. They both grow well from simple cuttings 5½" (14cm's) inserted at 45degrees for two thirds of their length when the rains start, and will produce vast amounts of cutting material in the second and future years from planting. During the rains they grow rapidly and as they hang down and cover the soil they root into the banks, holding and protecting the soil. When the dry season comes they stand drought amazingly well, in an exceptionally dry year in Taiwan when even the native trees were suffering, the only green was the Desmodium.

As mentioned elsewhere, in our experience and those of others, Napier grass/Elephant grass (*Pennisetum purpureum*) is not to be recommended for this purpose, though it is useful for grassed seasonal waterways where it receives nutrients washed down with the water. The reason is that Napier is a very greedy plant. It draws moisture and nutrients from at least as far from it as the height it grows which may be 6 ft (2 metres) or more. It grows well till it exhausts the soil and then becomes very poor, Setaria grass (*Setaria sphacelata*) does not seem to have the same problem. Napier grass has another problem in that rats love it, they like to burrow under its roots which are sweet. They go out to steal from surrounding crops, and their burrowing undermines the slope and can start new landslides.

5.11 Suitable grasses and legumes for use in soil conservation.

This is an important subject, it is not enough to simply say 'divert run-off-water to a grassed waterway? See Appendix 5.

5.12 The importance of vegetative cover

Vegetation is very important for protecting the soil.

a). The raindrops fall from a great height and so they batter and loosen the soil. Then the loosened soil is battered into the pores of the soil surface, blocking them and so reducing the soils ability to soak up the rainwater. This results in run-off which causes more spoiling of the soil. When plants cover the soil the raindrops hit the plants and break up into sprays which fall gently through the foliage to the soil.

However, trees can also cause soil loss due to their leaves concentrating the rainfall towards the dripline above their most active roots. In a natural situation the soil is protected by vegetation so the water percolates gently to the soil surface. If there is no vegetation beneath the trees the concentrated drip area is eroded, particularly when broad leaves drip very heavy drops or even if there is a constant flow. Leafy vegetables such as Chinese cabbage may be battered to death under such trees. So more suitable vegetation is needed covering the soil surface.

b). The plants are barriers to run-off, slowing it down so that it causes less damage.

c). The roots of the plants (some much more than others) bind the soil so that it is not so easily moved when wet or by the force of run-off water.

d). The roots and other plant remains add organic matter to the soil, thus improving the soil structure so that it is more porous and less easily eroded; they also improve the fertility of the soil so that the plants grown are more vigorous resulting in better protection to the soil.

I believe it is important to keep the soil covered with vegetation if possible, it is often better to have the soil covered with weeds than to leave it bare, but they should be cut before their seeds develop. Of course the weeds which are difficult to control should not be encouraged. Weeds often serve another useful purpose in that they are a natural way of restoring the fertility of soil. The weeds which grow are an indication of what the land is needing, the weeds which grow are those which can extract the needed nutrients from the soil. As they die the nutrients so extracted are added to the soil thus improving it for later crops. See Chapter B4.

5.13 Better road construction

There is much harm done by careless road construction in hill areas. Spoil from the road buries fields and orchards below the road. It washes down blocking streams and rivers so that they burst their banks destroying fields, road, bridges etc.

Very often there is no plan for proper drainage, so runoff from the road pours down hillsides, sometimes causing landslides and other problems. Proper protected waterways should be made and the road drains flow to them.

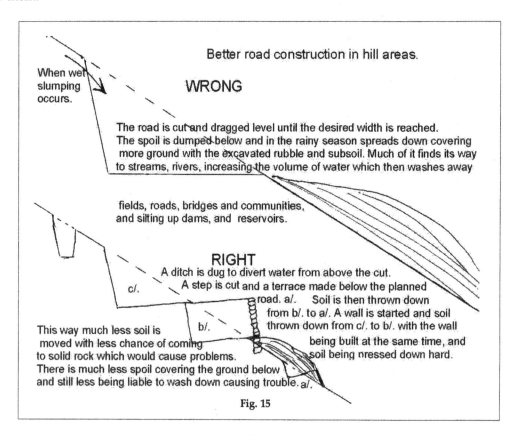

Better road construction in hill areas.

WRONG

When wet slumping occurs.

The road is cut and dragged level until the desired width is reached. The spoil is dumped below and in the rainy season spreads down covering more ground with the excavated rubble and subsoil. Much of it finds its way to streams, rivers, increasing the volume of water which then washes away

fields, roads, bridges and communities, and silting up dams, and reservoirs.

RIGHT

A ditch is dug to divert water from above the cut. A step is cut and a terrace made below the planned road. a/.   Soil is then thrown down from b/. to a/. A wall is started and soil thrown down from c/. to b/. with the wall being built at the same time, and soil being pressed down hard.

This way much less soil is moved with less chance of coming to solid rock which would cause problems. There is much less spoil covering the ground below and still less being liable to wash down causing trouble. a/.

Fig. 15

CHAPTER 6

SOIL AND FERTILITY CONSERVATION
SECTION TWO: TERRACE MAKING

6.0 Introduction

The requirements for lasting soil conservation on cultivated slope land.

Before starting conservation work we should consider the following requirements:

1. We need to finish with as much soil as possible where it belongs. Though occasionally we may move soil from where there is more to where there is less, to even up the field.
2. We need to finish with as much as possible of the top soil (the living soil) on the top.
3. We need to finish with terraces which are level across the slope, with a slight backslope into the hillside, or in situations where the rainwater does not soak in easily, and where the rain is heavy, terraces with an even slope to allow excess water to flow to protected waterways.
4. We need to have terraces which will be stable and will need little or no maintenance.
5. We need to have a method which uses the least labour to get (1) (2) (3) and (4).

In some places (2) will be the most important need. If we are going to plant vegetables, or if there is not much topsoil, or, if the land we have is not as much as we would like, or if labour is very cheap or perhaps there is no other work.

If the land will be ploughed, particularly by small tractor, or if rainfall is very heavy and long lasting (3) is important.

In other cases such as where the slope is steep and the rain heavy, or the soil is very unstable (4) will be the most important factor.

In places where labour is difficult to get or too expensive, or erosion is rapid and erosion of soil and soil nutrients must be stopped quickly, we will need to choose methods which put (5) as most important. This may mean that some of the topsoil is buried and some of the subsoil will be on the top. The alternatives would be a few terraces made very carefully with all the top soil kept on top, but the rest of the land being too badly eroded to be worth saving; or, the use of a small bulldozer. This may not be as unfeasible as it sounds, the Good News Bulldozer is modification to walking or two-wheeled tractors such as paddifield tractors. The wheels are converted into spud wheels and the good news bulldozer frame fixed onto the tractor. First a pass is made the rotary cultivator down to loosen the soil, then the loose soil is bulldozed to where needed.

A good way of terracing with a bulldozer, such as the Good News Bulldozer, is to push the topsoil from a strip onto the land below, then level off the subsoil with a back slope and repeat on the strip above. When this process is finished move down levelling off the topsoil; this method is illustrated in 6.3 Throwing down method.

There are terracing systems on sloping land which have been producing crops for 3,000 years. Properly made terraces are the only dunable way to be able to crop steeply sloping land.

T intro1 The 3,000-year-old Ifugao Rice Terraces, Banaue, may be rightfully counted among the wonders of the ancient world. Carved out of mountain sides and employing an ingenious irrigation system, this particular group covers an area of 400 sq. klms. If placed end to end the terrace walls would cover an estimated distance of 14,000 miles, or more than half-way round the world.

By contrast to the pictures on the left, observe T intro 2, 5 and 6. Land which was contour stripped.

These south facing slopes were once very fertile, without conservation they are now only able to grow occasional poor crops. However terracing can still make a huge and lasting difference. This land can grow good crops again.

T intro 2.

T intro 3. From the Philippines   Ref 67 p133.

As the top of the strips are infertile manure is added to compensable, but monsoon rains will wash out the plant foods.

T intro 4.

This hillside was contour stripped, the end result is it's fertility has been stripped and it is eroded. **T intro 5.**

This is what happens with contour strips in time, now the area hardly grows grass. Even the land in the foreground was cropland. **T intro 6**

Trials were constructed on the non contour stripped part of the slope of T intro 2. Equal areas of slope were measured out to compare terracing with slope cropping. Equal amounts of farmyard manure were applied to each treatment and maize was sown. The terraces produced three times the yield of maize of the slopeland.

It may seem hard to believe why should there be such a difference.

1. In volume two we learn how it was discovered over 2,000 years ago that plants grew better on loose soil. That is why double dug plots (plots where the soil is loosened for two spades deep) produces better yields of vegetables than single dug plots. The plant roots can spread deeper into the soil, reaching minerals beyond the reach of plants where the under soil is hard to penetrate, air and rainwater can also penetrate deeper. The roots are healthier as they are not waterlogged when it rains, the soil does not dry out quickly and there is air in the soil. (Roots do not like to go where there is no air).

2. When the early rains came it soaked down into the terraces, this was not a problem as the terraces were made properly and so keyed into the hillside. On the slopeland was ploughed soil on top of hard subsoil, the rain could not penetrate deeply. Much of the rain ran off the slopeland carrying some of the plant foods from the manure with it.

3. The seeds germinated and started to grow. The early rains ended and in the hot sun the slopeland soon dried out, and the plants started to wither. Most of the moisture in the terrace plots was lower in the soil so did not quickly dry out. The maize plants were able to push their roots deeper to where there was still moisture and continued growing into strong plants, their leaves partially shading the soil.

4. When the main rains came the plants on the terraces were able to respond and grow, quickly covering the soil, protecting it from the sun and from the battering of the raindrops. The small withered plants on the slopeland took longer to recover and start to grow again. The rain battered the soil on the slopeland and much of it ran off carrying precious plant food with it. The flowing muddy water tended to seal the surface so even less water penetrated the soil.

5. On the slopeland the battered surface soil crusted and so there was less air in the soil. With less plant food available and a poor environment for the maize plants to grow in they grew slowly. Weeds sprang up in the bare soil between the plants, competing for what plant food there was.

On the terraces the water soaked in, washing plant foods near the surface down into the lower root zone. The maize plants grew strong and tall shading the soil and choking out the weeds.

6. When the rains ended the slopeland quickly dried up, the soil became hard, and the growth of the maize slowed and stopped.

On the terraces the combination of the strong deep roots and the moisture stored deep in the soil kept the maize growing strongly, producing many more, much heavier, cobs.

The result, the huge difference in yield between the two treatments. The slopeland had less soil than before, it had degraded over the season and would not be able to produce a crop for a few years.

The soil on the terraces was better than before and if a green manure crop such as Velvet bean had been planted under the maize when the main rains had come, so that it would climb and cover the maize stalks after the harvest the soil would be much improved. The velvet beans would produce a mass of organic matter, which could have been used as a mulch for subsequent crops as well as manuring the soil. In addition the root nodules would be providing nitrogen for following crops.

Providing we can make terraces which will not slide, keeping most of the topsoil on top, terraces which can be made by farmers with a reasonable amount of labour for the perceived benefit, it should be clear that the only sensible way to conserve crop land is by making terraces. I believe that what is taught in this chapter should enable people working in a range of conditions to terrace their land.

This shows part of a soil conservation trial from 1965-1968. It showed that only terracing was a durable method.

This stony land was terraced in 1967, a visitor in 2002 found the land was still as fertile.

When I started work at the Yu Shan Mountain Farmers Agricultural Training Centre, East Taiwan, in 1965, I initiated a soil conservation trial with Terraces, Half Terraces, Grass bund contour strips and stone bund contour strips. The trial ran to 1968 and it became increasingly clear that the terraces were the only durable form of conservation on such slopes. The yield of crops on the terraces improved over the period, while those on the half terraces declined, and those on the contour strips declined greatly. All treatments were given the same manuring, and grew the same crops sown at the same time. The trial board showed each treatment with the labour spent working on each, the dates sown and harvested, and the crop yields. Students and the many visitors were very interested in the progress of the trial and started to build terraces on their own land. This greatly surprised government workers who had been trying to encourage soil conservation with little effect.

When I became farm manager we cleared and terraced the hillside land. In 2002 a person who had visited the site told me that she had seen the terraces that year and that she was told that they were as fertile as ever.

In the years since 1968 in both Taiwan and Nepal I have conducted similar unbiased trials and observed other schemes, nothing has contradicted the evidence of that first trial. Only terracing has been shown to be durable, of course the terraces have to be made correctly and carefully, keeping the top soil on the top. And they have to be made in an efficiently appropriate way. This chapter explains a range of methods for a wide range of conditions.

6.1 Problems with wrong terracing

While terracing is often regarded as unsuitable or unpractical because of the way it is usually carried out, I know from experience that they can and should be both suitable, practical and economical. The correct criticism of terraces is that:

(A) they are most often made by simply dragging the soil down hill, forming a level wedge of loose soil. In the rainy season this loose soil may become waterlogged, heavy and slippery, and start a landslide.

(B) The first soil loosened and moved is the fertile topsoil which formed the bottom of the wedge. Next is the poorer half fertile undersoil; and lastly the infertile subsoil which becomes the infertile surface of the terrace. Even with heavy manuring this soil gives poor crops, while the rich topsoil is wasted underneath.

(C) Such terraces are usually made as projects rather than by the farmers themselves. So the work is done poorly and there is often little interest in maintaining them.

(D) The terracing is often highly labour intensive, or by using expensive earthmoving equipment, both methods being uneconomical.

6.2 A flexible approach needed

As in other aspects of rural development we should be slow to criticize traditional methods even though they may appear wrong by the standards we have been taught or read about. Some examples should show what I mean. It would seem that the ideal fields are flat. 1. Easy to cultivate and work on. 2. The rainfall will soak in and so the soil moisture level will last for a longer period and crop growth will continue for longer. 3. The manure and fertiliser which is applied will not be washed down the slope when it rains. 4. The soil will not be eroded and plant nutrients washed out. 5. If the terraces are properly made the crop yields will be higher and with good management the fields will be as good or better in 10, 50 or 100's years' time. Cultivating steep slopes is hard work and it is not sustainable. I am firmly of the opinion that generally speaking terraces should:

A. be made in an economically viable way.

B. be made in a way that keeps all or most of the fertile topsoil on the top.

C. be almost flat, with a slight slope away from the edge of the terrace.

However, we need to consider the particular circumstances faced by farmers in each place.

If the subsoil is very soft and loose it could be that the farmers have found that if they have flat terraces the water quickly soaks down into the subsoil. When there is steady rain the subsoil becomes liquid and slips down the slope. In such a situation farmers may have learned that the terraces should be made sloping, so that much of heavy rainfall runs off. see T8 and T9.

In other situations they may make flat terraces but make them narrower so that the terraces are not dug so deep into the subsoil, and there is less chance of excessive accumulation of water soaking into one place and causing a slide.

In another area though the subsoil was firm the terraces were made gently sloping. On reflection I think it might be so that when there was a light to medium rain it would soak in and crops sown at the time of the light rains would be able to establish and start to grow even though the rains may only fall for a few days. Then when torrential rain came excess could run off rather than causing damage.

Terracing needs to be done according to the conditions as they are seen after we have started the work.

Very often it cannot be done by following a fixed plan. On a typical hillside conditions will vary, and careful office planning for graded or level terraces will often fail due to some sections having more or less topsoil or stone than other sections.

Where topsoil is deeper some of it may need to be taken to areas short of topsoil. Extra stone in other areas may be used for walls in areas with less stone. (Providing terracing is started at the bottom of the slope some compensation can be made by shaving more soil from the back wall of the terrace to level the terrace up.)

If the land is terraced properly, a stony hillside will seldom or never need special drainage, except to stop water from other areas adding to its own rainwater. In my experience a hillside which prior to terracing would have mud sliding down and blocking the gate when there was heavy rain, in one case burying it, had after terracing, negligible run off during the worst typhoon for 40 years. At the same time there were very many landslides and much serious flooding in the neighbourhood.

If a track goes diagonally up the slope, the track ditch will usually be sufficient to carry excess water from the terraces, provided that they are properly made.

If hillside rundown ditches have to be constructed and the slope is more than 7% check dams should be built at appropriate intervals to break the speed of the water. If more than 40%, steps should be made for this purpose. See Chapter 5.8 figure 14.

Rocky Areas - Some areas may have large rocks just below or even at the surface. It may not be possible or feasible to move or break up the rocks in which case alterative treatment is needed. The soil should be collected into the hollows and spaces between the rocks and levelled; then trees or bamboo planted. Creeping legumes such as Greenleaf Desmodium should be planted under the trees, they will spread over the rocks, utilizing the space and reducing the drying effect of sun and wind. The legumes can be cut for animal fodder or used to make compost for manuring the trees.

The Good News Bulldozer (See illustrations 25(a) and 25(b) and appendix 4.)

At first when I was in Taiwan labour was very cheap and we made terraces using only human labour. After a few years the situation changed and it was difficult and expensive to get labour for such work.

There are several ways of constructing terraces but few are used for dry conditions, most are used for the making of paddy fields.
The reasons are that :-
1/.Paddy fields usually grow more valuable crops than dry land crops, so the labour of making them is considered to be more worthwhile.

2/.When terraces have been made for paddy fields, water is run into the field and mixed with the soil to make mud. The mud is trodden by bullocks and or people, which settles it and makes the earth firm. It is again stirred up and trodden down until the mud has sealed the terrace to make it into a paddy field. After this water does not soak through into the soil underneath. See T1/. Because of this terraces used as paddy fields are usually successful.

When terraces are not treated in that way the soil which has been moved is soft and loose. When heavy rains come the water quickly soaks into the loose soil, it becomes softer and heavier. If (as is normally done) the soil has just been dragged down till a level terrace is made, there is a wedge of soft heavy wet soil on top of the firm and sloping surface of the hill. See T2/.

If the rain continues the terrace will slide down the hill and so the work is wasted and the field is worse than it was before. The precious topsoil has been removed. In fact a landslide may develop. T3/.and T4/.

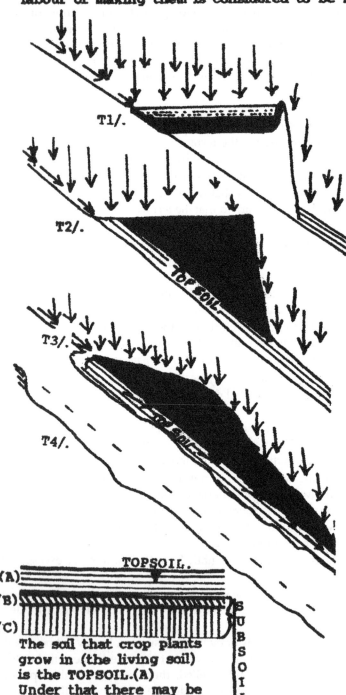

The soil that crop plants grow in (the living soil) is the TOPSOIL.(A) Under that there may be UNDERSOIL.

It will be a different colour and more fertile than the subsoil soil and may be very thin. Under that is the subsoil (C) which may be poor soil, sand, stones, shale (soft crumbly stone), or rock. Unless named specially both the undersoil and the subsoil are called subsoil.

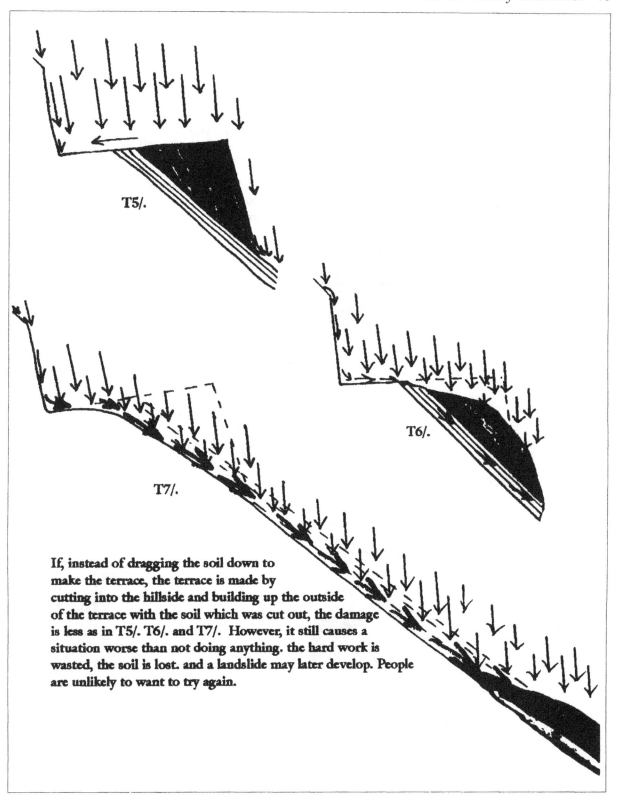

T5/.

T6/.

T7/.

If, instead of dragging the soil down to make the terrace, the terrace is made by cutting into the hillside and building up the outside of the terrace with the soil which was cut out, the damage is less as in T5/. T6/. and T7/. However, it still causes a situation worse than not doing anything. the hard work is wasted, the soil is lost. and a landslide may later develop. People are unlikely to want to try again.

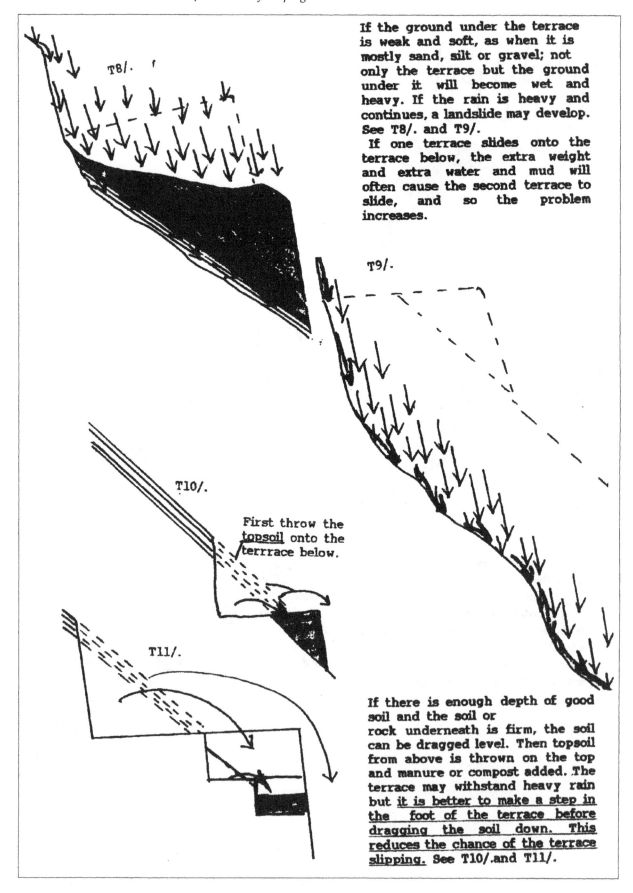

T8/.

If the ground under the terrace is weak and soft, as when it is mostly sand, silt or gravel; not only the terrace but the ground under it will become wet and heavy. If the rain is heavy and continues, a landslide may develop. See T8/. and T9/.

If one terrace slides onto the terrace below, the extra weight and extra water and mud will often cause the second terrace to slide, and so the problem increases.

T9/.

T10/.

First throw the topsoil onto the terrace below.

T11/.

If there is enough depth of good soil and the soil or
rock underneath is firm, the soil can be dragged level. Then topsoil from above is thrown on the top and manure or compost added. The terrace may withstand heavy rain but it is better to make a step in the foot of the terrace before dragging the soil down. This reduces the chance of the terrace slipping. See T10/.and T11/.

In places where labour is difficult to get or too expensive, or erosion is rapid and erosion of soil and soil nutrients must be stopped quickly, we will need to choose methods which put 4/ as most important. This may mean that some of the topsoil is buried and some of the subsoil will be on the top. The alternative would be a few terraces made very carefully with all the top soil kept on top, but the rest of the land being too badly eroded to be worth saving.

When planning to make terraces it is a good idea to first make a test terrace of about three metres long, or at least a trench in the shape of a terrace. That will help to decide what kind of land it is.
- How deep is the good soil?
- Is it rocky underneath?
- Is it firm shale or strong soil?
- Or is it weak with much loose shale, small stones or gravel, silt or sand?

INDIRECT METHODS OF TERRACING:-

PLOUGHING DOWN TO CONTOUR LINES. This can be appropriate where there is enough depth of good soil, and where erosion is very slow. If considering this method it is important to check that when the strips have become terraces, there will still be enough good soil at the top of the strips to grow good crops. If there is not sufficient good soil, the top of the strips will become steadily poorer because the poorer crops will provide less protection from sun wind and rain. Due to the uneven fertility harvest will be difficult, as, due to the uneven distribution of water and plant nutrients, the plants will mature at different times.

For this method contour lines are marked out, (A). (For the method of doing this see Appendix 7). Then a shallow furrow is ploughed. The next furrow is ploughed uphill onto the first furrow, the next is ploughed downhill. The rest of the strip is ploughed downhill. When ploughing, plough shallow at the bottom of the strip increasing the depth towards the top so that more soil is moved from the top. See T12,13,14/.

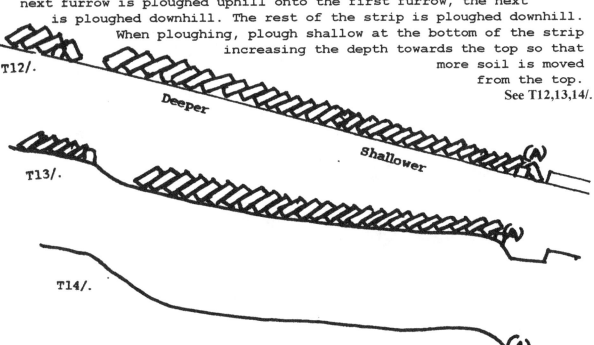

T12/.

Deeper

Shallower

T13/.

T14/.

CONTOUR STRIPS.
Another method of indirect terracing is by <u>planting contour strips</u>. Strips of grasses and / or other plants such as pineapples or sisal, are planted along the contour lines. The idea is to check runoff and to hold back some of the soil so that it builds up and eventually forms terraces. This of course takes longer than the plough down method.   See T15/.-T19/. While contour strips can be appropriate in some situations, in this authors experience they are really lazy terracing and in most cases worse than doing nothing.

<u>Wrong contour stripping.</u>

See chapter 23 on wrong methods, and 5.1 and 5.4

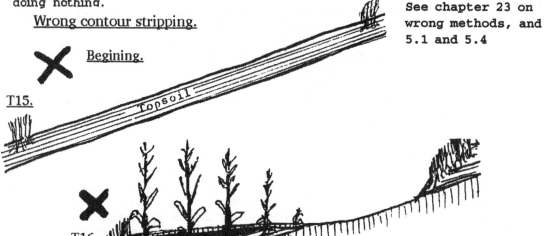

✗ <u>Begining.</u>

T15.

Topsoil

✗

T16.

When grasses are used to make contour strips Napier grass should not be used. T17/. is an example of a very unsuitable contour strip. The relative space between contour strips is often much greater than this illustration, so that unless the topsoil is unusually deep, a large part of the field will be bare of topsoil when or before the terraces have formed. Water will flow over the bare land and erode the terraces. Though the first soil washed down is the top soil, it is later covered by the poorerunder soil. Eventually that is covered by washed down subsoil, and so neither the crops nor the contour strip plants can grow.

Steep slope: early and fast run off.

<u>Later</u>

The wider the spacing between the contour lines the wider the area of useless land.

Poor plant cover.

✗

Run off still occuring.

T17.

In a wet season this may slide down

Napier grass takes nutrients from surrounding soil and starves nearby crops.

Rats like to live among the roots of Napier grass, their burrowing causes collapse of the contour banks, they also rob the crops.

The difference in height between the contour strips should be limited; so that when the land between the strips becomes a level terrace, the part which was at the top of the slope will still have at least 5" (12.5 cm's) of topsoil.

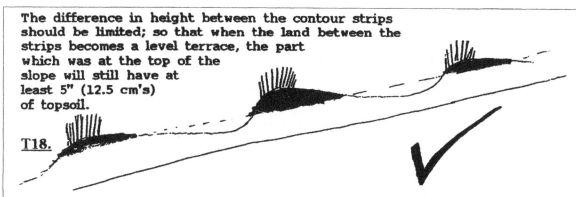

T18.

This means that on steep slopes or where the soil is very shallow the contour strips would be so close that little cropland would be left. This method may still be useful in order to have mini terraces develop naturally. After these are complete the grasses may be taken up and such crops as tea, pineapple, banana, papaya, sweet potatoes, peanuts, ginger, asparagus, or mulberry bushes for fruit or feeding silk worms, grown. Mini terraces should be made where the underneath soil and subsoil is soft and weak. See mini terraces.6.8

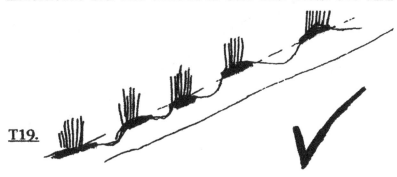

T19.

HALF TERRACES:
Conventional half terraces are made by digging a ditch and throwing the soil downhill to make a bank. In theory they stop the water flowing down the hill. Some half terraces are sloped sideways so that excess water runs to waterways. If rainfall is light they can be useful, the runoff water does not increase to a level which causes erosion.

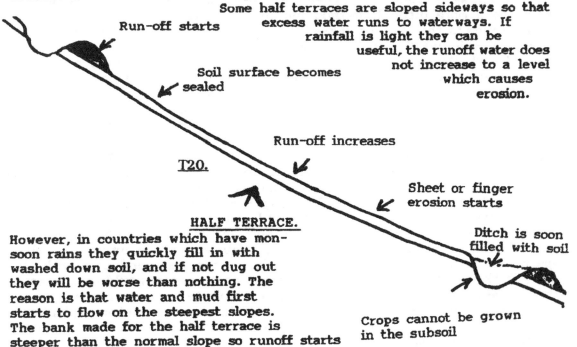

Run-off starts

Soil surface becomes sealed

Run-off increases

T20.

Sheet or finger erosion starts

Ditch is soon filled with soil

HALF TERRACE.

However, in countries which have monsoon rains they quickly fill in with washed down soil, and if not dug out they will be worse than nothing. The reason is that water and mud first starts to flow on the steepest slopes. The bank made for the half terrace is steeper than the normal slope so runoff starts

Crops cannot be grown in the subsoil

earlier on the banks. The water and mud runs down and over the surface of the field sealing the soil so that it cannot absorb the rainfall as easily. This results in more runoff which runs down over the now filled in ditch. It runs quickly over the bank adding to the water already running off the bank and gaining speed and the amount of the water and mud. If the field had been left there would not have been the early runoff caused by the steeper slope on the half terrace banks and in more of the rainfalls the rain which fell could have been absorbed where it fell.

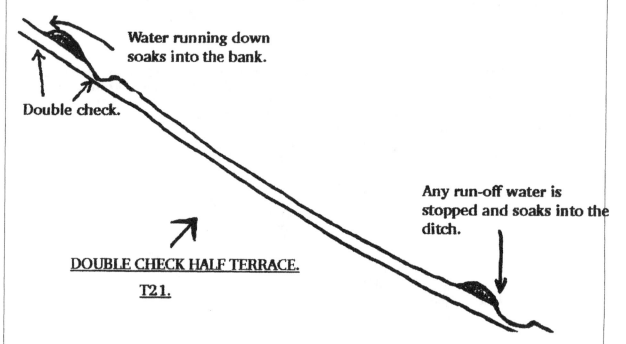

Water running down soaks into the bank.

Double check.

Any run-off water is stopped and soaks into the ditch.

DOUBLE CHECK HALF TERRACE.

T21.

DOUBLE CHECK HALF TERRACES:

To avoid the problems with conventional half terraces, Tansen or double check half terraces were tried, and proved to be much superior. At first sight it looks to be a more difficult method but in fact it is not.

The difference is that two or three furrows are ploughed along the contour; turning the soil uphill if possible, loosening a strip of soil 2-3 feet (60-90 cm's) wide. The loose soil is then thrown to the top side making a bank, so it is the reverse of normal half terraces. Although it seems harder to move the soil uphill, in fact there is little difference as with both methods the soil has to be lifted; when it is already lifted there is not much difference which side it is put on. In fact as will be seen less soil needs to be moved than with the normal method so there is less work than with normal half terraces. A key difference with this method is that the bank which as explained is the place where water run off first starts, is followed by the ditch. This means that there is less run off and so less erosion and less filling in of the ditch. The absorption area instead of only being the ditch is first the top side of the bank, then if rain is very heavy and water flows over the bank it is stopped by the ditch. Because of less runoff and the double check the ditch does not need to be as deep and so it can still be planted either by fodder crops or by field crops. In trials over five years we found that with this system if the double checks were as close as terraces would be, terraces partly formed naturally, so they were then converted into terraces. (To convert double check half terraces into full terraces see the explantation following throwing over terracing.) However this method does slowly convert to terraces; for explanation see T34 and T35.

NEITHER HALF TERRACES OR DOUBLE CHECK HALF TERRACES CAN NORMALLY BE USED ON VERY STEEP SLOPES.

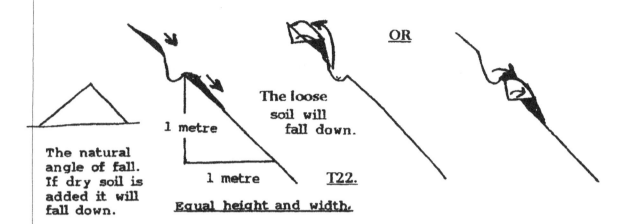

OR

The natural angle of fall. If dry soil is added it will fall down.

1 metre

1 metre

The loose soil will fall down.

T22.

Equal height and width.

Soil conservation work is normally done in the dry season when the land is empty and the soil is dry and lighter to move. The natural angle of slope of soil is approximately 60˚ or 133%, loose dry soil less. If we put soil on slopes of this slope or greater slope it will fall down the hill. If we do use one of these methods we must make a level base to build the bank on.

A problem with all these indirect terrace making methods is that it takes several years, perhaps 20 or more, for the terrace to develop, during that time plant foods, soil and water are being lost, except with the double check half terrace system, if it is maintained. It is not always realised that plant foods have to be soluble like salt or sugar so as to pass into the plant roots; when they are dissolved in water they are without colour, and so water running off land which looks clear may contain plant foods.
It is harder to cultivate, and harder to work on sloping land. Even though the soil at the top of the slope may still be deep enough when the strip has worked level; the underneath soil will be shallower. This will mean that the soil will dry out quicker, and there will be uneven growth over the terrace, with better crops grown where the topsoil is deeper. As the developing terrace was built on a slope it could also slide in a very wet season, or if there was an earthquake. Landslides would however be less likely to happen with indirect terrace making methods than with direct dragged down terraces, as the soil on a terrace which builds up over years is steadily consolidated (becoming firm) and joined to the soil under it.

FOR WAYS TO CONVERT THE INDIRECT SYSTEMS INTO CORRECT TERRACES WITH THE TOPSOIL SPREAD OVER ALL THE TERRACES SEE BELOW THE DIRECT TERRACE MAKING SECTION. T33-T36

DIRECT TERRACE MAKING.
THROWING DOWN METHOD.
If all the land is to be terraced at the one time, we normally start at the bottom of the land. After marking out the contour lines, (see chapter on surveying) the top soil from the strip between the bottom two contour lines is thrown to the land below. Or, it can be carried to above the strip of land which will be the top terrace or to another suitable place. Next a step is made in the foot of the terrace see T23; then the rest of the strip is levelled with a back slope which is set as explained in the Appendix 7, surveying.

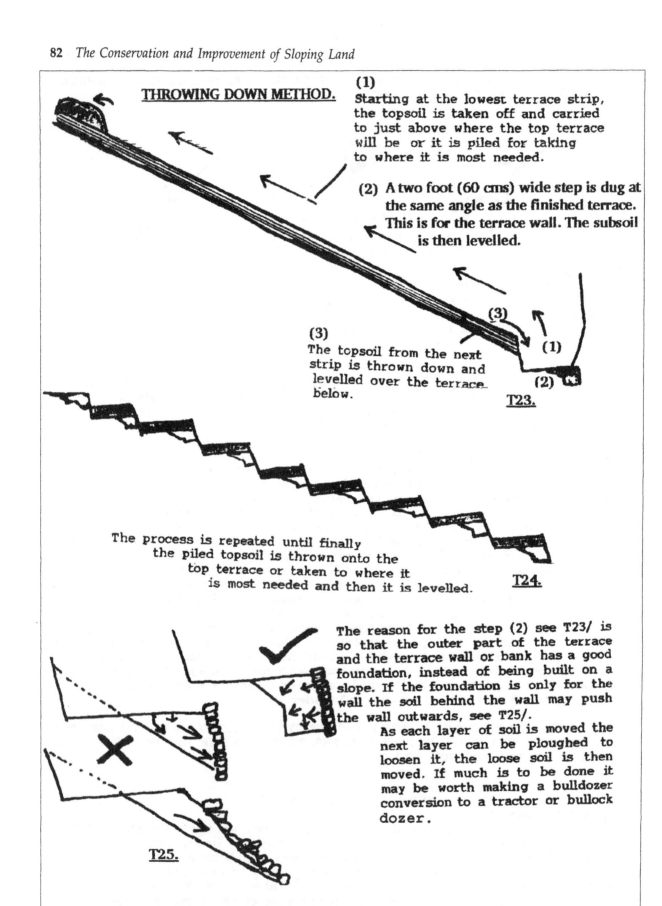

THROWING DOWN METHOD.

(1) Starting at the lowest terrace strip, the topsoil is taken off and carried to just above where the top terrace will be or it is piled for taking to where it is most needed.

(2) A two foot (60 cms) wide step is dug at the same angle as the finished terrace. This is for the terrace wall. The subsoil is then levelled.

(3) The topsoil from the next strip is thrown down and levelled over the terrace below.

T23.

The process is repeated until finally the piled topsoil is thrown onto the top terrace or taken to where it is most needed and then it is levelled.

T24.

The reason for the step (2) see T23/ is so that the outer part of the terrace and the terrace wall or bank has a good foundation, instead of being built on a slope. If the foundation is only for the wall the soil behind the wall may push the wall outwards, see T25/.

As each layer of soil is moved the next layer can be ploughed to loosen it, the loose soil is then moved. If much is to be done it may be worth making a bulldozer conversion to a tractor or bullock dozer.

T25.

The reason for the step.

B1. The bulldozer is raised as the rotovator is lowered to break up the soil ready for the bulldozer to push it to where it is needed.

B2. The Good News Bulldozer Conversion. After the rotovator has dug up the soil it is raised and the bulldozer lowered to push the loose soil to where needed.

B3. The topsoil from this strip was pushed to the terrace below, then the subsoil levelled. Next the topsoil from the strip above is pushed down and spread over this terrace.

B4. While the first terrace are growing crops, during a slack time more slope is being converted to a lasting more fertile future.

T25(a)

With the shortage of labour in rural areas there were now walking tractors for cultivating the rice fields and other work in most villages. I realised that in many areas in developing countries, there are walking tractors with rotary cultivators used for rice field cultivation. Some are normally used for pulling trailers, but the cultivators are available for fitting to them. I wondered if we could make a bulldozer conversion which people could hire and fit on their local tractor and use it as a bulldozer; first using the cultivator to break up the soil, then reversing, lowering the dozer blade, and pushing the loose soil to where needed. Such a bulldozer could be used for levelling land, making terraces and local roads or tracks.

After the bulldozer has made the terraces the land can be cultivated by tractor and all work is easier.

I drew up a simple plan and sent it to an organisation for agricultural engineering research, for their advice and opinion. They replied that they thought it should work, they would be interested to know if it did.

However, when I had it made, I found that it did not work. I had thought that when the depth wheel was lowered to raise the cultivator the bulldozer blade would come down into the loose soil, then as the tractor moved forward the soil could be pushed to where needed. The reason why it did not work was that there were now three weight bearing points, the depth wheel, the drive wheels, and the bulldozer. As the tractor tried to push the soil the drive wheels slipped and turned without being able to move the tractor.

I found that the solution was to make cranked push bars, they were fastened to the casing for the drive shaft of the cultivator. When the tractor moved forward and the bulldozer blade pushed into the soil the push arms pushed down on the axle, and so the more the soil was pushed the more the pressure on the wheels. Instead of the wheels slipping they gripped better and so if the bulldozer depth wheels were too high it was possible to stall the engine; or if the depth wheel on the tractor was set too low.

The bulldozer parts were made by a local welding shop, the wheels were paddy field wheels, with spuds welded on instead of the normal paddles.

The conversion kit was designed to be able to fit a range of walking tractors, so that it could be hired out and fitted to any local tractor.

B5.   The Good News Bulldozer conversion can fit on most two wheel rice tractors. First the cultivator is lowered to loosen the soil, then the loose soil is dozed. The picture shows a rock being dug out of the ground.

We found that it was better not to use the depth wheel on the tractor to control the bulldozer action. Instead we used the depth wheels on the bulldozer to control the depth, and adjusting the angle of the blade till it easily entered the soil. We used scrap broken truck springs as weights attached below the tractor handle bars to act as counterweights to the bulldozer when we wanted to lift the bulldozer out of work. The bulldozer worked very well as can be seen from the pictures.

We were very pleased with the performance particularly when starting to make terraces when we were driving across the slope at the start of a terrace. We found that unlike a normal bulldozer the Good News Bulldozer could go crabwise (diagonally) across the slope using the steering clutches to control the direction, so that while the tractor wanted to slip down the slope we could aim it slightly up the slope to compensate and still angle doze the soil from the top of the slope down to where it was needed. Only once did we have it roll over, it was when the lower wheel slipped into a hole where a rock had been. However the two of us easily rolled it back onto it's wheels.

The bulldozer was designed to push the soil either forward, or to either side. It was pivoted in the centre and the straight parts of the push bars (which were made of tubular steel used to make tripods for lifting logs of wood) telescoped with pins through them to lock each position.

The bulldozer depth wheels and their screws were scrap from something else but could be simply wheels attached to a sliding upright bar with ways of fixing it at different positions, perhaps by welding a bolt to the top which could be raised or lowered and locked by two lock nuts.

Soil conditions are often uneven, such as where some patches have very little top soil while others have more than is needed; or where some patches have many rocks which are taken to make terrace walls, and so the remaining soil is less than in other parts.

In those area's the terrace becomes lower and so run-off concentrates there; in heavy rain this may flow over the terrace and cause a gully to develop. For that reason we need to be able to level the subsoil of the terrace; we do this by taking subsoil from the higher parts to the lower parts and making the terrace level but with the correct backslope. See Appendix 7 Surveying for the correct back-slope.

The process is repeated with the terrace above, with the topsoil being thrown down on to the level terrace below it, and spread evenly. When finished the topsoil which was thrown down from the first strip is taken to where it is needed. This may be to the top terrace, or to add to terraces which do not have much good soil in order to make the terraces more evenly fertile. The top terrace may have other soil added or be planted with soil improving crops such as Sunnhemp or Adzuki beans; see the chapter on soil improvement in volume 2. Or it may be planted with trees or bamboo.

It may seem a big job to carry the topsoil of the bottom terrace somewhere else, but we should consider that each year manure is carried up to the fields and yet much of the manure on steep slopeland is washed away in the runoff water and so wasted. When the fields are properly terraced all the manure and plant food stays in the soil so the labour of making it and carrying it from the farm to the fields is not being wasted; the same amount of manure will grow more food. Also, with this method all the top soil (the fertile living soil), is saved and is kept on the top.

Trials carried out in Nepal comparing two strips of hillside next to each other and of the same size, and with the same amount of manure on each; gave the following yields of maize the first year after one was terraced using this methods.

The traditional sloping field gave 7.8 kg of cobs and 12.5 kg of stover (maize straw). The terraced strip gave *31.3 kg of cobs 29 kg of stover. See 6.0.*

Soil loss could easily be seen on the slopeland but was absent on the terraced land. The terraced land would continue to produce crops, the slopeland would continue to lose soil and plant foods, and soon be uneconomic to use for crops. For relatively small areas of land where there is sufficient labour available or where some mechanization can be used, this method can transform the future in a sustainable way, provided of course that good management is used, as explained in volume 2.

This method is good in that it keeps all the living topsoil on the top and except for the bottom terrace topsoil, soil or subsoil is only moved once apart from levelling off. It can also be used with a bulldozer or bullock dozer.

Throwing Over Method.

This is useful for three kinds of situations:

1. When terracing a single strip of land or a small area, where the throwing down method is unsuitable.

2. Where a hillside while not soft is not very stable (firm).

3. It can be used as part of the Good News Method.

(a) If a series of terraces are made the topsoil from the lower one third of the strip is ploughed or loosened and thrown onto the middle one third. The topsoil from the top one third of the strip is thrown onto the land above. The subsoil from the top or inner one third is ploughed or loosened and thrown over to the lower one third. This is repeated until the lower/outer one third is higher than the inner one third. The height difference needs to allow for the loosened soil settling, when the outer one third should still be slightly higher so that excess water will run towards the inside of the terrace. See T26. T27. and T28.

(b) If for some reason the topsoil cannot be thrown onto the land above it is thrown on top of the topsoil from the lower one third which has been put onto the middle third.

With all terracing systems it is best to:-

A) Include a step in the terrace wall into which trees for green manure or fodder, and creeping

legumes for fodder, green manure, and for protecting and binding the soil, are planted, see T44.

B) Check that any surface water from above will be diverted, and not run onto the area; unless systems for trapping water which are explained in the next chapter on water conservation are being used.

C) Check that there is a back-slope from the outer edge of the terrace to the inside edge. This should normally be 5% which means that the inside edge should be 5 cm's (2") lower for each metre (6' 7") of distance across the terrace. So if' the terrace is 2 metres wide the inside edge should be 2 x 5= 10 cm's lower than the outer edge. This slope can easily be set using the Good News Level, see the Surveying. Appendix 7.

The reasons for this slope are:-

1. The soil at the outer half of the terrace is loose soil, and it will settle down over the next year or more. So without a backslope when the terrace is first made, there would be a forward slope after the soil had settled down. This would run excess water over the edge and onto the terrace below. After the soil has settled there should still be a slight back-slope, so that any surface water will flow inwards to where the soil is firmest

As mentioned before there should be a slight variation for different soil. If the soil packs down easily, such as with sandy soil, slightly less slope should be used. If it is very lumpy, spongy, or loose and so does not pack

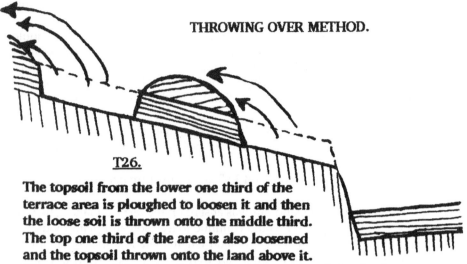

THROWING OVER METHOD.

T26.

The topsoil from the lower one third of the terrace area is ploughed to loosen it and then the loose soil is thrown onto the middle third. The top one third of the area is also loosened and the topsoil thrown onto the land above it.

The subsoil soil is then ploughed and a step made in the lower strip as shown for the throwing down method, see T25 and T27, the rest of the lower strip is levelled. The loosened subsoil from the top one third is thrown over onto the lower one third. More subsoil is loosened and thrown over until the lower one third is slightly higher. The aim is that when the looser soil on the lower one third has settled down it will still be higher so that there will be a slope in towards the hill.

NOTE. If the subsoil is very stony make the lower part less than one third, and the upper more than one third, so that the upper part does not need to be dug our as deeply.

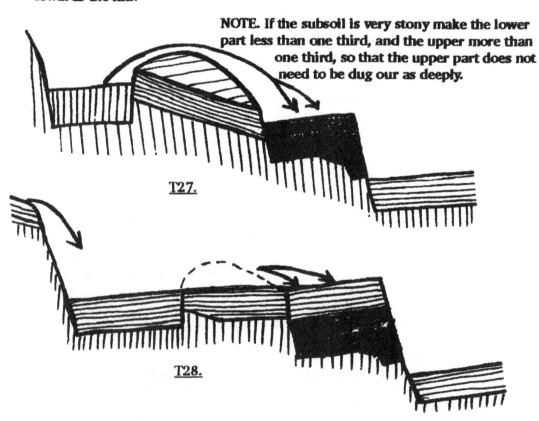

T27.

T28.

Next the heap of soil on the middle one third is thrown onto the lower one third, and the soil from the lower one third of the strip above thrown onto the top one third. See T28/.

The throwing over method has an advantage when working on land which could possibly slide in a very wet season. As the middle strip of soil is undisturbed it acts as a stabiliser to resist any sliding.

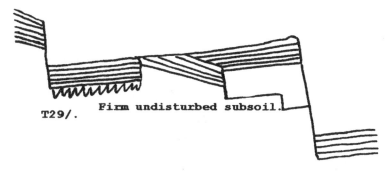

T29/.   Firm undisturbed subsoil.

With this method some of the undersoil will be exposed on the top in the middle strip, see below, it can be mixed into the surrounding topsoil, but if it is important for all the soil to be of even fertility the good soil can be moved from the middle strip, and the underneath soil levelled. Then the topsoil which was moved is spread back over the middle strip. See T30/. This still leaves the deeper soil firm.

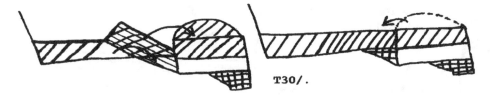

T30/.

SIDEWAYS METHOD.

This method is sometimes suitable for terraces where the soil depth is fairly uniform. The soil is only moved once, and there is not as much topsoil to carry from the first place levelled as with the throw down method. It is suitable for wider terraces where not all the soil can be thrown to the next strip in one throw. At the end the topsoil from the first strip is put on the last strip, or where most needed. There are two disadvantages with this system.

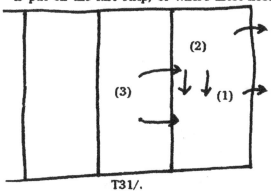

T31/.

(1) The topsoil is thrown to the side.
(2) Level the subsoil.
(3) The topsoil is thrown from the next strip onto the levelled undersoil and subsoil.

A/. If the strip of land has patches of different soil, some areas with more rocks which are used to make the terrace walls and so that area becomes lower. Or perhaps there are areas with not much topsoil, and others with deep topsoil some of which could be taken to the poor areas. With the other methods it is easy to balance up the subsoil before the topsoil is spread over the terrace. Normally that cannot be done with this method. Sometimes it is possible to make the terrace level by shaving more soil from the back bank of the terrace to spread on the low places. For this reason terracing should start from the lower part of the field.

B/. With this method it is not possible to use a plough to loosen up the soil before moving it, except for initially ploughing the top soil, though it may be possible to use a small rotovator.

First, the topsoil is loosened and moved from the end strip and the subsoil levelled.
Next, the topsoil on the second strip is loosened and moved onto the end strip
The subsoil on the second strip is levelled and the process repeated until finally the topsoil from the first strip is spread over the last strip or where it is most needed.

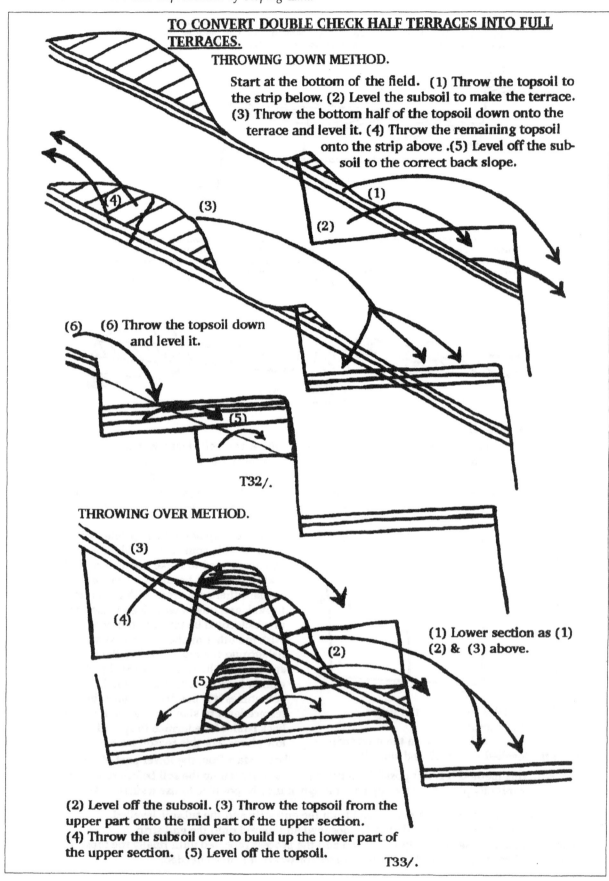

TO CONVERT DOUBLE CHECK HALF TERRACES INTO FULL TERRACES.

THROWING DOWN METHOD.

Start at the bottom of the field. (1) Throw the topsoil to the strip below. (2) Level the subsoil to make the terrace. (3) Throw the bottom half of the topsoil down onto the terrace and level it. (4) Throw the remaining topsoil onto the strip above .(5) Level off the subsoil to the correct back slope.

(1)

(2)

(3)

(4)

(6) (6) Throw the topsoil down and level it.

(5)

T32/.

THROWING OVER METHOD.

(3)

(4)

(2)

(5)

(1) Lower section as (1)
(2) & (3) above.

(2) Level off the subsoil. (3) Throw the topsoil from the upper part onto the mid part of the upper section.
(4) Throw the subsoil over to build up the lower part of the upper section. (5) Level off the topsoil.

T33/.

HOW TO CONVERT HALF TERRACES AND CONTOUR STRIPS TO SUSTAINABLE TERRACES.

As explained earlier, half terraces and contour strips should not be used unless the slope is gentle, the rainfall is not heavy, and the good soil is deep. Even then it is better to use double check half terraces.

Typically the end result of contour strips and half terraces is as shown below. If the land had been terraced first the volume of topsoil would be greater as it would have all been kept on the hillside. It is often claimed that contour strips and half terraces stop all loss of soil and plant nutrients; in my experience this is not correct. While some soil is kept and builds up, other soil and plant food continues to flow down the slope,

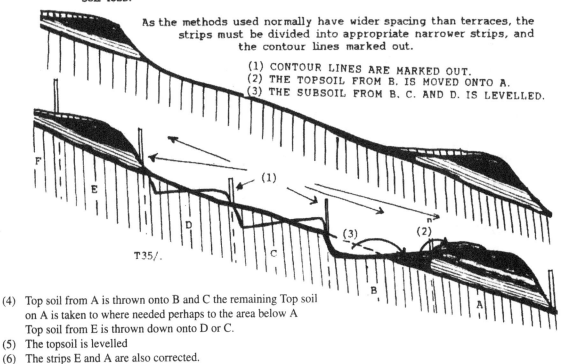

particularly where the slope is long. These methods can even increase overall soil loss.

As the methods used normally have wider spacing than terraces, the strips must be divided into appropriate narrower strips, and the contour lines marked out.

(1) CONTOUR LINES ARE MARKED OUT.
(2) THE TOPSOIL FROM B. IS MOVED ONTO A.
(3) THE SUBSOIL FROM B. C. AND D. IS LEVELLED.

(4) Top soil from A is thrown onto B and C the remaining Top soil on A is taken to where needed perhaps to the area below A Top soil from E is thrown down onto D or C.

(5) The topsoil is levelled

(6) The strips E and A are also corrected.

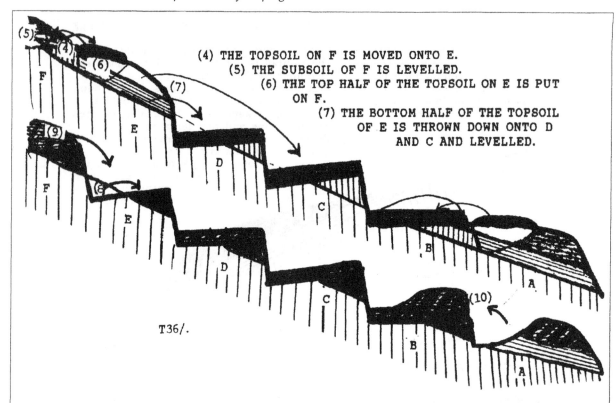

(4) THE TOPSOIL ON F IS MOVED ONTO E.
(5) THE SUBSOIL OF F IS LEVELLED.
(6) THE TOP HALF OF THE TOPSOIL ON E IS PUT ON F.
(7) THE BOTTOM HALF OF THE TOPSOIL OF E IS THROWN DOWN ONTO D AND C AND LEVELLED.

T 36/.

(8) THE SUBSOIL OF E IS LEVELLED.
(9) HALF OF THE TOPSOIL OF F IS THROWN ONTO E AND THE TOPSOIL OF F AND E IS LEVELLED.
(10) THE TOP HALF OF THE TOPSOIL ON A IS PUT ONTO B.
(11) THE SAME PROCESS IS REPEATED ON THE LOWER CONTOUR STRIP AND THE OTHER CONTOUR STRIPS OR HALF TERRACES.

FROM WHAT WAS SAID BEFORE IT WILL BE SEEN THAT UNLESS THE GOOD SOIL IS DEEP AND THE STRIPS ARE NARROW, NEITHER CONTOUR STRIPS NOR ORDINARY HALF TERRACES ARE SUSTAINABLE SYSTEMS, THEY WILL RESULT IN AT LEAST SOME OF THE SLOPELAND BECOMING INFERTILE. IT WILL ALSO BE SEEN THAT TO CONVERT THEM LATER TO REASONABLE TERRACES WITH SOME TOP SOIL, IS COMPLICATED, IT TAKES MUCH MORE WORK, AND SO IS SELDOM APPROPRIATE.

down well the slope should be slightly steeper.

D) The next bunds should be made along the outer edge of the terrace and at intervals of 5 - 10 metres along the terrace, according to whether the soil, is very open or not.

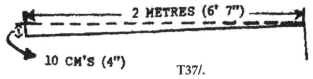

T37/.

The reason for this is that normal level soil can absorb the rainwater which falls on it, particularly if it has already been wetted, but it can't absorb much extra water. Even if finished terraces were exactly flat it is unlikely that they will stay absolutely flat, some areas will settle (sink lower) more than others. In heavy rain water will move to the lowest places, these will be unable to absorb it. The water may then increase until the new terrace may break, or the water will flow over onto the terrace below, adding more water to it. That would increase the chance for the same thing to happen to the lower terraces; a landslide or gully may develop.

If the soil is very open it may absorb the extra water until it becomes heavy and soft and may then slump or slide.

By limiting surface water to it's own limited area it will be absorbed and the build up of water will not occur.

When ploughing simply lift the plough over the bunds. The following year the bunds should be repaired. After that the soil will normally have become firm enough, and if the terraces are fairly level the bunds will not be as important. Remember that this repair of the bunds should be done before the rains start, as dry dusty soil easily seals and causes run-off. After the soil has become damp the water usually soaks into it easier.

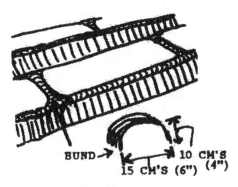

T38/.

NOTE a) If the land is unstable unsuitable contour farming or broad based terraces can cause landslides.

NOTE b) It is important to realise that occasionally there is unusually heavy rainfall. Preparation should be made for such an occurrence. If terraces or contour bunds are too long, water accumulating in particular areas could burst the terraces or bunds, perhaps causing gullies or landslides. If the soil is unable to absorb a sudden heavy rainfall contour lines should slope to waterways. It may be that the soil can as is usually the case, absorb the water falling on it, but is not able to absorb extra water from neighbouring areas. To prevent that occurring, cross bunds sometimes called basining should be made to keep the water to the area in which it falls.

NOTE c). The horizontal height difference between terraces should vary according to varying slope, the type of land, and the depth of the soil. Some land has very shallow soil and so the height between terraces should be less or too much digging out and moving of hard and perhaps rocky subsoil will be needed. Similarly stony soil will probably have rock fairly near the surface, so it will be difficult or impossible to make deep terraces. The water holding capacity of the soil should also be considered, if the soil becomes too wet and heavy it is more likely to slide. Softer soil will more easily become too heavy and slide if the terrace is too big. Gravelly soil will also be more likely to slide when wet.

6.6 What is practical and sustainable?

If the soil is fairly deep double check half terraces briefly described earlier, can be practical and sustainable, but will take time to develop during which soil may be lost in prolonged heavy rainfall.

Apart from being a sustainable way of land management, does terracing make much differences to the potential yield of crops grown on them?

Since first working with comparing different soil conservation practices in 1965-68, properly made terraces always gives better crop yields. One example I can give is of a trial made on a piece of fairly uniform slopeland. Two strips were measured off with identical area, one was terraced manually by the throwing down method, the other left as the rest of the area, cultivated and sown still sloping, a common practice in the area. Both strips had identical manuring and were sown with maize at the same time, yet the difference in crop yield was clear. The slopeland strip giving less than one third of the yield of the terraced strip. On analysis I came to the conclusion that the reasons were:

• When the early rains fell, though they were fairly heavy they only lasted for two or three days, followed

WHAT IS PRACTICAL AND SUSTAINABLE:

The trouble with the methods explained up to this point is that either they are slow and do not really save the soil, or they need a lot of labour.

As mentioned earlier it normally takes about 600 years or more to develop 25 mm's (1") of topsoil from rock. Perhaps 500 years to develop 25 mm's (1") of subsoil. Yet it is very common for 6.5 mm's (¼") of soil <u>150 years of soil development</u> to be lost in one year on a sloping field. The average loss of soil on steep maize fields during the annual monsoon, on the land facing this writers office in Nepal was 65 mm's (2½")<u>1,500 years of soil development lost in one year.</u>

From the trials carried out in Nepal two methods were shown to be practical for the average farmer.

THEY WERE THE DOUBLE CHECK HALF TERRACE, (SUITABLE FOR DEEP SOIL.) AND THE GOOD NEWS OR STEP BY STEP COMPLETE TERRACING SYSTEM.

Normally I would suggest that the double check system described on T21 is used as a first step before the Good news system, however as mentioned before our experience was surprisingly good and with a short distance between the double checks they will turn into terraces. There will not be too much loss of topsoil if managed properly. The explanation seems to be as shown in the illustrations which follow. The final stage as with the Good News system is to dig out a step in the terrace bank throwing the good soil onto the inside part of the terrace below. Plant fodder of green manure trees such as Leucaena, Sesbania sesban, or Acacia albido, along the step. The trees will cause little shade, their roots will help to hold and stabilise the soil, their leaves are used for fodder to make more manure for the soil, or, used directly as green manure. Their branches are used for firewood, while straight trees may be left at intervals and allowed to grow taller to be used for building materials.

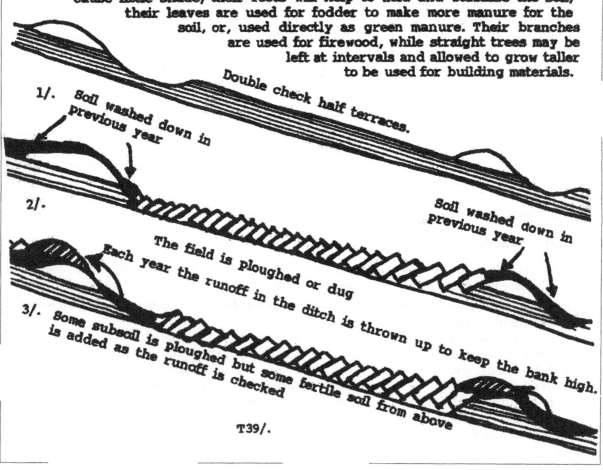

Double check half terraces.

1/. Soil washed down in previous year

2/.

Soil washed down in previous year

The field is ploughed or dug

Each year the runoff in the ditch is thrown up to keep the bank high.

3/. Some subsoil is ploughed but some fertile soil from above is added as the runoff is checked

T39/.

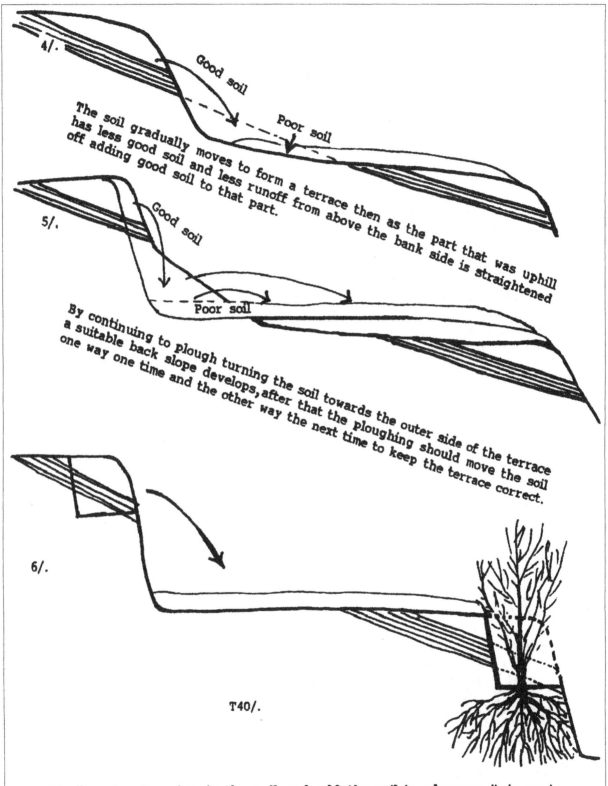

4/.

Good soil

Poor soil

The soil gradually moves to form a terrace then as the part that was uphill has less good soil and less runoff from above the bank side is straightened off adding good soil to that part.

5/.

Good soil

Poor soil

By continuing to plough turning the soil towards the outer side of the terrace a suitable back slope develops, after that the ploughing should move the soil one way one time and the other way the next time to keep the terrace correct.

6/.

T40/.

Finally cut out a step in the wall and add the soil to wherever it is most needed. Plant green manure or fodder trees, grasses and creeping legumes in the step. The trees are kept cut low so that they do not cause too much shade, or Acacia albido which sheds its leaves in the rainy season is used.

by the usual hot dry weather for a few weeks until the main rains started. On the slopeland much of the rain ran off, and what soaked into the soil did not soak in as deeply as on the terraces. A) because the land was sloping. B) Because the subsoil was hard and dense. On the terraces all the rainfall had soaked into the soil and because the soil was more open the water had penetrated deeper.

- After the rain stopped the slopeland quickly dried up. The plant roots had not been able to develop much before growth stopped due to lack of moisture, it could only continue after the main rains came. The plants on the terraces continued to grow, and as the moisture in the top of the soil dried up the roots penetrated deeper to where there was still moisture, their root systems became well developed, ready for strong growth when the rain came.

- When the main rains came the plants on the slopeland were poor and stunded, and as the roots were shallow growth was slow. Even when the plants grew larger the roots could not develop well due to the shallow soil and hard dense subsoil. When the rain was heavy, the shallow soil became too wet so the shallow roots could not breathe. Much of the manure in the soil was washed out and down the slope in the run-off water.

- The plants on the terraces grew away quickly and soon shaded the soil, protecting it from the battering of the monsoon rains and UV light, so there was less leaching (loss) of plant foods. There was no run-off, so the manure stayed in the soil to feed the plants. Because their roots could penetrate deeply into the soil they were better developed and could reach more plant foods. Because the soil was open and deep the heavy rain could soak through and not flood the soil, so the roots could still breath and feed. The plants grew tall and strong and so produced more and much heavier cobs.

The land used was in fact south facing in the northern hemisphere so subshine was not a factor. However, we can realise that on colder north facing sloped terraces will produce better crops as can be seen in T40/a.

I expected the farmer to be keen to make more terraces but as he was also a shopkeeper and the land was not his he considered the labour involved was too much for him to bother.

This prompted me to consider how we could develop a more practical way to stop the soil, water and plant food loss quickly and develop the slopeland into a sustainable and improving system, the Good News Terracing System was the result.

Why not use these direct or indirect terracing systems?

The problem with all normal direct and indirect making methods is that it takes several years, perhaps 20 or more, for the land to be terraced or the indirect terraces to develop. During that time plant foods, soil and

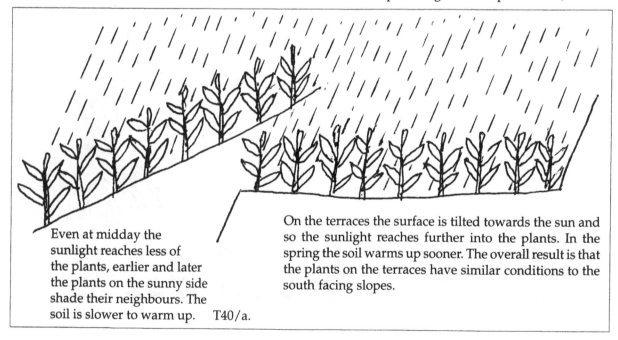

Even at midday the sunlight reaches less of the plants, earlier and later the plants on the sunny side shade their neighbours. The soil is slower to warm up. T40/a.

On the terraces the surface is tilted towards the sun and so the sunlight reaches further into the plants. In the spring the soil warms up sooner. The overall result is that the plants on the terraces have similar conditions to the south facing slopes.

water are being lost, (except with the double check half terrace system if it is maintained and the half terraces are close together). It is not always realised that plant foods are soluble like salt or sugar so as to pass into the plant roots; when they are dissolved in water they are without colour, and so water running off land which looks clear may contain plant foods. As the developing indirect terraces are built on a slope they could also slide in a very wet season, or if there was an earthquake.

It is harder to cultivate and work on sloping land. Even though the soil at the top of the slope may still be deep enough when the strip has worked level; the underneath soil will be shallower. This will mean that the soil will dry out quicker, and there will be uneven growth over the terrace, with better crops grown where the topsoil is deeper.

As mentioned, when using a small bulldozer, the throwing down method is a good method in that except for the topsoil on the first strip - the soil is only moved once, and it is thrown downhill. All the topsoil ends up on top and the crop yields are increased dramatically compared with those from sloping fields.

This method can also be done manually unless the terraces are very wide.

However, in many situations it needs more labour than is available or people are willing to give or pay for, and it is a slow method. It may take an average farmer in a country like Nepal 10-20 years to terrace his farm. During that time his land would have lost much soil and plant food.

It would be better to have a system which although it did not keep all the topsoil on the top would stop the loss of soil and plant foods and water quickly. Even though some of the topsoil was buried the total topsoil on the top would be at least as much as it would be if it took several years before the land could be terraced.

The BIG QUESTION was how to have a system which was quick and easy and practical for normal farmers to use?

The Good News step by step terracing system is the answer. I have proved that it works. I believe that two or three neighbouring farmers can get together and by spending two or three weeks a year, will be able to stop soil and plant food loss in two or three years. In 9 - 12 years all their land will be terraced and the soil fertility stabilised or improving. If this method was used widely there would be a dramatic change in the future of many third world hill farming communities; with hope and security, where they now face no hope, and no security, when in future years the soil can no longer produce worthwhile crops.

6.7 The Good News Step by Step Terracing. See T41

The Good News terrace system was developed in response to the need to develop credibility quickly, with a method which is feasible for ordinary small farmers to manage with the help of a neighbour.

The concept behind it is to stop the loss of soil and plant nutrients as quickly, simply and effectively as possible, and to show that the effort will be economically worthwhile.

The basic idea is that rather than try to terrace whole slopes a staged process is used. In the first stage terraces are only made at sufficient intervals of 2,3 or 4 strip distances, to stop the run-off of water, and with it soil and plant foods. As farmers see that the terraced land produces better crops they are encouraged and, so as they have time and resources they convert all their cropped slopeland by stages into sustainable fully terraced fields.

When all the land has been covered with the first step, the process is repeated with the strips in between the first terraces. When this is completed the terraces are widened to join up with those above and below; finally fodder and or green manure trees and planted in steps in the terrace walls; and Green leaf Desmodium is planted to cover the terrace walls and banks.

The Good News Step by step system of terracing can be used with whatever method is locally appropriate and results in a) stable terraces, and b) terraces where most of the topsoil will be kept on top.

The terraces will initially need to be wide enough for bullocks to turn with the ploughs, later they may be widened. In our experience making such terraces using the method described, will on average hillside land with two people and a team of bullocks on average create 10 metres of terrace per hour. In seven hours work 70 meters, six days = 420 metres of terrace; even at less than half the speed say 200 metres per week. If two farmers get together for a month in the slack season they could each have at least 400 metres of terraces made. See T40/d.

This terrace admittedly easier than most, has all the topsoil on the top, and yet was built averaging 10 meters per hour for two men, so for a six hour day 60 metres. Two farmers working for one week of six days 360 metres, the following week another 360 metres on the other farm would stop the loss of soil and nutrients on the areas worked. The crops would be so much better on the terraces, and encourage the two farms to do more terracing the following year till all was terraced. T40/b.

Good news terracing the first step. Soil and water moving down the slope is stopped by the terrace below it, so there is no loss. The terraces produce better crops which encourage more terracing until the first step is finished, then the slope land is also terraced. T40/c.

Good News Terracing.

1. What is needed is a method which quickly stops the loss of plant foods, soil, and water, and which is feasible for ordinary farmers to use. A method which shows quick and profitable results, and is efficient in terms of labour expended, is sustainable and results in increasing fertility.
The Good News Terracing method does that.

The method does not terrace all the land at once but uses a step by step approach. Contour lines are marked out and interceptor ditches are dug to divert outside water from running onto the area 1. Next one in two, three or four contour. strips are made into terraces, see 2 and 3. The terraces are not made the full width of the strips, this means that the area can be covered fairly quickly, the runoff water, plant foods and eroded soil are checked by the terraces and so erosion stops. The crops grown on the terraces will be much better than those grown on the slope, this encourages the farmers to make more terraces the following year, see 4.
The terraces are not built on a sloping surface but are keyed into the hillside and so if made properly cannot slip, see 5.
When all the fields have been treated in this way the process is repeated on the inbetween strips, see 6.
After all the terraces are made they are widened, see 7. and steps cut out into which suitable legume tree seedlings and cuttings of Greenleaf Desmodium are planted, see 7 and 8. The Desmodium will hang down and root into the terrace wall, see 8. It will hold and protect the terrace wall and bank and produce high quality livestock feed or it is cut and used as mulch.

In countries like Nepal most farmers have less than 0.5 hectares of land, assuming that it is all cropland it could be represented as below. 83.3 metres across x 60 metres down = 4,998 (5,000) square metres =0.5 hectare.

60m

83.3m

If the farmer made pilot terraces at the rate of one out of three final terraces it would take 83.3 x 8 =666.4 metres of terracing. In my experience with this system two men with bullocks to plough to loosen the soil, can average 10 metres per hour on fairly good land when making terraces two metres wide. If they work for seven hours per day =70 metres, 666 divide by 70 metres = 9.5 days. Of course on bad land it could take 20 days or even more. However if two farmers get together and work for 10 days on each farm during the slack time of the year they will have checked erosion on at least half of their slopeland fields. As the crops will grow much beter on the properly made terraces the farmers will be encouraged to continue the work each year. Terracing the land in between and if they manage the soil carefully building a sustainable future.

T40/d

Good News Method 1/.

First use a Good News Level to mark out the contour lines, then dig the interceptor ditch if needed. (See surveying, Appendix 7)

1/. Plough the lower 2/3rds of the strip to be terraced.

2/. Throw the loose soil from the top 1/3rd of the ploughed area onto the unploughed land above it, and throw the soil from the bottom 1/3rd onto the middle 1/3rd. Repeat the ploughing and throwing until all the fertile topsoil on the top and the bottom of the strip has been moved.

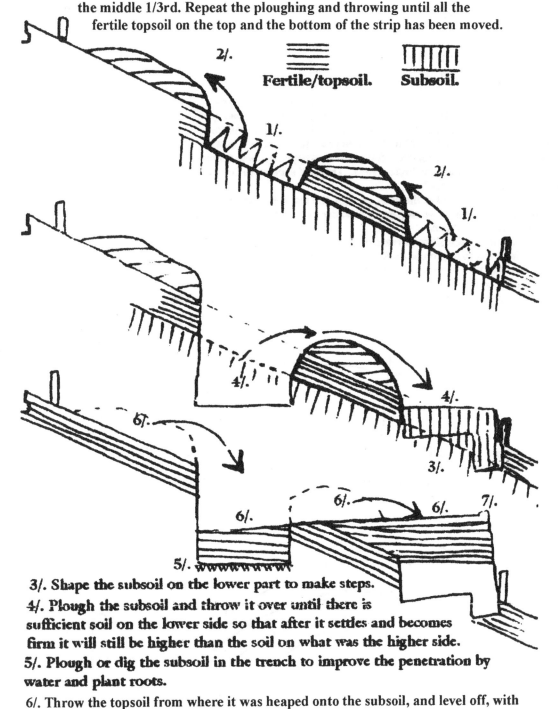

Fertile/topsoil.   Subsoil.

3/. Shape the subsoil on the lower part to make steps.

4/. Plough the subsoil and throw it over until there is sufficient soil on the lower side so that after it settles and becomes firm it will still be higher than the soil on what was the higher side.

5/. Plough or dig the subsoil in the trench to improve the penetration by water and plant roots.

6/. Throw the topsoil from where it was heaped onto the subsoil, and level off, with a suitable slope back into the hill, and with a bound-7/- along the edge.

T42/.

Good News Method 2.
Where the subsoil is of fairly good quality, is also firm, and where the rainfall is not high and so landslip before the terraces become stable will not occur. If in doubt shape the subsoil as in T42 3.

Stage 1. Dig an interception ditch along the top of the area to be terraced. Mark out contour lines where the terraces will be. The Good News Level explained in Appendix 7 **is good for making contour lines.**

Just as with method 1/. the first
need is to stop the loss of moisture, plant
food, and soil from the land as quickly as possible,
so only one in two to one in four terraces will be made at
first depending on the expected amount of rainfall. If it will be heavy
one in two should be made. also for speed, these first terraces will be
approximately 3/4 of their final width. The plant foods washed down
to the terraces from the slopeland above will be stopped by these terraces
and so the terraces will grow better crops. This will encourage future work,
until all the land is terraced.

Stage 2/. leave the top 1/4 of the strip, plough the top half of the remainder.
Then drag or push the loose soil down onto the bottom half, and repeat
until the terrace is formed.

There should be a back slope to run excess water to the back part of the
terrace which has firm ground underneath. There should be sufficient slope
so that after the moved soil has settled there will still be a back slope.
Basins should be formed so that excess water cannot flow sideways to a low
place and accumulate, perhaps breaking the terrace. See the illustration T38/.

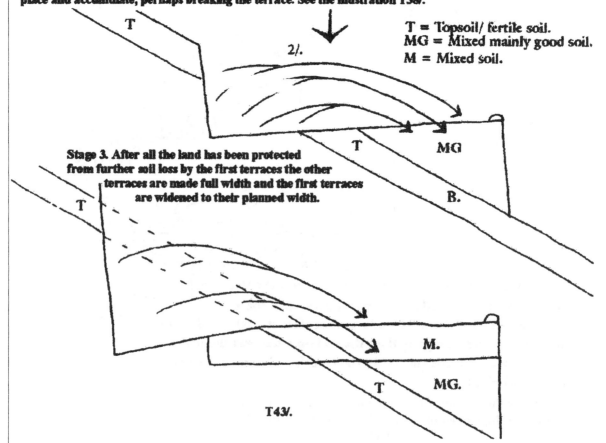

T = Topsoil/ fertile soil.
MG = Mixed mainly good soil.
M = Mixed soil.

Stage 3. After all the land has been protected
from further soil loss by the first terraces the other
terraces are made full width and the first terraces
are widened to their planned width.

T43/.

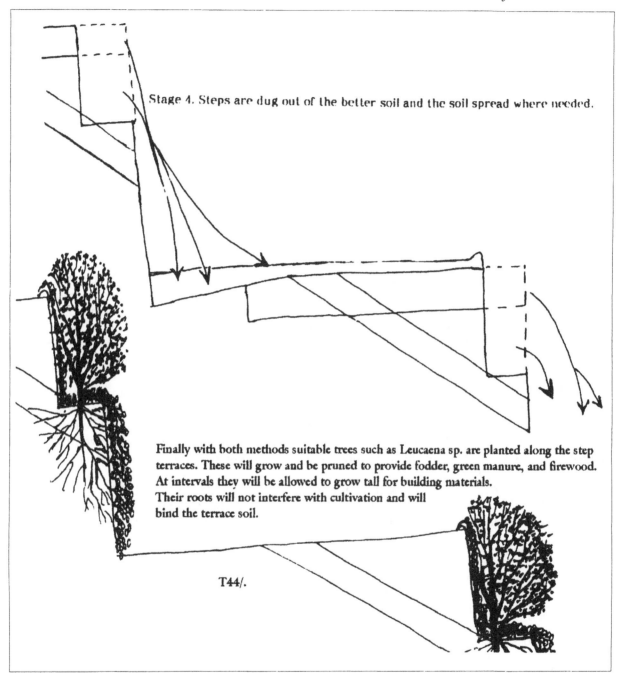

Stage 4. Steps are dug out of the better soil and the soil spread where needed.

Finally with both methods suitable trees such as Leucaena sp. are planted along the step terraces. These will grow and be pruned to provide fodder, green manure, and firewood. At intervals they will be allowed to grow tall for building materials. Their roots will not interfere with cultivation and will bind the terrace soil.

T44/.

It is difficult for people who do not have spades or shovels, it is sometimes harder to throw soil over than drag it down, it may be worthwhile supplying shovels.

We have used the throwing over methods in Good News Terracing and with two men moving the soil and building a pressed earth wall and an old man ploughing as needed, the following rate was achieved. The terrace was two metres wide and 79 metres long and took 5½ hours = 14.36 metres (47 feet) per hour. As the old man only ploughed, if the men had to do the ploughing as well as the other work we could say they would average 10 metres per hour for two men or 5 metres per man hour. If a farmer and his neighbour joined forces marked the contour lines and then worked for 8 hours a day on similar terracing they would be able to do 400 metres in five days. If they spent 10 days terracing = 800 metres terraces, on each of their farms during the slack season in two or three years they would be able to cover their sloping cropland with the first stage; stopping the erosion and loss of water and plant food. They would be getting better crops and it would be easier working on the terraces than on the slope land. Their future would be much more hopeful. In perhaps ten years time the terracing and planting of the green manure/fodder trees and legumes would be completed. They would have a sustainable future if they managed the soil well, in fact the crops would improve so that it would be an improving future instead of a future with at best poorer crops and harder work to produce them.

In another area where the farmer could not get bullocks for ploughing and so all the soil had to be loosened by hand the rate of work was only 3.66 metres per man per hour.

Of course if the bullocks were weak and unable to plough the harder subsoil, or there were many rocks to break up there would have been much slower progress.

If the rocks had to be broken up and built up into the terrace walls it would be much slower, with experience the work will become easier and quicker. It has to be compared with the alternative, the longer nothing is done the harder it will be and the poorer the results. If it is left too long it will be too late; there will be no future.

NOTE a) If the land is unstable (If the underneath structure is itself soft, shale or other loose material, instead of firm solid material or rock, it is unstable,) unsuitable contour farming or broad based terraces can cause landslides. (See T49.)

NOTE b) it is important to realise that occasionally there is unusually heavy rainfall. Preparation should be made for such an occurrence. If terraces, contour bunds or ditches are too long, water accumulating in particular areas could burst the terraces or bunds, perhaps causing gullies or landslides. If the soil is unable to absorb a sudden heavy rainfall, contour lines should slope to waterways. It may be that the soil can as is usually the case, absorb the water falling on it, but is not able to absorb extra water from neighbouring areas. To prevent that occurring, cross bunds sometimes called basining should be made to keep the water to the area in which it falls. See 6.5 T38.

NOTE c) The horizontal height difference between terraces should vary according to varying slope, the type of land, and the depth of soil. Some land has very shallow soil and so the height difference should be less or too much digging out and moving of hard subsoil, stones and perhaps rocks will be needed. The water holding capacity of the soil should also be considered, if the soil becomes too wet, heavy and lubricated by the water it is more likely to slide. Softer soil and gravelly soil will more easily become too heavy and slide if the terrace is too big.

IN SOME AREAS WITH SUCH SOIL THE FARMERS PLOUGH THE LAND UP AND DOWN THE SLOPE. THIS WOULD SEEM TO ENCOURAGE EROSION BUT I SUSPECT THEY DO IT BECAUSE THEY REALISE THAT IT IS BETTER TO RUN THE WATER OFF QUICKLY THAN TO HAVE A LANDSLIDE.

6.8 Mini terraces

There are several situations where mini terraces are most suitable, they should be made when:

A. it is unsuitable to make normal terraces.

B. the slope is very steep and the work which would be needed to make normal terraces would not be economic.

C. when rock is near the surface.

D. when there is little labour available and so close set contour strips can be made which will in a few years turn the land into mini terraces after which the plants in the strips can be taken out and more important plants planted.

E. when the plants to be grown will grow better with more light. Crops such as vines, mulberry bushes for fruit and fodder, and most crops where there is not much sunshine during the peak growing period grow better with more light. They will receive more light when grown on mini terraces. Plants which are easily attacked by disease may grow better on mini terraces as there is better air movement.

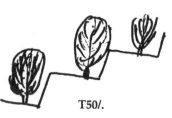
T50/.

F. When the terraces are small, less of the soil is moved and so less disturbance and opening up; also there is much less water at the inside of each terrace and so there is much less risk of problems developing.

G. Mini terraces are suitable for such crops as tea, pineapple, banana, papaya, sweet potatoes, peanuts, ginger, asparagus, strawberries, or mulberry bushes for fruit or feeding silkworms. We grew mulberries with 1.2 metres

(47") between the rows and 40 cm (16") between plants. Other recommendations are 0.75 m - 0.9 m (29.5" - 35.5") between rows, 0.45 m - 0.9 m (18" - 35.5") between plants; and 1m (39") between rows and 0.4 m (16") between plants. From this it can be seen that planting distance can be flexible to fit the space on the terraces. If the rows are closer the plants should be further apart. The outside of the terrace at least should allow space for a path.

One way to make mini terraces on soft/shaly soil without rocks for the walls with one metre risers (the height between terraces) is to consolidate the terrace banks by treading along the edge. The slope will be less steep than for normal conditions, but after one or two years when the terraces have become firm enough, the slope of the walls can be trimmed to make steeper.

H. For an illustration of how close contour strips can develop into mini terraces see T19.

6.9 Keyline Terraces

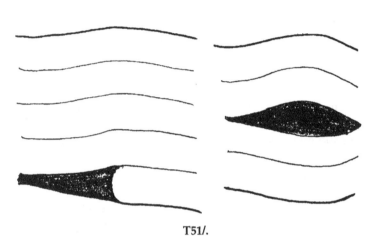

T51/.

Sometimes it is important to have the terraces or contour strips of parallel (equal) width. This makes sowing and cultivation quicker and easier. However if the slope is uneven the distance between contour lines will also be uneven and it will not be practical on very uneven slopes. With slopes which are fairly even, soil may be trimmed from banks of wider areas and added to narrow areas. While most of the terraces may be made of equal width some will have to be uneven with narrow ends which cannot be ploughed. It may be better to leave the most difficult areas planted to grass and trees. See T44/.

6.10. Improving terraces which are already made but sloping

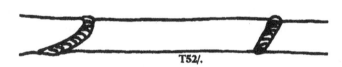

T52/.

Terraces which are sloping to one side.
We need to make bunds across as in paddy fields and lift the plough over the bunds in the same way as going between paddy fields. The good news level or a water level should be used to decide where to put the bunds. As the area between the bunds will gradually become level the distance in height should be set so that when the soil is level the part which was the highest will still have sufficient topsoil.

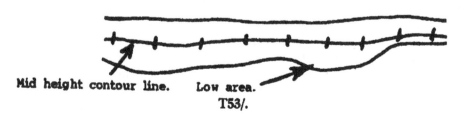

Mid height contour line. **Low area.**
T53/.

Terraces which are sloping to the front and irregular shape.

Contour lines should be marked out at suitable levels.

Example:- See T53/. This was an old contour strip which had developed into an irregular and sloping

and eroding piece of land with deeper soil at the lower side and shallow soil at the upper side above which was another contour bank with deeper soil.

How we corrected this field. The top soil having been moved to just above the top line. As the lower line and the median line joined at the right hand side there was nowhere to put the surplus (extra) subsoil from the top side; so it was carried to fill in the low area. The terrace was made level and then the topsoil replaced.

6.11 Drainage of terraces

A complicating factor when people talk about terrace systems, is the understanding that terraces must be made with a side slope, to drain away surface water when heavy rainfall occurs.

In my experience though this can be an important factor, the problem is often much less than is commonly thought. When we have paddy fields we seal them so that the water cannot soak into the ground; and so we have to provide drainage for rainwater which is more than is needed by the paddy.

If the rainwater can soak into the ground there is little which cannot be absorbed. The problem comes when an area of soil has not only the rain which falls on it, but also water from the surrounding area. If the fields have an open surface and do not have a hard plough pan or are otherwise compacted, they will normally be able to absorb their own rainfall.

In the worst typhoon in 30 years in an area on the east of Taiwan; 50″ (1.27 metres) of rain fell over two days. During the typhoon this author went several times through a 1,000 square metres area of newly terraced land. It was reddish laterite soil with some stone, when wet that soil acts like sticky clay. Though we had a road zig zagging up the hillside with a ditch to intercept side flow from the terraces, I was amazed to find that there was almost no run off at the bottom of the area. Prior to the terracing, even with normal heavy rainfall there would be so much run off that the gate at the bottom of the area would be blocked with mud.

The important point is that land should only need to absorb its own rainfall. Graded (sloping) terraces are more likely to have problems as muddy water running across a soil surface seals the pores (spaces and holes in the soil surface) so that it becomes more and more like a paddy field, and so there is an increasing rate of run-off.

The water running along the terrace accumulates (comes together) in any lower area. The extra wetness and weight of the water tends to cause the area to sink lower and so more water accumulates on it. The ground becomes softer, and if the rain continues may cause a small slip; so that the water pours through and down the terrace wall and bank onto the terrace below. This extra water increases the chances of a similar problem on the terraces below, which have even more water on them than the top terraces.

Graded terraces do not have a quick rate of flow but they have an increased amount of flow.

If cross barriers keep the water in its own area it is normally less likely to cause serious damage.

If surface drainage is needed, it is best to have a back slope to the inside corner where the underneath soil is firmer; if sideways slope to waterways is needed there should be flow slowing bunds across the shallow ditch along the inside corner of the terrace.

There may be a point in having graded terraces where the soil seals very quickly, but they need to be made carefully or that can make the situation worse; the Good News Level can also be used to make graded terraces. See Appendix 7. There should be regular maintenance to level up any low places.

When graded terraces or ditches are made the slope should not be too great, or the sideways flow will erode the soil. However if the slope is not enough the water will not flow far but collect at low places or stop at places where the soil is not quite level and is blocking its path.

Normally the slope is 1% or 0.5% on VERY even terraces. If the soil is rich in organic matter the slope can be slightly higher, as such soil does not erode as easily. On sandy soil the slope should be less. If too much water collects the distance between waterways should be made less. Local experience is the best guide as amount, duration, and type of rain; the % slope and length of slope, all vary too much to make rules about this.

It could be said that land which is weak underneath should have graded terraces to take the water away so that it does not soak in and soak the underneath soil or shale. My feeling is that in such situations only mini terraces should be used. See Mini terraces above.

Many people remove the stones from soil but it should be remembered that stones in soil act as a mulch, protecting the soil from being battered, and loosened, or sealed by the raindrops. The sones keep the surface open and slow down side flow. Another important factor if that if the soil is not very deep and it contains 50% stones, if we remove the stones the remaining soil depth will be only 50% of what it was, which my be too shallow for good plant growth.

6.12 A final important point

Soil conservation is best done on a linked basis.

It is of no use for some farmers to work hard at soil conservation if other farmers above are not being careful. Excess water causes gullies to form, and excess water from the land above flows down over their neighbours land and bursts the terraces. Or, the land above turns into landslides which bury the land below.

Landslides commonly occur when high land which previously was forested becomes degraded (poor almost bare land); or it is cleared, cultivated, and sown to such crops as maize. When the rains come the soil which is bare and loose, perhaps recently ploughed or hoed, becomes wet and heavy and starts to slide. Or water which previously fell harmlessly onto forest land now falls onto bare ground and so runs off onto neighbouring maize fields which become wet and heavy and start to slide. What may start as a small slide easily becomes much bigger.

If good farmers cannot encourage their neighbour on the land above to practice soil conservation, they should plant a bamboo grove at least two or three metres thick along the top of their land, to block, or reduce, the effect of a landslide. However there may still be water running from the landslide and so the interception ditch should be big enough to divert such extra water.

6.13 Terrace walls

A vital part of terrace making is the correct building of the walls.

There are three common misunderstandings.

1) Many people think that terrace walls are the same as garden walls but that is incorrect.

2) Other people think that the walls have to be very thick to support the thrust (pushing out pressure) of the soil. For this reason many people consider that they cannot afford the expense, or the labour of moving and building up the large quantity of stone which would be needed.

We will consider how the minimum of stone can be used to build suitable walls.

3) Another, though which may stop people from building terrace, is that they think the walls must be made of stones, and they do into have sufficient stones to use for that work.

We will consider how the lack of suitable stone need not be a problem.

The principles of terrace wall building

a. We need to consider the **natural angle of slope of the soil** we are going to use for the terrace. This affects the way we build the wall.

Sand about 45°(100%) Medium soil about 60°(133%) Packed clay soil 75°(167%)

The principle is that if sand or very loose dry sandy soil is poured on the ground it naturally settles at a slope of about 45° from the ground. If we do the same with normal soil the angle is steeper. If we put soil extra to the natural angle it will fall, we can call it the **fall soil.** So when we build the wall we need to consider the natural angle of slope of the soil which will be used in order to decide how to make the wall to support the fall soil.

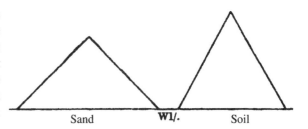

Sand   **W1/.**   Soil

b. If there is a layer of loose bare stones or gravel which will be exposed when terraces are made, the stones or gravel will become slippery when wet and fall out, leaving a space which may cause the soil above the space to collapse. We should also call such stones or gravel fall soil. A wall should be built to stop the loose material falling out.

c. We build the wall so that it supports the **fall soil.** At the same time we pack the soil to make it firmer and so less likely to fall. The wall does not have to support all the soil but only what would fall -the fall soil.

With sandy soil there will be more soil which would fall so the wall must lie back more to reduce the amount of fall soil and so that more of the weight of the wall is pushed against it.

d. Another factor which can affect terrace walls is **thrust,** this is a pushing force. Soil which is built on a slope will tend to move in the direction of the slope particularly when it is wet and heavy. When the soil on the slope behind the wall is dry it has a light thrust force which may be supported by the wall. If there is heavy treading of the soil or the soil becomes wet and heavy the thrust increases, the lower soil is compressed and then the thrust turns outwards and may push the wall out. W4.

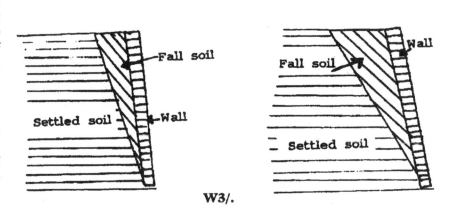

e. The angle of the wall should vary according to 1) the materials used for making the wall, 2) the amount of **fall soil,** 3) the type of soil behind the wall, and 4) thrust which may occur. The Good News Level see Appendix 7 sets guide marks for the angle of the walls according to the materials used. These are of course only guides as in some places the rocks or stones or clods making up the walls will be larger, and of better shape, than in others.

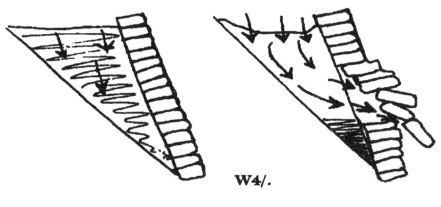

The guide angles of slope of the wall are for sandy soil 65° = 145% slope. For normal soil 67.5° = 150% slope. For clod walls 70° = 155% slope. For stone walls 82.5° = 183.35% slope. W5/-

f. The sloping soil behind the wall should be cut back to a distance which is equal to half the distance between the top of the sloping soil and the wall. Then the main thrust will be downwards instead of pushing out against the wall. W6.

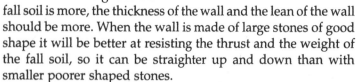

g. The thickness of the wall and the angle of the wall can be varied according to the conditions. The more straight up and down the wall is - the more land is available for crops, but less if the weight of the fall soil is more, the thickness of the wall and the lean of the wall should be more. When the wall is made of large stones of good shape it will be better at resisting the thrust and the weight of the fall soil, so it can be straighter up and down than with smaller poorer shaped stones.

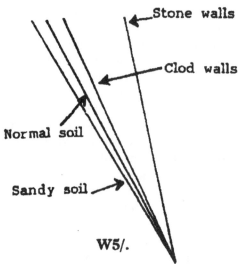

h. One of the important differences between a normal wall and a terrace wall is *the angle of the stones in the wall.* With a normal wall they should be laid flat. With a terrace wall they should be laid at a right angle (the normal corner angle) to the face of the wall, so they slope downwards towards the soil. W7.

This makes them lean against the soil and makes the wall stronger. The fall soil pushes down on the stones making them less easily pushed outwards. To push them outwards they have to be pushed upwards against the weight of the stones above. If the stones are laid as in a normal wall they are pushed out much more easily.

The picture W7a. shows a wall which had been built with the stones laid flat as in a normal wall. After heavy rain the soil had pushed the wall out. We relaid the wall as in the photograph with the stones sloping downwards and there was no more trouble. The wall was still sound when I saw it 17 years later.

i. If a high wall needs to be built it is better to build it in 2 or more sections instead of one high wall. The division of the sections could be planted with legume plants such as Greenleaf Desmodium, and fodder and green manure trees such a Leucaena. W8.

j. If there is a situation with an unusually big height difference between two terraces or very unstable (loose or weak) soil, it may be better to build it in two stages with a year or more before building the second stage to allow for the soil to settle and become firm.

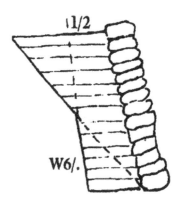

k. A terrace bank normally has two parts. The bottom part is where the lower terrace sub soil was cut out, and so it is firm, hard undisturbed subsoil. If it is only firm soil without sandy or stony areas, there is no need to support it with a wall; but it should be protected from erosion by being covered by creeping legumes such as Greenleaf Desmodium, or grasses such as Molasses grass or Signal grass.

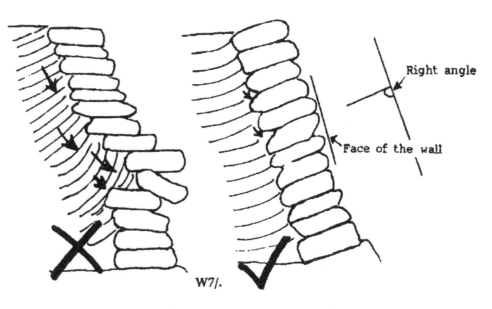

The top part will either be a wall, or shaped and tramped down soil. It also should have creeping legumes such as Greenleaf Desmodium planted along the top. This will grow down and in the rainy season will root into any soil the stems reach. It will in time cover the banks protecting them from erosion, holding the soil together, and providing either very good animal fodder, or useful plant material, to be added to the soil as a very good green manure.

w7a/.

l. If there are not sufficient stones to make all the walls of stone and so clods will also be used it is best to put the stones at the bottom and the clods at the top of the wall. Another possibility is to make double use of the stones, see the end of this chapter.

m. If there are not enough stones and clods to build the wall the wall is made as a split earth wall. W10.

The bottom part (A) which is the undisturbed subsoil from where the subsoil was cut out for the terrace below; should be trimmed to be as upright as possible. If it is firm it is better for it to be upright so that it is less easily eroded in the rains; the water running off, not softening and cutting into the soil.

The top part (B) which is made of loose soil should be less steep. If there are any hard lumps or turves (the roots of plants together with soil) they should be used for the outside of the wall. As the terrace soil is built up the soil clods or turves that are available should be placed on the outside edge first, then just behind and overlapping more clods for at least 1 foot (30 cm's) then spadefuls of soil so that the inner soil holds the outer soil, see the illustration. W11/. Each layer of soil at the edge and two feet (60 cm's) from the edge should have loose soil packed in the spaces and then be treaded firmly *not jumping* and with the feet together, before the next layer is added. When complete it is good to plant grasses or legumes to grow on both (B) and (A) to hold and protect the soil.

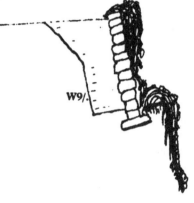

n. With all methods it is important to have a firm foundation for the wall and it must have the correct slope, see W9 and W5.

DIFFERENT STAGES IN MAKING TERRACE WALLS, **not all are used for every situation.**

- Stage 1) Decide on the materials and the method to be used.

- Stage 2) Decide on the most suitable angle of the walls.

- Stage 3) If there are no stones or rocks for the terrace walls dig down to see if the subsoil is hard firm and can be used to make clod walls.

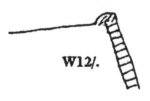

- Stage 4) If there is not enough hard soil to make a clod wall prepare to make the terrace as explained in m.

- Stage 5) Prepare the foundation; see n.

- Stage 6) Dig out the space behind the wall as explained in f and W6.

- Stage 7) Build the walls keeping the space behind the walls filled in and packed down up to the walls.

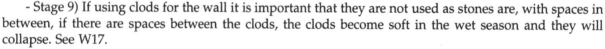

- Stage 8) If using stones clods or turves for the wall the top surface of the stones must be at the sloping back angle or slightly more level; but not slope towards the front of the wall. See W7 and W19.

- Stage 9) If using clods for the wall it is important that they are not used as stones are, with spaces in between, if there are spaces between the clods, the clods become soft in the wet season and they will collapse. See W17.

- Stage 10) When the walls are finished pile soil on the top to make a bund so that water from the terrace cannot flow over the wall and erode it. W12.

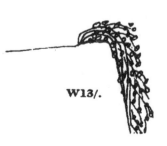

- Stage 11) **Plant plants** or cuttings of a creeping legume such as Greenleaf Desmodium where it can grow along the top of the terrace (not on top of the stones). It will grow over the wall protecting it, rooting in where there is soil and so helping to bind the surface of the soil. It will reduce the effect of summer heat reflecting from the walls, and can be cut and used to improve the soil or fed to animals to produce more income and more manure for the soil. W13.

HOW TO KNOW WHAT KIND OF WALL TO MAKE

Earth wall

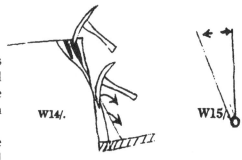

If there are few or no stones available for building the walls dig down to see if the subsoil is hard and so clods of subsoil can be used instead of stones. Turves (Soil held together by the roots of plants) may also be used. If there are not enough stones or stones 2) 5) 6) 7) 10) 11) see above.

(1) If the soil is very soft it may be best to build up the terraces in stages of about one metre of height. Tread the soil carefully and make the wall firm by thumping with the back of a spade. Then when the soil has become firmer and is held and protected by plant roots and tops another metre is developed. This is repeated until the needed size of terrace to complete. In such a situation it would be even better to have a step in the middle planted with green manure or fodder trees. They will help to stabilise such soft soil.

(2) Another method-split earth walls, see W14/. and W15/.

When the soil has become firm and grasses and other plants are binding the soil together, the top part of the wall can be prised out and soil packed in to straighten the wall. The lower part may be trimmed back to make more room for crops. W14. and W15.

Stone wall

If there are enough suitable stones available (they must be fairly flat or square, without more than one fourth being round or a difficult shape (such as three sided), a stone wall can be built. The stones can be built in different ways according to the type of stones. Some examples are given below. As much as possible the stones on top must bridge the gap of the stones below them. As different stones come to hand the stones can be laid in different ways. It is important to remember that speed, strength and cheapness is more important than looks. If the person building the

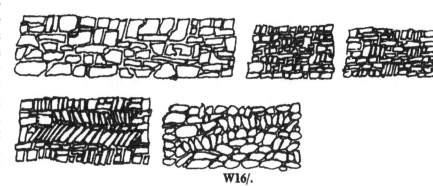

W16/.

wall is an experienced stone worker used to building houses or normal walls, he may chip almost every stone; but if he does, although the wall will look nice, it will be too expensive, so not many terraces will be built. Just as working clothes would not be suitable to wear when visiting an important person, the clothes for visiting an important person would not be suitable for farm work. So the most suitable walls for terraces are different from the most suitable walls for a palace garden. See W16.

Use the principles a, b, c, d, e, f, g, h, i, j and n.

In addition follow the stage 1) 2) 5) 6) 7) 8) 10) an 11) which follow.

Clod and / or turve wall. If there are not enough stones but there are plenty of strong clods which can be used, a clod wall should be built, see W11, W17 & W18.

Use the principles a, b, c, d, e, g, h, i, j, k, l, and n.

In addition follow the stage 2) 5) 6) 7) 8) 9) 10) and 11) explained above.

After making the firm foundation level trim the bottom of the clods flat with a heavy knife and lay them in position. Then put stones or small clods at the outer edge and in other holes; pack the rest of the hollow with soil to block any gaps between the clods and to make level. Leave a little loose soil on the top for bedding in the next layer of flat bottomed clods. See W17.

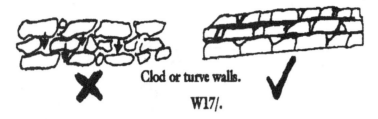

Clod or turve walls.

W17/.

When three layers of clods are in position the face of the wall can be trimmed off with the heavy knife. Remember that the wall should be built at the correct angle which is different from a stone wall see e/. Or measure by hanging a stone on a piece of string 3 ft (91.5 cm) long, so that the stone just touches the outer edge of the foundation. Measure in 12½" (32cm) from the top of the string and that is the correct angle see the drawing.

Another method for making clod walls. It is a little more complicated but the wall may be a little stronger. This method may be more suitable if the clods break easily when there is too much shaping.

W18/.

W19/.

Instead of trimming the bottom of the each clod, most of the clods can be left with their basic shape. The top surface of the clods when laid in position should be flat or leaning inwards or slightly cupped, see W18. and W19.

As with the other method all the spaces between the clods must be tightly packed with stones or soil.

Walls which are properly made will not need much attention but should be checked and if there is damage it is best to repair it quickly before it spreads and causes more work.

When there are not enough stones for all the terrace walls we can have *Double use of stones for walls.*

When using the good news system of terrace building we first terrace alternate strips of the field. After doing this to all the land we come back and terrace the remaining strips. As this will be at least two years later, the soil will have settled and become firm. If it was difficult to make terraces without stones the stones used for the walls of the first terraces could be taken to use for the later terraces. Suitable grasses and legumes should be planted to protect the terrace banks; and a step dug out and trees planted in the step, the tree roots and the other plants will stabilise hold and protect the soil bank.

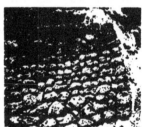

A wall of subsoil clods.   W20/.

The Overall Plan

The plan is to have the terraces as level as possible while moving the smallest amount of soil to enable this. In fact due to the fact that fields are seldom uniform (the soil all the same) in depth of topsoil; in the amount of stones or firm clods which can be used for the walls; or in the soil profile (curves and slope of the strip); so the finished terrace will seldom be level unless we adjust the soil. We can do that if after moving the topsoil and loosening the subsoil we take subsoil from the higher areas to level up the subsoil before replacing the topsoil.

Level terraces normally have a surprising ability to absorb rainfall, though they may not be able to absorb additional water from elsewhere.

If terracing cannot be done from some reason contour strips or Tansen half terraces should be made with contour ditches to carry excess water away. On steeper slopes the risk of erosion is greater and so the ditches need to be placed closer horizontally (across the slope).

To calculate the vertical (upright) distance between contour strips or Tansen half terraces measure the depth of soil which could - perhaps with a little - help grow crops.

Then set the vertical (upright) distance as that depth minus 4" (10 cm's). This is so that as erosion moves soil down the slope and a rough terrace develops there will still be a minimum of 4"(10 cm's) of reasonable soil at what was originally the top of the slope.

WATER CONSERVATION

7.1 Introduction

'In her book "The last Oasis" (Earthscan Publications), Sandra Postel maintains that the world is entering an era of extreme water shortage which could be disastrous for agriculture---. One reason for the slow down in growth of irrigated areas is that these projects have been getting increasingly expensive. **The good dam sites have been used and it is becoming harder to find economically viable new water projects---**. It is often cheaper to put in systems to save and conserve water than it is to build dams and put in new irrigation systems.'[108.]

As more land is denuded or badly managed more water

Fig. 1. Instead of the rain water flowing down and eroding the slopes it should be slowed or harvested in some way.

runs off the land, instead of replenishing the soil and the below soil water storage. As the water runs off it causes damage and flooding. Streams and water springs which used to supply water for drinking and household use, or rice fields, dry up.

In more and more countries such as in India more water is pumped from underground reserves than is replenished. Each year the water level lowers and so deeper and deeper boreholes have to be made. Like oil, fossil supplies of water (water stored for thousands of years)are being used. It is not only becoming more expensive to obtain the water but the water is increasingly being effected by salty water infiltrating the vacuum left in the water bearing layers. The water is less and less good for the land, increasing the tendency for salinity spoiling the soil, and

Fig. 2

reducing the yield in irrigated areas. There is also the problem of finding drinking water for people and their livestock; less livestock means less manure for the soil, and so poorer soil and poorer crops.

From an article in Appropriate Technology. Vol 29. No 1 March 2002.

> Grain production in China is expected to fall, as water shortages begin to bite. Consequently, grain imports may have to rise.
>
> This scenario emerges from a survey which reveals that China's water situation is far worse than realised, writes Lester Brown of the Earth-Policy Institute. The water table under the North China Plain, which produces over half of China's wheat and a third of it's maize, is falling faster than thought---.
>
> The study, conducted by the Geological Monitoring Institute (GEM) in Beijing, reported that Heibei Province in the heart of the North China Plain, the average level of the deep aquifer dropped 2.9 meters (nearly 10 feet) in 2000. Around some cities in the province it fell by 6 meters.---.
>
> As thousands of wells run dry, so the three rivers that flow eastward into the North China Plain - the Hai, the Yellow, and the Huai - are drying up during the dry season. Lakes are disappearing too.---.
> Whatever it does, China will almost certainly have to turn to the world market for grain imports. If it imports even 10 percent of it's grain supply - 40 million tons - it will become overnight the largest grain importer, putting intense pressure on exportable grain supplies and driving up world prices.

The full article and additional information from the Earth Policy Institute, 1350 Connecticut Av. NW; Washington DC 200-36, USA www/earth-policy.org[210p58].

I have seen television programs and advertisements urging people to support drinking water projects. The program said that for £5 per person a water system can be supplied giving those persons clean water for life. Sadly that is not often the case. They will not have clean drinking water for life as the aquifers are not being replenished, and so springs dry up and the water level in wells falls, and supplies fail. Very often new water supplies are required because the ones previously used have now dried up, so that merely providing new supplies is treating the symptoms rather than the cause. It is time that the cause was realised and treated, otherwise the problem will continue to worsen till there are no more sources to tap.

Water and soil conservation are closely related, in most cases what effects one also effects the other. When we conserve soil we reduce run-off and have better plants and so have better soil protection, in turn there is less surface sealing and more absorption of rainwater.

As mentioned earlier the author first had experience of this as the farm manager of an agricultural training centre farm in Taiwan. I was appalled to find that in heavy rain, mud and water poured down the sloping fields, blocking the field gates.

Fig. 3 The village committee showed me this pond, it is all they have now for people and animals as the other water sources have dried up, they asked for a pump to supply water from a spring down the hill.

We then commenced soil conservation work, making terraces wherever it was practical to do so; with side slope to ditches at the side of the track which zigzagged up the slope.

After this work was completed the worst typhoon for 60 years caused terrible damage and flooding to the area.

I was anxious to see how the terraces were standing and was amazed to see at the height of the rainfall only a trickle of runoff water in the ditches. The only terraces damaged were made by two particularly inept students.

I learned then and later that if terraces are properly made in fairly open soil, even if the soil contains clay the soil can absorb all the rainfall *which falls on them.*

There will be situations where the soil seals and does not absorb, or where the terraces have been used as paddy fields before, where a graded side slope is needed, in such situations it is important to make sure that the water channels are kept clear. *The problems arise when water from elsewhere is added to that which falls on a given area,* such as where terraces are constructed below an area of slope land, or a road, and there is nothing to stop water running onto the terraces.

Problems will also occur when terraces are not level and water accumulates at one place. This can occur when carefully graded terraces have been made; the water which was planned to flow sideways into waterways is stopped at a low point. This could be because after completion the terrace has settled lower at that point, or because the grading was faulty, or because the water is blocked by rubbish or by soil or rocks fallen from the terrace banks above or during the cultivation of the field.

Fields which were paddy fields prior to the drying up of the water source may still hold water in the rainy season. They may be opened up by deeper than normal ploughing to break up the waterproof pan formed when the field was a paddy, this will both help the growth of non-paddy crops grown on them and allow more water to be absorbed. In such a situation there should not be the problem of the land becoming soft, waterlogged and slipping, as the soil will have had years to settle and become firm and stable. Alternatively if there is still a reduced water supply the paddy may be converted into a fish pond or a water holding area to supply water after the rains have ended.

7.2 Watershed conservation

7.2a Catchment areas

Some projects have the mistaken idea that protecting the local watershed means simply planting a few trees round the area of the local spring. This is simply protecting the water head, not the water shed which is a much larger area. To protect the watershed the area above the spring for at least 15 acres (6 hectares)

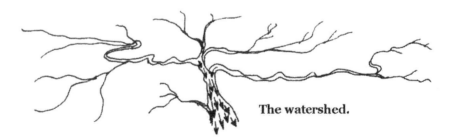

The watershed.

Fig. 4 Water conservation within a district should if possible start from where the run-off commences, with as much interception of the flow as possible at each stage.

for a small spring, should be managed to conserve the water. The water supplied by the spring or well can only be as much as the water which percolated to the water storage area which supplies the spring or well.

7.2b Replenishing the underground water

Why do streams and springs dry up?

There are main aquifers (water holding layers) and underground streams and rivers. Above these are secondary aquifers which hold a certain amount of water and supply some surface springs but dry up in the dry seasons.

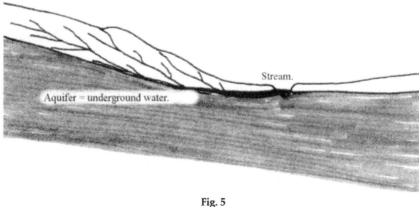

Fig. 5

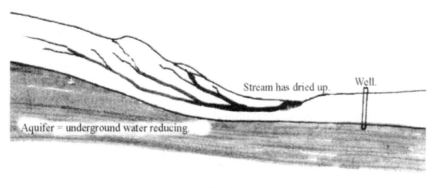

Fig. 6

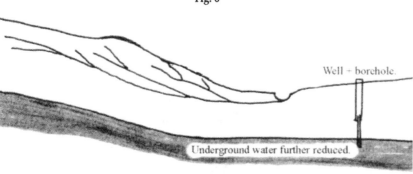

Fig. 7

Stage 1. The trees shrubs and undergrowth of the forest cover and protect the soil, the rainwater soaks into the ground and down to the aquifers. The run-off water is clear and there is not a very great amount. The rivers and streams continue to flow all through the year without very great variation in the amount of flow; there are many springs of water.

Stage 2. Trees are cut and land is cleared and cropped.

Less of the rainwater is absorbed so more of the water runs off the land. The rivers become too full and may overflow and cause flooding. In the dry season the rivers have less water, some of the springs and streams dry up and deep wells have to be dug to supply water on the flatter land. Fig. 6.

Stage 3. More land has been cleared and water conservation methods have not been used, so even more water runs off, the level of the aquifers is lower and so the wells are not deep enough and bore holes are needed to deepen them. Fig. 7.

Stage 4. Even less water soaks into the soil, so it soon dries up. The crops are poor and the people are unhappy.

More water is used than is able to reach the aquifers so the water level in them falls even lower. In the hilly area the springs have dried up, or only run for a short time after the rains. Paddy fields dry up and people have to travel longer distances to find water. There may even be fighting over the water that there is. On the flatter areas only the wealthy can afford to deepen the bore holes to reach the water, and as they continue to pump at greater expense the water is used up; there is no future.

IF INSTEAD OF DESTROYING THE FUTURE THE PEOPLE USE SOIL AND WATER CONSERVATION AND WATER HARVESTING. THE RUN-OFF IS REDUCED AND WATER IS FED BACK TO THE UNDERGROUND STORAGE OF AQUIFERS AND UNDERGROUND STREAMS AND RIVERS. THERE IS WATER FOR PADDY FIELDS AND FOR HOUSEHOLD USE; THE CROPS GROW BETTER AND THE PEOPLE ARE HAPPY. FIGS. 8, 9 & 10.

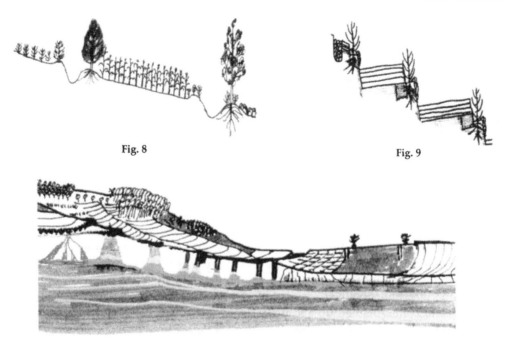

Fig. 8   Fig. 9

Fig. 10

7.3 Managing run-off

When rain falls on the bare surface of the soil it often seals the pores; particularly after a long dry period when the soil is hard and dust is washed into the cracks. At such times although the land needs water, much of what falls runs away.

Fig. 11

By providing interceptor/absorption ditches and allowing the runoff to enter the subsoil we are enabling the land to have extra absorbing capacity and be better able to sustain good crop growth for a longer period.

To avoid the problem of run-off from terraces where the soil is either not very porous (not able to absorb the water quickly) or, as the terrace is newly made the soil is loose and soft and so liable to slide when wet, the terraces should have a back slope as explained in chapter 6. Cross bunds should also be made to divide the terraces into basins, with the top of the bunds lower than the edge of the terrace so that if an extremely heavy rainfall occurred and the water could not soak into the soil, it could flow sideways to an emergency waterway, to interception areas see below.

Another way to increase absorption of rainfall and reduce runoff is to make half moons or mini terraces below fruit or other trees or even in crop and vegetable fields where rainfall is less than best or may cause run-off. Figs. 12&13.

Fig. 12

In crop lands simple bunds of stones or soil and/or branches, reduce run-off, also Tansen double check half terracing explained in chapter 6. The latter is made by ploughing or digging along contour lines. Then the loose soil is thrown onto the topside of the furrow, not downwards as in normal half terracing. The result is that water moving down the field is blocked by the ridge of soil and soaks in. If the runoff is very heavy and flows over the ridge it is again checked by the furrow and soaks in or flows to a more absorbent place or waterway.

If possible sloping fields should be terraced as explained in chapter 6, where this is not possible or is delayed, ploughing and sowing should be done across the slope. Subsequent hoeing of the crop should also be done in a way which keeps ridges across the slope. This will hold most of the rainwater and plant nutrients in the field, instead of them being washed away.

Fig. 13

Absorption ditch.

Fig. 14

Landslides commonly occur because water running from one or more places concentrates into a soft area; soaking into the ground, softening and lubricating the soil and increasing its weight. If land is terraced properly with interception ditches to stop outside water from running into the area, water should not concentrate but be absorbed.

If the ground does not easily absorb water and the underlying soil is stable, provision should be made for it to run into interception/absorption ditches and soak into the ground. See figure 14.

If there is a danger that such water soaking into the ground could possibly cause a land slump, or a landslide, normal ditches should be made to carry the run-off water to firm areas, where there are absorption ditches, a soak-away pit or a storage area, where it can be absorbed, stored, or can flow without causing harm; this will:-

1. Avoid heavy run-off which causes erosion and flooding.

2. Replenish the ground water and the water bearing layers underground; these supply the wells and springs needed for use in the dry season.

3. This will even up the flow of rivers and streams so that irrigation and water for drinking and other use is easier and more reliably obtained.

Forest trees are too often planted up and down the slopes so encouraging water runoff. Research in Taiwan has shown conclusively that when the trees are planted across the slope not only does it improve soil and water conservation, the trees grow much better and harvest is sooner.

7.4 Waterways. See also 7.10d

A common fault in books about soil and water conservation is that though they say that excess water should be diverted to grassed waterways, this author has not found any which recommend suitable grasses and legumes for planting in such waterways. Very often the grasses are unsuitable, they may die during wet or dry conditions, or become eroded during the period of heavy rains, and so instead of holding back water and sediment, they make things worse. To help with this appendix 5 describes suitable grasses and legumes for waterways and for pastures.

To protect the sides of the bigger waterways, to hold the soil and maintain the sides of the watercourse tree or bamboo species which are resistant to water logging should be planted at the sides.

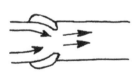

Fig. 15

In some cases breakwaters should be made to divert the current from the sides. These can be earth and rocks planted with bamboo or other suitable plants, or be of rocks or gabions.

If such methods are not used or if the current is very erosive such as on sandy areas or the outside of bends in the watercourse, walls may be needed. Without such protection the sides will erode, collapsing into the watercourse and possibly diverting the water causing erosion and landslides. Landslides eventually expose the water bearing rocks so that the added water which emerges causes more damage, at the same time draining away the water reserves. Because of this, methods to prevent or check landslides explained elsewhere in this book, should be used for the same reason.

Baskets of bamboo, saplings, split branches, creepers or wire, are filled with stones.

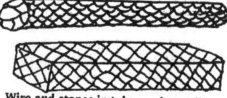

Wire and stones in tubes or boxes like cages are used to make gabions.

Fig. 16

When the waterways slope more than 30%, steps should be built to break the force of the water. If the flow of water is more than 70 cm (27½") wide, barriers should be made across the flow.

Footings of stones or wood should also be made on the lower side to prevent the overflowing water washing out the base of the barrier or eroding the bed of the waterway. Figs. 18 & 19.

These barriers hold back stones sand and silt which build up till eventually terraces are formed which when soil and water conservation is improved and so runoff is reduced, can be used to grow crops or trees.

As mentioned earlier in 5.2, 6.11and 7.3, diversion or interceptor ditches should be dug above fields to prevent water from above damaging the fields. If the soil is soft weak soil which may collapse if too wet, the ditch should be open, smooth and sloping; it must be kept clear, so that water falling into it can be quickly diverted. If the soil of the field has firm subsoil or rock and the topsoil free draining and is not likely to slide, the ditch can be mainly an interceptor ditch; it can be flat and filled with stones so that it does not need to be kept clean. While water will move through the stones it will move slower and more will be absorbed into the ground. The moisture will help the crops and also replenish ground water and reduce or remove the problem of getting rid of excess water.

7.5 Mulches

As mentioned elsewhere in this book, mulches not only help plant growth and preserve soil fertility

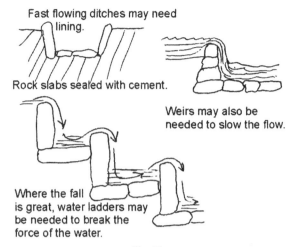

Fast flowing ditches may need lining.

Rock slabs sealed with cement.

Weirs may also be needed to slow the flow.

Where the fall is great, water ladders may be needed to break the force of the water.

Fig. 17

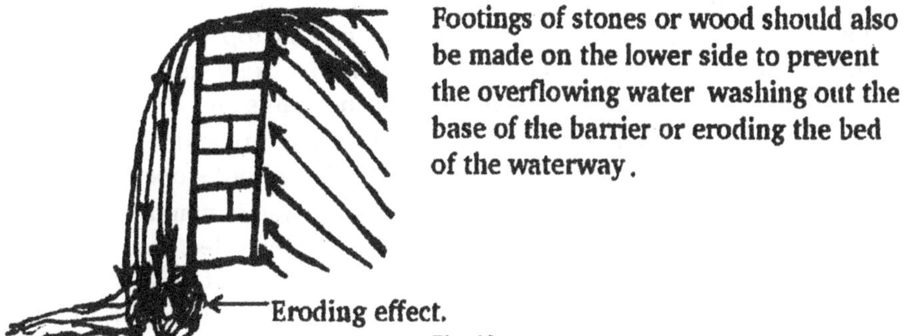

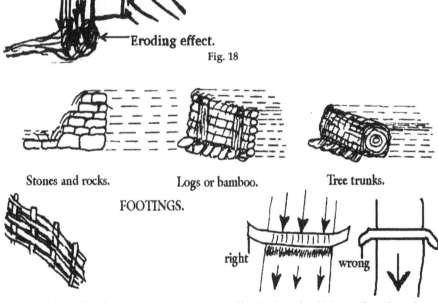

Footings of stones or wood should also be made on the lower side to prevent the overflowing water washing out the base of the barrier or eroding the bed of the waterway .

Eroding effect.

Fig. 18

Stones and rocks.   Logs or bamboo.   Tree trunks.

FOOTINGS.

right   wrong

Woven sticks or bamboo perhaps two barriers with stones or grass in betweeen.

The barriers should be well set into the sides of the waterway and swept backwards

Fig. 19

but they also conserve moisture, they reduce or prevent surface sealing so that most or all rainwater is absorbed. They also obstruct the flow of surface water, so decreasing the speed and amount of the surface runoff, and increase the amount of water penetration. Mulches reduce surface evaporation or drying by wind, and also, because mulching improves the fertility and texture of the soil this indirectly helps water conservation, as the soil structure is better able to absorb water and produces better plants, which in turn reduce runoff.

Stone mulches can be used to reduce evaporation especially in dry or desert areas. Richard St Barbe-Baker the "Man of the Trees" has told how this technique can be particularly beneficial to sapling trees in desert areas.

Stones are of benefit to plants in the following ways:
• By providing shade from intense day heat;
• by releasing stored heat to the soil at night;
• by preventing poultry or small animal damage to roots;
• by preventing wind lifting of roots;
• by providing shelter for worms and small soil organisms;
• and on very cool nights, by causing water to condense on their surfaces.

I would add that stones seem to encourage soil microorganisms, they also stimulate some plants. Bamboo for instance always seems to grow better when some of the roots are in contact with rocks.

Other important effects of a stone mulch are to reduce evaporation, to protect the soil from erosion by both wind, rain, and degradation by UV (in sunlight) light, to reduce the growth of weeds which compete with the plants for water and nutrients; and to help in the absorption of rainwater.

The soil beneath stones has a more even temperature and humidity than other soil.

A common mistake of people who farm stony soil is to remove the stones and rocks. Not only does this remove the stone mulch, but it reduces the depth of the top soil which in stony land is often shallow to start with. So the result is a very shallow top soil over poor stony subsoil. It is better to leave the stones in the top soil and have a greater depth of soil useful to the plants and able to absorb and to retain more water.

In poor growth conditions with a shortage of manure and a shortage of soil moisture, plants in stony soil do better than those on soil of the same depth without stones.

The reason is that if we have a given amount of water or manure applied to a square metre of land which contains 50% of stones, the water or manure is concentrated in only half the volume of soil, so that the amount of water or manure in each cup of soil is twice that of land which is 100% soil.

As the plant roots grow between the stones they grow in richer, moister soil than that in shallow soil which has had the stones removed. The stony soil allows deeper penetration of rain, and in addition there are the other advantages of the mulching effect of the stones which are on the surface.

Mulches of plastic and of stones compared with no mulch for growing vegetables in dry areas.

Growing vegetables in arid lands is often difficult, if not impossible, because water for growing the crop under traditional cultural practices is not readily available. Annual rainfall is less than ten inches (25 cm), with much of it falling during a few short periods. Run-off is great, resulting in much loss of this water. Furthermore high temperatures and low humidity that exist in arid lands cause much water loss through evaporation. Another problem found in arid lands is the poor quality water.

Mulches conserve soil moisture and provide an environment around the plant more favourable for growth and production. The introduction of plastic films prompted much work on mulching in recent years. Many procedures for mulching with plastic, paper and other materials have been developed. Gravel has showed promise as a mulch in dryland production. —.

The object of the study was develop a technique for growing vegetables with a minimum amount of water under arid conditions.

The plastic aprons were made of vinyl, 6 ml thick and approximately 1m². at the centre of each apron was a 2" (5 cm) hole with flap edges which allowed easy seeding yet covered the soil surface next to the plant. Within a 6 " (15 cm) radius of the plant stem many ¼" (7 mm) holes in the plastic served as entry to the soil and root zone for run-off water from the surface of the apron. These holes are covered with an attached piece of plastic in such a way that the rainwater is funnelled through the holes to beneath the plastic apron, but evaporation is inhibited due to the cover over the holes. (This author would suggest for practical use punching the holes with a V or X shaped punch allowing the water to penetrate, yet leaving the flap to retain moisture).

These aprons were anchored by covering the edges with soil at the rim of the basin. The aprons were a light green colour which faded to a dull white.

The gra.el was light grey and varied in size from 3/16" to 5/8" (4 mm - 17 mm) diameter. The depth of the gravel was about 1½" (40 mm) and covered an area equal to that of the plastic."

The cucurbit Yellow Straightneck summer squash was grown.

For the mulched plots the water used was about 2" (50 mm) in 1969 and about 5" (130 cm) in 1970 is significantly low, and indicates that water for crop production can be used much more efficiently than normal practice.

Both years the mulched plots produced the best yields with plastic aprons doing the better job. Yields though limited by frost, were equal to more than 350 bushels per acre, a good average yield.

The plastic, and to a lesser extent the gravel, reduced moisture loss as indicated by water added and the yields obtained. Also the mulches reduced crusting over germinating seed, which was a problem in the bare plots.

Conclusions. Vegetables can be grown under arid conditions with a minimal supply of water. Mulching, especially with the plastic apron, makes this type of growing possible. A complete system utilizing water harvesting, mulching and trickle irrigation shows much promise for growing vegetables where normally it is very difficult to do so. [95.]

7.6 Windbreaks. See also 3.9

These also help conserve moisture, particularly in areas which suffer from typhoons or hurricanes, or hot drying winds. When the rain is lashed at the soil by the force of the wind, surface sealing may be increased. Windbreaks also reduce soil drying and hardening and dust formation, and so improve the ability of soil to absorb rainwater, particularly on sloping land.

On many soils the first rain of the rainy season causes most run-off and erosion and landslides, the surface of hard dry soil seals easily, particularly if there is soil dust on the surface. Water runs to more porous areas, soaks in, in quantity, and causes a landslide; or as a high proportion of the rainfall runs off, gullies are formed. The vegetation is sparse and dried up and so gives little protection.

Windbreaks, like mulches, help to keep the soil from becoming too dry. There is better vegetative cover, and the soil being less dry and baked is more able to absorb the rain when first it comes.

7.7 Difficult areas

Areas which are unable to absorb water such as areas of rock or areas with shallow soil over rock become hot, and the air above them becomes hot. Air flowing over them then dries surrounding areas which then become hot themselves. We can convert such situations by such means as planting bamboo around and among the rocks. Bamboo seems to thrive in such situations. We can also plant plants such as Greenleaf Desmodium cuttings, or passion fruit, at the edge of such areas to grow over them and shade them; at the same time producing fruit or fodder. Passion fruit can also be planted to grow up the bamboo culms.

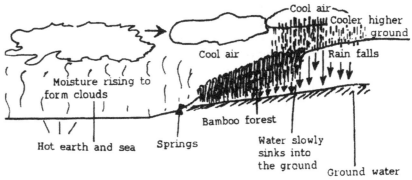

Fig. 20

Fig. 21

The areas become green, cool, and useful, or we can collect the water which runs off these impervious areas to give extra water to gardens, fruit trees, etc.

Where there is thin soil overlying rock or clay pans it may be practical to move the soil into hollows so that trees can then be planted to utilise the rain and condensation from the rock. Of course if a given amount of land receives water from an equal size impervious area it will have double the normal rainfall. This will make an important difference to what can be grown. Land which is bare and hard or crusted from which most of the rainwater runs off can be changed into an asset. See fig. 21.

Bamboo forests are a quick and good way to reforest, as harvest can start in 4 years from planting cuttings or root splits, see the chapter on Bamboo. B2.

In Taiwan it was made compulsory for areas of poor, steep or erosion susceptible land to be planted with bamboo. After the bamboo forests were established there was less and less run-off, more water in the springs and streams in the dry season and rainfall increased in the areas with larger bamboo forests. To understand why, we should understand what causes rain to fall.

When clouds move over a cooler area the droplets of moisture in the clouds come closer and join to form bigger droplets, when the droplets become a certain size they are so heavy that they start to fall as rain. Walking through a bamboo forest it is surprisingly cold, so we can understand that if clouds pass over or near a cold bamboo forest they will produce rain more often than if the same clouds pass over a hot area. If it rains one day the area will be cooler the next day, so clouds passing over are even more likely to produce rain. See Fig. 20.

In the evening, in a bamboo forest during humid weather it can appear to be raining, as moisture condenses on the leaves and drips down. In misty weather bamboo leaves absorb moisture and transfer some of it into the root zone so on digging down we can find that soil round the roots is moist. The daytime micro climate is affected by the moisture absorbed at night which may account for the coolness within and above a bamboo forest. More on bamboo in Chapter B2.

7.8 Converting unused or underused water into useful water

This is related to the point above when it concerns areas such as domes of rock, rocky areas or roads. The author was stationed at a mission hospital compound on a hillside. Complaints were made about the drainage from the hospital laundry, and also rainwater from the compound; during the rainy season the water was flooding staff homes below the compound. In between the homes and the compound was a bare, rocky, stony hillside of unused waste land. I asked for permission to turn the area into a bamboo forest and water utilisation area, so that as well as turning a hot bare unsightly patch of land into a beautiful and income generating bamboo forest, the drainage problem of the hospital would be solved. This was agreed to and was very successful. Side ditches leading from the drainage water ditch were made to distribute the water so that it soaked away among some of the bamboo. The area was transformed in three years. Comparison trials were also carried out with a range of species of bamboo, each species being grown in both very dry rocky land, and rocky land which was occasionally slightly damp by irrigation from the ditch. It is now a demonstration area of 24 species of bamboo growing in different conditions. The sale of the bamboo adds to hospital funds, local people have the leaves for animal fodder and the drainage problem is solved.

Rainwater tanks. It is not often realised that in parts of rural Australia household water supply is not piped but is rainwater from the roof stored in large circular galvanised tanks. Where piped water is supplied the rainwater is used for washing and bathing.

Utilising household waste water for growing vegetables. The waste water can be used for watering the garden. Washing powder contains phosphate and bar soap contains potash, both of which are needed by many third world soils.

The author's family lived in a water deficit area of Nepal where for much of the year water had to be carried and was limited. An oil drum was made into a filter and supply tank supplying trickle irrigation pipes for the vegetable garden. See 7.12 fig. 52. All the used household and washing water was poured into the drum. The filter removed materials which would clog the trickle irrigation holes and we had no problem of blocking, the water emerging looked clean. The vegetables grew well under this system for the last 2½ years we were in that project. Nepali people often divert water from the public bathing and laundry area to their vegetable gardens; I have not heard of any problem. See also 7.10a

7.9 Using water more efficiently

Growing rice more efficiently is an important way to conserve water.

As deforestation takes place there is more run-off and so less water soaks down to replenish springs and other water sources. Irrigation systems become short of water, this can lead to serious disputes with fighting, bribery and even murder in some cases.

In Taiwan a large irrigation system was affected in this way and so there had to be strict rationing of the canal water. Experience showed that contrary to previous belief, if paddy fields were not irrigated until they just started to crack the yields remained the same or even increased. The reason is that 1) Poisonous gases are released and replaced by oxygen. 2) Sunlight shining on the base of the plants stimulates tiller growth (the production of side shoots). These early tillers (side shoots) will produce full heads of grain, unlike any later tillers which can only produce poor heads with many unripe grains at harvest.

When this regime was part of a system of direct seeding of rice by direct drilling of the seed in clumps, the yields can increase by 25% using 20% less water. The author has used this system in both Taiwan and Nepal with up to 28% more grain, of higher quality than with transplanted rice.

Cultivating for optimum water penetration.

Where the soil surface has dried out completely and is hard and fairly smooth, heavy and sudden water application can result in poor absorption, run-off and erosion. In dry weather, it is best to cultivate so as to form clods. With such clod formation:-

- the surface area for absorbing water is increased,
- the water can quickly penetrate through air pockets in the soil, and,
- the rate of evaporation is slowed down.

(Where it is possible to cover the soil with mulch (see mulches 7.5) and zero tillage this is not needed, however in many situations there is no mulch available.)

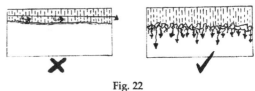

Fig. 22

7.10 Water harvesting. (Collecting and using water)

7.10a The courtyard system

Some notes from an article in the July 2000 issue of 'Enable' the newsletter of the Association for Better Land Husbandry.

Chinese farmers in the drought prone loess region in Northwest China have made their own experiments to harvest as much as possible of the scarce local rainfall resources to develop what is now popularly known as 'Court-Yard Production Places.'---.

In 1995 Tens of thousands of people and livestock suffered from a dire shortage of drinking water and people had to get up early in the morning and travel miles to fetch water. Large areas of crops ended in total failure that year and many factories were temporarily shut down in Gansu as no water was available to the production lines. In places, schools were also forced to close as the teachers and pupils had to look for drinking water elsewhere---.

The Chinese experience draws heavily on indigenous cistern building, but extends it much further by using modern materials and modern design technology to construct a network of various cisterns, which then form a steady source of water for supplementary irrigation both to the small family courtyards and the farmland where a variety of vegetables, fruit trees and cash crops are grown. In recent years, both with drought and good rainfall alike, such cisterns have contributed much to increase farmland productivity and household income, and farmers like them very much.---.

Certain creative characteristics are particularly worthy of attention.

1.The rainwater catchment plots are carefully positioned in such a way that (a) they occupy unused land, for instance roof surfaces, road surfaces and bare slopes; (b) they are in higher places than the water cisterns to ensure automatic inflow of the collected rain; (c) they can catch as much rainfall as possible. Linking the catchment plots and the cisterns are fairly sophisticated underground pipes, at the end of which a series of de-silting tanks are built to prevent the sediment from entering the cisterns. To further improve water quality, farmers normally clean the catchment plots before the rain starts, and sometimes disinfectants and used to purify the water.

To save cropland and to ensure better rainwater-catching, some inventive Gansu farmers have begun to experiment using plastic films to collect water in the farm fields. Because local rainfall normally occurs is several heavy storms in the post-harvest autumn season, the use of such plastic films which can easily be taken away immediately after the rain and kept for re-use in the future, affects virtually no crop planting in the next season at all. As far as one could see from the impressive water harvesting sites, farmers are in fact very skilful at making their own decisions, and as long as their productivity can be reasonably increased they are more enthusiastic than anyone else to adopt new technologies.

The courtyard production system is very much one of concentrated farming; surplus labour in the region now finds better usage on the small courtyards, which tend to become sites of experimentation and innovation. As water no longer poses a problem and the courtyards are just outside the living quarters, farmers can spend more time observing, caring for, and improving farming right by their own house premises. As a result, productivity is high in the courtyards and crop diversification also becomes normal practice, thus effectively turning the courtyards into a mosaic of vegetables, fruit trees and new crop varieties. With this, the time of no drinking water and inadequate food is gone. As the water collected by one farmer in his water cisterns has to come from land surfaces above which may belong to a number of other farmers. So, the local farming community usually comes together to see how best to plan their rainwater harvesting arrangements for the whole tract of land in which each one of them has a stake. In the process, farmer to farmer exchange of ideas and sharing of experiences are fostered very naturally. While conflicts do arise on occasion, it is solidarity which eventually wins the day.[184]

7.10b Harvesting from flash flood gullies and stream beds

The following method will be suitable for land with a comparatively gentle slope, not for hillsides.

A method traditionally used in the Yemen by local farmers and taken up by other farmers in the region; was introduced to Somalia in the late 1940's. The aim is to maximise absorption even intercepting water from above the farm or flowing at the sides of the farm.

A series of low earth banks are built across the areas which would otherwise erode into gullies. The aim is to collect and hold as much water as possible while allowing for large storms to pass without causing much damage.

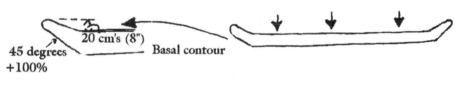

45 degrees
+100%

20 cm's (8")   Basal contour

There is no complete blocking so that if very heavy floods occur the water will overflow at the arms instead of bursting the banks. Fig. 23.

Fig. 23

The banks are about 3 ft (91 cm) high with a base of 6-8 feet (1.83-2.84 m) and an even slope on both sides.

The earthbanks are up to 100 metres long (the author would feel 30-50 m maximum would be safer, I would imagine the banks are raised by digging out a ditch along the top side, this will increase the water holding capacity). The earth banks are usually made at 1.5-2 feet (46-61 cm) vertical intervals (height difference between banks). There should be at least 10 metres between the end of one bank and the end of the next otherwise the overflow water from the two banks could create a gully. In areas where heavy storm rains are expected a crescent shaped bank should be built to block and spread the flow as it flows towards the next bank. Fig. 24.

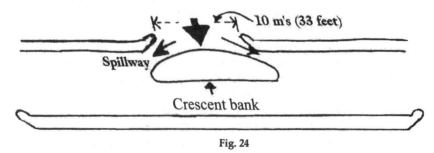

10 m's (33 feet)

Spillway

Crescent bank

If the land is stable, ditches can be dug to absorb water into the subsoil. These are dug across the spillways or at appropriate intervals across the slope, where the spillway water accumulates more water than can be used in the strips and the banks below it. Fig. 24.

Fig. 24

A personal letter from David Sanders explained that he had seen some of the water spreading structures in Yemen about 12 years ago. His impression was that they were very extensive and built on flat or very gently sloping land in the wadi (storm water streams) beds. Water is diverted out of the main streams when they are in flood by diversion banks. The banks that he saw were protected at the spill end with rocks. Water was diverted into a number of ponds, one after the other. The ponds had banks about three to five feet high. Once a pond filled, the water overflowed, over a stone spillway, into the next pond. As soon as the ponds had dried up enough, they were planted to sorghum which grew on the conserved moisture and probably saw very little or no rain. He had seen similar systems being installed in what is now Eritrea.

This writer very much regrets that he has lost the source of the base of the above article and is very willing to acknowledge the source if informed.

7.10c Waterway interception ditches and soak away's

At the sides of the waterways interception ditches divert the water back into the fields. These slope inwards but may have open ends to allow excess water in times of flood rains to flow into grassed or stepped waterways. Fig. 25.

Interception ditches for protecting the surface yet harvesting runoff. When we were preparing to create a soil conservation trial on stony land in Nepal, we dug an interception ditch to divert water from the slope above; we filled it with large stones. Later we noticed that in the plots below the point where most water entered the ditch, the crops were growing much better; the water was seeping down to

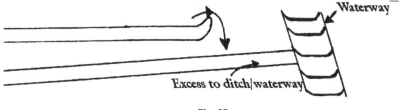

Fig. 25

the plots instead of down the ditch. See the illustrations T41 in chapter 6 and Figure 14 in this chapter.

This principle could well be used in other similar conditions. If the water was running over the surface it would cause erosion; by being diverted to slow soak underneath the surface the water was being conserved without damage or loss.

In cases where terraces were wrongly made or there was no firm absorbent subsoil or underlying material such soak away trenches could be unsuitable as the top layers of soil would become soft and porridge like, and a landslide may result, particularly if the underneath layers were sloping.

Common sense is needed when considering if methods will be suitable in a given situation. Digging a pit to understand the under the surface material and structure will help.

7.10d Water absorption barriers. See also 7.4

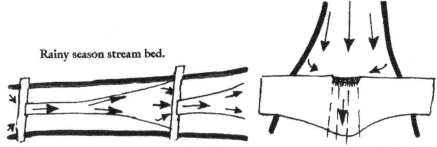

Water is checked, and some is absorbed, over time the areas between the check barriers will silt up and crops can be grown in the silt. By this time other conservation measures should have reduced the flow in the channels so that most water will be absorbed and used by the crops growing in the built up soil.

Fig. 26

Fig. 27

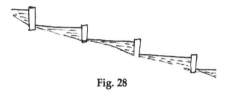

Fig. 28

Another way of harvesting water and at the same time developing more cropping land is by making a series of barriers down a gully or wet season stream bed.

The end result is a series of terraces. When we block off gullies we reduce the speed of flow so reducing erosion, increasing the amount of water absorbed and also enabling the silt to settle out. This reduces the volume of the runoff so reducing flooding and damage of land downstream, at the same time keeping the silt in the area. It is of no value to Nepal to have the island which is forming in the Bay of Bengal from soil originating in Nepal. As the silt builds up behind the barriers the absorptive area is increased and so there is less run-off. The field which develops can be planted to crops which themselves both use water and retard the flow. By this time soil and water conservation should have reduced the flow in the channels so that most will be absorbed or used by the soil and the crops growing in the built up soil.

7.10e Swales

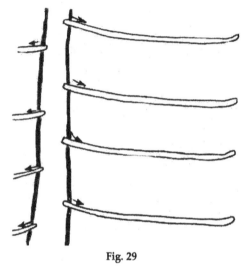

Fig. 29

A swale (Fig. 29) is a barrier of stone, soil or plant matter or a ditch made across the slope with a slow slope to a ditch to carry excess water away; or it is doubled back in a zigzag manner till all the water is absorbed. The idea is to divert and spread the water flowing down a slope so that it can soak into the soil, it is very useful where rainfall is not regular, but when it does come is heavy, and normally quickly runs away with little penetration. It also reduces the erosion of the water pathways.[44] p480.

Water is diverted from the watercourse and spread over the land to one or both sides of it.

If the water is muddy and of greater quantity, barriers to divert the water will be more suitable. Barren land can become more and more productive over time as silt builds up and crops are grown.

There are special crops adapted to this kind of system which have been developed by local people in areas such as dry parts of Mexico and Australia. Seeds are quickly grown after the water has come.

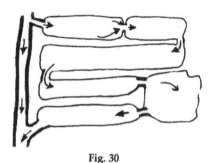

Fig. 30

Sometimes the area flooded by the swales diverting the water flow is shaped into basins for more effective use of the water. The basins hold and absorb more water and plants such as melons and pumpkins planted in the moister soil can grow outwards over the dryer areas utilising the space and reducing the evaporation from the area.

7.10f Ridge and furrow

An alternative way of using the water from swales, or where there is no watercourse utilising rainfall only, would be to shape the area into ridges and furrows. The water from the hard uncultivated ridges would then run into the cultivated absorbent furrows, so that the furrows which are planted receive 2, 3 or more times the average rainfall, depending on the area of water harvesting ridge left between the cultivated land, decided according to the amount of water needed.

7.10g Tied ridge or ponding system

Construction of tied ridges has been found to result in striking yield increases for cotton, cowpeas, millet, and sorghum in the semi-arid tropical areas of Africa. This two year study compared tied ridges to simple contour ridges, quantifying their effects on the soil water regime, crop water use, and growth patterns of cowpeas in Burkina Faso's Sudan Savanna.---.

Experiments conducted in sandy loam topsoil with a sandy clay subsoil suggest that tied ridges may present an alternative for semi-arid region farmers faced with unpredictable rains. In ridges — soil water content increased by an average of 30.5 and 24.6 mm per week in 1985 and 1986 respectively over simple ridges. Root growth was increased by tied ridging. The tied ridges showed a higher level of vegetative growth which was positively linked to increased water availability during rainless growing periods. Grain yields were 51% higher for tied ridges in 1985, a relatively dry year, but not significantly in 1986, a high rainfall year.

However, the study also showed that waterlogging induced plant stress is more likely to occur in cowpeas planted with tied ridging in years of above average rainfall. Cowpeas planted in both ridging systems during 1986 showed signs of waterlogging stress but, because the rains did not occur during the sensitive flowering and pod formation phases, grain yields were unaffected. It should be noted that cowpeas are particularly sensitive to waterlogging.[100.]

With some tied ridge systems the width of the ridge is equal to the width of the furrow and the main crop is grown on the ridges. This is where the rainfall is heavy and constant so that the furrow would sometimes be too wet for crops. In drier areas the main crops are planted in the furrows though sometimes spreading over the ridges. Cross ridges link the main ridges which are on the contour. This keeps the rainwater where it falls so that it can soak in. The cross ridges are slightly lower than the main ridges, so that if there is a very heavy rainfall which brings more water than can be instantly absorbed in the least absorbent places; excess water can move sideways rather than running down hill.

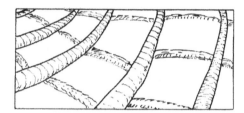

Fig. 31

This system cannot be used where the rainfall is greater than the soil can normally absorb. However if the land is ploughed, dug, or cultivated before the heavy rains the soil can absorb much more and under this system soil will not be removed from the ridges.

7.10h The planting pit and ridge system

With this method shallow pits are dug with the soil being mounded into ridges between the pits. Fig. 32. The pits are about 20 cm (8") across and 30-35 cm (12-14")apart. This method normally prevents run off and concentrates what water there is so that it penetrates deeper and so is not quickly dried out and can keep the plants planted in the pits growing for a longer time.

Fig. 32

7.10i Zai holes

With this method hard crusted earth from which most rainfall would run-off has holes or hollows dug in it at intervals. Available plant remains, trash, leaves and available compost is placed in the bottom of the holes. When the rains come the water runs off the crusted earth into the zai holes. Much or all of the water soaks in through the trash and into the loose soil below the trash. Seeds are sown through the trash and crop plants grown. After the rain has finished the trash acts as a mulch so the moisture remains much longer, some of the moisture which soaked into the underlying soil also rises with nutrients to nourish the plants. After a year or two of better crops there may be more trash available than before, and other holes may be dug, so that gradually the hard crusted area is transformed.

According to the description of Zai holes in the book *Amaranth to Zai holes* published by Echo (Ref 185) the holes are 20 cm by 20 cm (8 inches by 8 inches) and 10 cm (4 inches) deep. Another advantage of the trash in the holes is that it encourages termites which make tunnels and so increase the rate of water infiltration.

The book quotes Tony Rinaudo in Niger as stating.

> Where farmers are using it, it is making a big impact on crop yields. Soils here are infertile and if farmers have manure at all they just broadcast it on top of their fields. Most if this is baked, blown

and washed away. If the manure and organic matter are placed in a zai hole, losses are minimised and nutrients are concentrated where the plans can use them. Crop plants have a competitive advantage over weeds that are not in the zai hole.---.

We convinced one farmer to try zai's on a small plot of barren land. He did and harvested100 kg of corn and 15 kg of sorghum. The next year farmers in 20 villages dug over 50,000 zai holes! We urged farmers to also try zai holes in their sandy soils. The results were so convincing that many are now digging holes on their own initiative.[185]

7.10j Spread over fields method sometimes incorrectly called Chinampas

This is a way of irrigating in the same way as that used with hill paddy fields but can be used for structures like swales or non paddy crop fields.

Water is diverted by a weir (a barrier of wood, rocks soil etc), or a jutting out pier of rocks or soil, or even a dam. There can also be catch trenches perhaps filled with rocks, these leading to pits or cisterns (tanks), or directly to the fields. There the water flows into the upper end of a terrace or ditch, spreading along slowly, wetting the land as it goes; when it reaches the other end it flows down to the strip below. When all the land is wetted any excess is diverted into a safe ditch or watercourse which may carry it to another similar system or to a water soakage or storage area. This is similar to paddy field systems but using narrow ponds to spread the water of over the land, so enabling the water to be absorbed for later use by crops; obviously a useful system where heavy but irregular rainfall occurs.

7.10k Chinampa system

This has been confused with the system explained above. The author recently came across an article which explained what the Chinampas system really was.

It was used by the Mayan people of central and south America and is still used in some places. It consists of raised beds surrounded by ditches. *"Trees such as willow and alder, which often grow in symbiosis with a nitrogen fixing actinomycete (Frankia sp.) are planted around the islands' perimeters (edges)".* [93]

7.10l Cajetes system

This system was developed among the Aztecs as early as 1,000 BC. ---. This includes the funnelling of rain run-off to a network of relatively small water tanks (cajetes) (this author would describe them as ditches) situated at the base of hillside terraces on cropland. The cajetes serve as catchments and compost pits for soil and organic debris carried by runoff water —. Nutrient laden soil and decomposed debris is returned to the fields while trapped water percolates (soaks through) to recharge the water table. In the meantime the tanks protect the terraces from structural damage due to runoff. The Aztecs had watercourses to channel run-off from very heavy storms and the mud which collected in the ditches was spread over the land. Other aspects of this system were intercropping, crop rotation, fallowing and the maintenance of border areas which provide space for high value plants such as fruit, fuel, and fodder trees and medicinal species. The trees stabilize the terraces act as windbreaks bring up trace elements and provide shade and shelter to bees and other beneficial organisms. Based on.[94]

7.10m Waru system

A similar system called the waru waru was used on flatter land on Peru's southern border with Bolivia on the altiplano, a vast plain 12,500 feet above sea level. On this land area-

are patches of corrugated land.

Each patch is divided into long narrow strips separated by furrows, some of which contain puddles of water.

Closer inspection reveals that the tops of the strips are populated by dry, hardy grasses whereas the vegetation in the furrows is lush and green. The local farmers call these waru or camellones. Until 1981, however, the local farmers had no idea that these represented persisting evidence of the remarkable engineering and agricultural skills of their ancestors. The fact that waru cover some

205,000 acres of land around Lake Titicaca suggests that the ancient inhabitants of the altiplano had hit upon a system that successfully tackled the considerable environmental constraints of farming the area. Archaeologists are now convinced that the waru were built specifically to protect crops from frost damage and floods. In 1981, Clark Erickson of the University of Illinois, recognized the archaeological significance of waru waru. But he also wondered if they might serve modern farmers as well as they did their ancestors. Erickson began to rebuild some of the raised fields. Using traditional Andean tools, local farmers planted an experimental field with potatoes, quinoa and canihua. Waru waru potato yields were more than 8 tons per hectare, compared with the average yield for the region of 2-3 tons per hectare. (Authors note these low yields indicate very poor growing conditions.) Today some 3,700 acres of raised fields have been reconstructed. The Peruvian Department of Agriculture is convinced of the value of waru waru in the region. The government now offers loans to farmers for rebuilding the fields.[183]

7.11 Water storage

Flatter land often has problems of water shortage. As we have seen in 7.2 wells which used to supply water for both domestic use and for irrigation dry up due to less water reaching the aquifers (the water storing places in the ground).

The common solution is to dig deeper wells or bore deeper tubewells. Much money is used for such projects but it is hiding not curing the problem, only curing the signs of the problem. The water level continues to fall. The fossil water reserves are being used and so fossil fuels are being used to pump up the fossil water, neither of which is sustainable or intelligent. It would be better to use the resources to conserve and harvest water.

Where water shortages are a problem in downstream areas it may be feasible for the people from such areas to donate labour or cash for water conservation measures to be created in the upstream area.

Water may be stored to extend the growing season.

Small dams or water holding areas . As previously pointed out, in some situations run-off water from bare or rocky areas, from roads and from buildings etc. increases the effective rainfall if diverted to swales chinampas or water holding areas. These may be ponds or small dammed gullies etc, lined with clay or banana leaves and mud; they can provide enough water for at least a fruit and vegetable garden.

Gabions, illustrated in 7.4 Fig. 16 are often used for making dams and weirs. A gabion is a basket; it can be made of split saplings or bamboo but is usually made of galvanised wire. The basket is filled with stones when it is used for water conservation, in other cases soil may also be used. Most gabions for use in streams and rivers are like woven tubes filled with stones and rocks, or, woven square or oblong cages filled with stones or rocks. In Taiwan I have seen short round upright gabions made of a bamboo cage filled with rocks and the gabions lashed together, this also seemed to work well particularly when using scrap steel cable from an old suspension bridge to lash in front of the gabions. At other times bamboo was used for this purpose but did not last as long.

The maximum pressure that must be taken into account when calculating the weight of a dam is always situated at a third of its height---.
In order to resist the energy of the water collecting above it, the weight of a gabion dam must be at least three times greater than the high water pressure to which it is exposed[90.]

When the dam is shallow less strengthening will be needed.

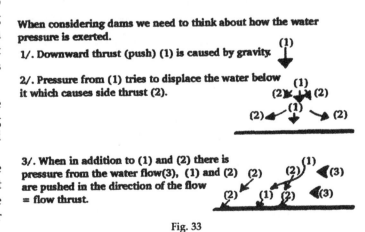

When considering dams we need to think about how the water pressure is exerted.

1/. Downward thrust (push) (1) is caused by gravity.

2/. Pressure from (1) tries to displace the water below it which causes side thrust (2).

3/. When in addition to (1) and (2) there is pressure from the water flow(3), (1) and (2) are pushed in the direction of the flow = flow thrust.

Fig. 33

4/.When the flow is blocked by an obstruction whether an object like a rock or a dam wall sideways thrust occurs in a similar way to 2/. , as the water at the wall which is blocked by the wall displaces away from the thrust.

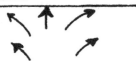

Fig. 34

5/. When the level of the water rises higher than the top of the dam and shoots over there is forward and downward thrust

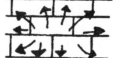

Fig. 35

6/. When making the wall we need to consider the strength of the wall and also the strength of the earth and rocks of the banks at the ends of the **wall, otherwise either** **the wall may be pushed down, or the water pressure will burst through the banks at the ends of the wall.**

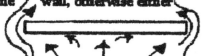

Fig. 36

7/. We also need to consider the downward thrust at the base of the wall, at the uper sides of the wall, and the forward and downward thrust at the base on the lower side. If care is not taken the base of the wall will be weakened and the wall may collapse.

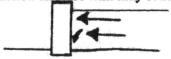

Fig. 37

8/. When planning the design of dams it is useful to consider other structures such as arched windows, tunnels or hump back bridges.
They are designed to take pressure from above and divert it to the side. At the same time the pressure compresses the curve or arch materials pushing them tighter together and so making them more solid and strong.

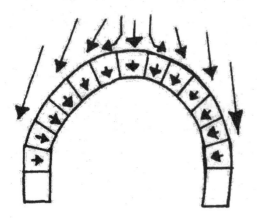
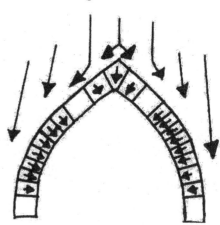

Fig. 38

9/.

In a situation where both sides of
the stream are solid rock and the
ends of the wall could be fixed
very strongly, the wall could be
curved towards the flow.

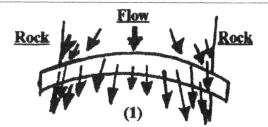

(1)

This would make a stronger wall, or, less material would be needed to
withstand the same flow pressure. It would also give a better spread of the
overflowing water.

If the banks are of earth and rocks but still strong (2) may be appropriate.

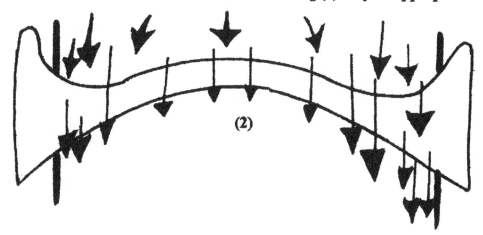

(2)

Fig. 39

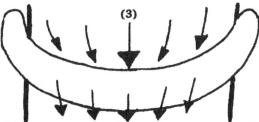

(3)

If the sides of the stream are weak the water pressure on the sides at
each end of the wall could cause the bank to collapse and the water to flow
around the ends. In such a situation the sides must be strongly reinforced,
and the shape of the wall should be either straight or curved away from the
flow (3), so that the thrust is directed away from the sides. This means that
the wall must be stronger particularly at the centre.
It may even need a two way curve to strengthen the wall (4).

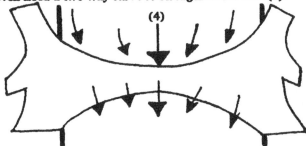

(4)

Fig. 40

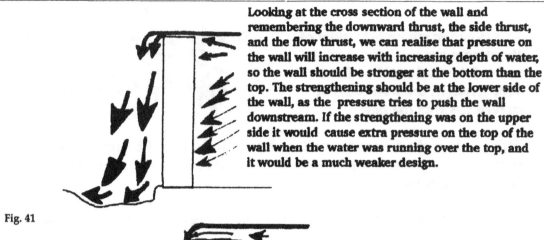

Looking at the cross section of the wall and remembering the downward thrust, the side thrust, and the flow thrust, we can realise that pressure on the wall will increase with increasing depth of water, so the wall should be stronger at the bottom than the top. The strengthening should be at the lower side of the wall, as the pressure tries to push the wall downstream. If the strengthening was on the upper side it would cause extra pressure on the top of the wall when the water was running over the top, and it would be a much weaker design.

Fig. 41

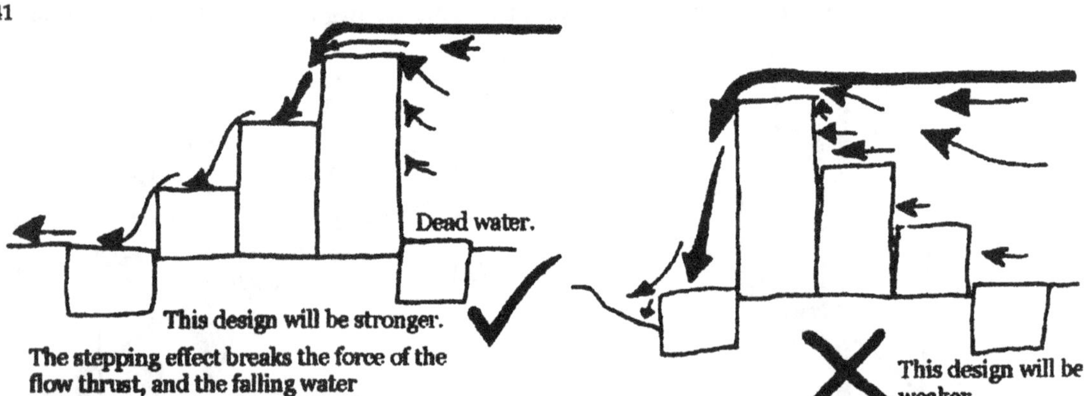

Dead water.

This design will be stronger. ✓

The stepping effect breaks the force of the flow thrust, and the falling water adds to the weight of the wall. The water pressure pushes the components of the wall tighter together.

Fig. 42

✗ This design will be weaker.

The stepping effect has a bad effect. The straight drop of water on the lower side is very erosive.

In a big dam it may be feasible to have pools in the step stages, giving a cushioning effect as the water flows into water, and adding extra stabilising weight of water. This will slightly reduce the weight of rock and concrete or other materials required to balance the pressure of the water above the dam.

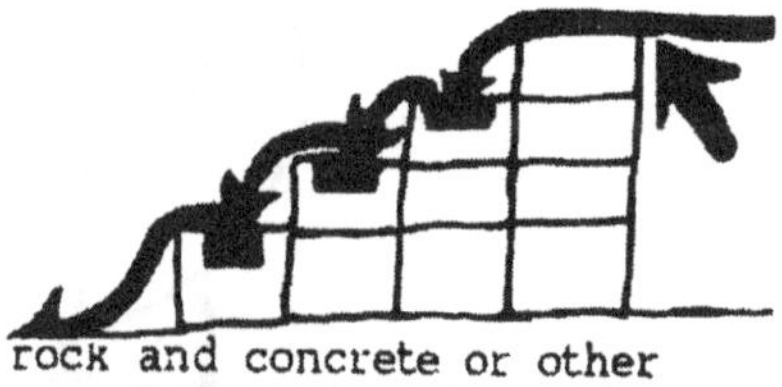

Fig. 43

Fig. 43/a

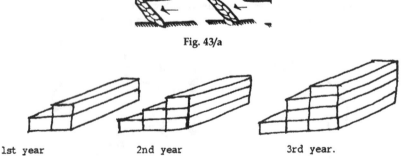

1st year   2nd year   3rd year.

Fig. 44

If there is no reliable engineer experienced in this kind of work who can calculate the strength and size of the dam, it is better to have a series of smaller dams rather than one big one. Or, the dam can be built up in stages, it's size increased each year according to experience.

Where the flow is already strong it is better to have better field conservation and small dams, than build a large dam. A series of conservation measures decreases the run off and increases the amount of silt that settles out in the area and remains for use in the district.

Gully dams may have pipes built into the dam walls at various levels to allow for different water levels and for silting up. Sometimes it is worth having watergates at the side of the dam to allow for flushing early in the rainy season so that silt does not fill the dam. Otherwise after the dam has silted up another dam is made above or below the first one and the area above the first dam used to grow crops in the silt.

As the effects of good soil conservation increase silting will be reduced.

Pools can be made and sealed in a similar way to paddy fields by ploughing, puddling and treading till the water does not leak away. Or, by sealing with clay if available or cow dung and mud plaster under banana leaves covered with mud. The author has no experience of this method although he has been told that it does work. Otherwise the bottom of the pool can be lined with plastic, covering the exposed plastic with leaves and mud.

Deeper pools are better than shallow ones which lose more by evaporation due to the greater surface area to volume but also because shallow water heats up quicker. If pools are dug out they should be deep and the soil removed should be used to build up the banks both to increase the capacity of the pools but also to reduce evaporation by hot air currents or wind passing over. The banks can be planted with grasses and other plants such as banana and papaya. The leaves and banana stems cut up and thrown in the water to feed fish; plants such as floating rice, water spinach may be grown in the ponds. Water hyacinth to be used for producing bio gas is another possibility, but care is needed to ensure that it does not get into streams and rivers where it can spread quickly, choke the flow and become a serious pest.

From an article in Appropriate Technology. Vol 29. No 1 March 2002.

> Grain production in China is expected to fall, as water shortages begin to bite. Consequently, grain imports may have to rise.
>
> This scenario emerges from a survey which reveals that China's water situation is far worse than realised, writes Lester Brown of the Earth-Policy Institute. The water table under the North China Plain, which produces over half of China's wheat and a third of it's maize, is falling faster than thought---..
> The study, conducted by the Geological Monitoring Institute (GEM) in Beijing, reported that Heibei Province in the heart of the North China Plain, the average level of the deep aquifer dropped 2.9 meters (nearly 10 feet) in 2000. Around some cities in the province it fell by 6 meters.---.
> As thousands of wells run dry, so the three rivers that flow eastward into the North China Plain - the Hai, the Yellow, and the Huai - are drying up during the dry season. Lakes are disappearing too.---.
> Whatever it does, China will almost certainly have to turn to the world market for grain imports. If it imports even 10 percent of it's grain supply - 40 million tons - it will become overnight the largest grain importer, putting intense pressure on exportable grain supplies and driving up world prices.

The full article and additional information from the Earth Policy Institute, 1350 Connecticut Av. NW; Washington DC 200-36, USA www/earth-policy.org

Soil and water conservation are closely linked. The value of water conservation is not only in the amount of water saved but:-
a) When water runs off land it usually takes soil and plant foods with it.
b) The more the water is retained in the field the longer the soil is covered and protected by growing crops.
c) A longer growing season results in better crops which means more organic matter is produced, which, if returned to the soil directly or as manure improves the soil structure and fertility. Better soil structure leads to better resistance to erosion and better water retention. Better fertility produces even better crops so even more organic matter and so an upward spiral. The greater crop residues can be used to mulch the soil, further reducing evaporation and blocking the harmful effects of UV light from the sun which destroys soil organic matter in the top 1-2 inches (2.5-5.0 cms) of soil.

Soil conservation reduces water run-off and maintains the depth of the soil and so the ability of the soil to retain rainwater. As topsoil is eroded whether by wind or water the crops are poorer and the growing season reduced. The underneath subsoil is usually less able to absorb the rainfall so water run-off is greater.

For more information on this subject see[90]

One of the alternatives in addition to others mentioned, is to divert water to recharging tubewells. Consideration must be made as to whether the recharging tubewells will affect the local drinking water, as the water will not be as clean as if it had soaked through the earth. If the drinking water comes from another source there is of course no problem. Experience in Gujarat state, India, recorded in [92] showed that:

> Constant pumping of groundwater and repeated droughts created a very precarious situation by 1988. The ground water tables lowered alarmingly. Many farming pockets started to become barren due to the inflow of saline water.
>
> In Rayan, the groundwater table which used to vary between 15 and 35 feet depth for the past 50 years, sank to a depth of 80-90 ft. Along with the low rainfall in 1985, 86 and 88 (it didn't rain a single mm in 87) the use of groundwater increased tremendously. Farmers went on deepening their wells and then made bore wells to go even deeper, until saline (salty) water at a depth of 250 feet warned them about the oncoming plight.---.
>
> The only option was to harness surplus rainwater which used to gush away in four small rivulets. Groundwater could be recharged by building appropriate structures at appropriate spots.---.
>
> Formalities were completed fast and the work for 15 masonry structures, 2 recharging tube wells and the deepening of one old pond began.---.
>
> The first 35 mm of rain caused all structures to overflow. Within 8 days all the harvested water disappeared into the ground. In some farmers wells which were located near the structures improvements were seen at an early stage. Others benefited a little later. During 1989, 310 mm (1 foot ¼ in) rain fell in 19 days---.
>
> The pre-and post-project monitoring of groundwater depth and water quality revealed eye catching improvements. There was a four feet rise in groundwater levels and a reduction of 300 parts per million in total dissolved solids in the water. For the first time in five years, farmers experienced some relief and hope.---. so before the monsoon of 1990, another sizable work was carried out. Now the Rayan project consisted of 18 masonry check dams, 3 ponds, one impressive underground check dam and 8 recharging tubewells.
>
> Although much water was recharged to the ground, the amount of water pumped out by 84 wells of Rayan farmers was still considerably more. There were only two alternatives, monsoons with well distributed rainfall, or development of proper water management techniques. The monsoon of 1991 seemed to have decided that Rayan farmers should learn the hard way about water management techniques; it only rained 43 mm. And thus despite impressive structures, very little water was harvested that year.
>
> Now farmers face dilemmas again. Policy makers used to feel self-satisfied when they sanctioned drought- relief works and subsidised fodder, (now one of the vice chairmen) took up the herculean challenge influencing those policy makers to turn to water harvesting structures instead of haphazard digging.
>
> At local level the outlook on farming is also beginning to change. The high-input and market oriented growing of groundnut and cotton is now being replaced by less water-consuming and salt tolerant sunflowers. Horticultural crops like pomegranate, sapota, ber (jujube) and custard apple, which perform well with less and lower quality water, are gaining popularity. Drip (trickle) irrigation systems are gradually being installed. It is circumstances such as the recent droughts which are finally leading Rayan farmers to sustainable agriculture.[92]

From the journal 'Appropriate Technology' September 2002.

> Though India is one of the wettest regions in the world with an average rainfall of 117 cubic metres over the plains, many parts of the country frequently experience severe water shortages. Radhakrishna Rao reports how some communities are overcoming this problem by reviving old and neglected techniques for harvesting and storing excess rainfall.
>
> In the tribal dominated Chhotanagapur belt of Jharkhand state the traditional system of conserving water is *ahar* or surface irrigation tanks which continue to be popular. It is essentially a hill technology that has been adapted to the gently sloping plains with the help of run off diversion channels called *pynes*.

An *ahar* is basically an earth filled check dam built across the natural drainage joining uplands for harvesting the run off. In the Palamua area of Chhotanagapur region, Dehra Dun based People's Science Institute (PSI) is involved in popularising *ahar* technology in a big way to boost water availablility for both domestic and farming use.

India's north western state of Rajasthan, a large part of which is covered by the formidable Thar desert, has a long and unbroken tradition of water conservation---.

In more than 700 villages of the state *johads* are now meeting water needs of the villagers without any hassels. Essentially *johads* are simple stone and mud barriers built across the contour of slope to arrest rainwater. They have high embankments on three sides while the fourth is left open for the rainwater to enter.

In the villages where *johads* have been revived water is shared judiciously among the villages but when water is scarce farmers are not allowed to grow water intensive crops.

A *johad* prevents rain water from running off, allowing it to percolate into the ground, recharging water aquifers and improve the water balance of the earth. Significantly, the engineering knowledge to make *johads* was entirely local and no outside expertise was utilised. Yet these *johads* have stood the test of time and admirably withstood the ravages of rainfall.

According to Rajendra Singh of TBS, whose contribution to the revival of *johads* earned him the Magsaysay award, there is not a single village in the country which cannot quench its thirst and that of the fields through the revival of traditional water harvesting techniques.

Ratakhurd village in the semiarid and undulating Alwar district of Rajasthan has become a sort of 'green paradise' following the efforts by NGO Professional Assistance for Development Action. (PRADAN) in the drought proofing of the village through the revival of traditional water conservation techniques revolving round baands. A series of baands put up along the hill slopes were found to arrest the rain water run off. Each baand has a spillway which passes on the excess water to the rest in the line.

Once checked, the run off percolates underground and increases the moisture content of the soil and recharges aquifers effectively. This makes water for irrigation available round the year. Starting with a single baand for demonstration of benefits, the organisation helped the local people rebuild as many as five baands at an estimated cost of Rs. 1,000,000 each.

In Bhunkara and 35 other villages of Amreli district of Gujarat state the traditional water harvesting technique revived by a voluntary organisation SKTGSM (Sri Kundla Taluk Seva Mandal) has given a big boost to farming activities. These villages known for their hard, rocky terrain on account of their peculiar geological features coiuld not conserve rain water. However SKTGSM changed the situation by building dykes to check rain water and feeding it onto check dams and percolation tanks. Water famine in these perpetually drought prone rural areas of Gujarat has become a thing of the past with flourishing green farms and fields everywhere.[211 p38.]

For more information contact the People's Science Institute, 252, Vasant Vihar, Phase-I, Dehra Doon - 248006 U.P. India. Fax 0135-620334; Email psi@nde.vsnl.net.in Web site www.indev.nic.in PRADAN, C/o GIVE Foundation, 613-615, JB Tower, Drive-In Road, Ahmedabad 380054, Gujarat. India. Fax: +91-79-685 5610; Email info@givefoundation.org

National water Harvesters Network, Centre for Science and Environment, 41, Tughlakabad Instiitutional Area, New Delhi-110062. India. Fax 91-11-6085879; Email cse@cseindia.org Website www.rainwaterharvesting.org

7.12 Watering and water management for vegetables

As mentioned in 7.9 where the soil surface has dried out completely and is hard and fairly smooth, heavy and sudden water application can result in poor absorption, run-off and erosion. In dry weather, it is best to cultivate so as to form clods. See Fig. 22. With such clod formation the surface area for absorbing water is increased, the water can quickly penetrate through air pockets in the soil, and, the rate of evaporation is slowed down. (Where it is possible to cover the soil with mulch (see mulches 7.5) and zero tillage this is not needed however in many situations there is no mulch available.)

Approximately 90% of a plant is water, so the maintenance of a constant water supply to a plant's roots and leaves is essential for its growth and survival. Just as a person will die after three to six days without water, so many plants will wilt and die in just a couple of dry days; a seedling deprived of moisture could die after just a few hours without water.

Under natural conditions, water is available to plants through rain and dew, through rivers, streams and other water courses, and through water surrounding the mineral fragments in the soil and absorbed in the sponge-like organic matter. In periods of dryness, as surface water dries out, so deep ground water is drawn up by capillary action into the subsoil/topsoil layers.

Methods of Watering

The best method of watering is one which applies water in a form as close as possible to rain i.e. in separate droplets falling almost vertically by its own weight, rather than a continuous stream hitting the soil with force.

The water falling forcefully tends to seal the surface so that there is little penetration and some run-off. As the water has not penetrated deeply it is quickly evaporated. With shallow watering the plants develop shallow rooting and so have contact with less soil and plant foods, they do not grow as well, and quickly wilt if watering is missed for some reason.

Some people use a watering can, a dipper or a hose to quickly apply a lot of water in a steady stream to a plant or a row. This method tends to produce a swamp:

- it wastes precious water;
- it dislodges the soil from around the plant's roots;
- the water fills up all the air pockets, preventing the plant from breathing;
- when it dries, the mud sets hard and dense, making it difficult for roots and air to penetrate the soil.

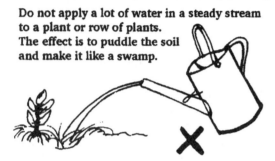

**Do not apply a lot of water in a steady stream to a plant or row of plants.
The effect is to puddle the soil and make it like a swamp.**

It then dries hard and compact

Fig. 46

Here are better ways of watering

An alternative used in small gardens is to bury an unglazed earthenware pot up to the neck in the ground between plants. Fill it with water and put a cap on it. Refill when needed, the water slowly passes through

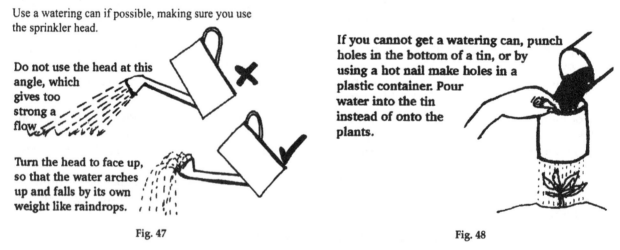

Use a watering can if possible, making sure you use the sprinkler head.

Do not use the head at this angle, which gives too strong a flow

Turn the head to face up, so that the water arches up and falls by its own weight like raindrops.

Fig. 47

If you cannot get a watering can, punch holes in the bottom of a tin, or by using a hot nail make holes in a plastic container. Pour water into the tin instead of onto the plants.

Fig. 48

the pot keeping the soil moist. Plastic soft drinks or water bottles if available can also be used, with the bottoms cut off. See Fig. 51.

Trickle irrigation

The most efficient method in the long-term is the trickle hose system. This method makes use of water pressure from an elevated tank to convey water along plastic supply hose to the individual beds, thence through narrow perforated hose laid beside the growing plants. The tiny perforations in this hose allow the water to seep directly to the growing roots, thus minimizing water loss by evaporation, misplacement or run-off.

Alternatively using a dipper broadcast the water in wide sweeps which disperse the water in droplets.

Fig. 49

Another method is to keep dipping a bushy branch into the bucket of water then shaking the branch over the plants. This is good for seedlings but does not give sufficient water in hot dry weather.

Fig. 50

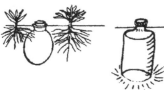

Fig. 51

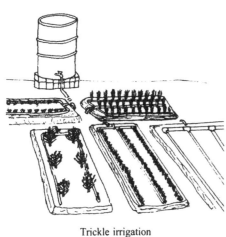

Trickle irrigation

Fig. 52

This system takes time, care and some expense to instal, but once set up it will save both water and labour. Once the system is installed, the only labour required is re-filling the tank, connecting up the perforated plot hose with the main supply hose and turning the tap on. After 30 minutes or so of trickle watering on one bed, the supply hose can be disconnected from that bed and reconnected to the next.

To instal the system, you will need:

* a tank made from a 44-gallon (200 litres) drum set upon a support (bricks, etc) at least one foot higher than the garden bed to be watered.
* The outlet tap should be located one inch above the bottom of the drum to allow sediment to settle. If a drum is not available, a very large clay pot with a tap could be used

The bigger the container, the better the pressure for watering.

* PVC plastic hose

For the main system, low-density, ½ inch cheap hose is easiest to use and has an effective life of four or more years, as long as you can bury the main supply system which otherwise is subject to damage as people walk along the paths. If it is not possible or convenient to bury this main supply hose, it is advisable to use high density ½ inch hose, which is much more expensive but is more durable.

For the perforated plot hose, use the low density type, for this is easiest and cheapest.

* Hose from tank buried under path, with outlet pieces coming up to be connected with perforated plot hose.
* To prepare the trickle hose. Using a fine nail, make holes in a line along the top of the hose. The holes should be about the thickness of the lead in a pencil. To test the efficiency of the holes, connect up the pipe to the water source with the holes turned upwards, turn on the tap, and observe the water jets. If any hole is too big, so the flow is more from it, you can fuse the plastic again by placing a hot match or nail against the hole; if it is too small, enlarge it just a little until water flows evenly from each hole. When satisfied that the flow is even, rotate the pipes so that the holes are downwards.
* Connect up individual plots about once in 5-8 days and allow the trickle to flow for about 15 minutes. Allow more or less time according to the water pressure and the needs of the plot. The soil will become moist in an inverted V below the hose. The plants are not splashed with muddy water which often carries or encourages disease. Weed germination is much less as only the soil surface beneath the hose is moistened. Much less evaporation takes place.
* The plot hoses arearranged according to the spacing of the rows of vegetables grown. They stay in place

for the growing time of that plot. After harvest, if necessary, the hoses are re-arranged to correspond to the spacing of the next crop in the rotation.

- To use the trickling water most efficiently, plant vegetables close to the plot hoses.
- Connect up only one plot at a time to allow good pressure from the tank. Make a bung of bamboo, maize or a turned over piece of hose to block the supply outlets that are not being used.
- In hot dry weather, a vegetable garden of 30 square metres will need approximately one drum (50 gallon/225 litres) of water per week if watering every second day. Note The plots are rotated, only some plots are watered at each watering. More water will be needed if the soil is shallow or gravelly, less if the soil is covered by mulching materials.

Fig. 53

When the author lived in Nepal where water was often in short supply and had to be carried to the house, the above system was modified to use all the waste water from the house not used for flushing the toilet. If a HASP compost toilet had been used (see volume 2) we could have used all the water for the garden. Buckets were placed under the sink and basin; when we used the home made solar heated shower, we stood in a large basin. The same basin was used for washing the clothes. Water not needed for the toilet was poured into a 44 gallon (200 litres) drum. The drum had old mosquito net over the

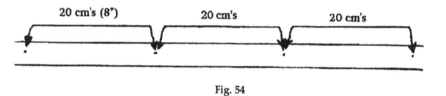

Fig. 54

exit pipe at the bottom over this were rocks up to near the top. Over the rocks was wire mesh with mosquito screen on top. Over that was a layer of sand with more old mosquito screen on top. The dirty water was poured onto the mosquito screen. It filtered through into the drum, and when needed the filtered dirty water flowed through the trickle pipes. Some gardens in the area had used household water like this for years for watering their vegetables, and there did not seem to be any problem.

Frequency of watering

The principles of good watering are:

- water before the plant is stressed by lack of moisture. If stress develops, there is setback in growth and increased susceptibility to disease.
- give less frequent deep watering rather than frequent shallow watering. With shallow watering, the roots remain close to the surface, whereas deep watering encourages the roots to penetrate further to where moisture and nutrients are more reliably available.
- In warm climates, it is generally best to water in the evenings to allow the water to soak into the ground rather than to be evaporated. However, in cold weather it is best to water in the morning so that plants are not cold and wet during the night. If this is not possible, water in the afternoon whilst the soil is still warm (about 3 pm). Seedbeds should always be watered in the morning.

Plants should not be watered on their leaves in strong sunlight, for this will cause the foliage to burn (water on leaves acts like a magnifying glass for the sun's rays). In conditions that encourage blight, downy mildew or other diseases, it is better not to apply water to the foliage, but directly to the soil under the plant's foliage.

As a general rule, it is better to give good deep watering one or two times a week rather than to give daily surface watering. By 'deep watering' we mean allowing a continuous fine sprinkle of water to flow through the topsoil and the subsoil for maybe thirty minutes so that the soil becomes thoroughly damp but not water-logged. We are trying to reproduce the situation in nature, where rain falls in sprinkling showers not just for two minutes or so, but for twenty minutes or more. Such good rain usually starts lightly, builds up in intensity and then eases off. This pattern is ideal, because the fine gentle rain at the beginning dampens the soil and allows better penetration from the later heavier flow.

By observation, you will establish the appropriate frequency of watering for your particular crops and conditions. Seedlings require frequent watering until established, whereas mature and hardy vegetables may need only 15-25 minutes of deep trickle watering per week.

Insert a spade into the soil half an ½ hour after watering. Pull it back to make a V-shaped slot in the soil so that it can be seen how deep the water has penetrated. The water should penetrate at least to the bottom of the main root of the plants.

7.13 Water requirements of a range of vegetables

Crop	Yield, in pounds per acre.	Gallons of water needed per pound of food
Onions	40,000	16+
Celery	60,000	22+
Lettuce	28,000	23+
Tomatoes	50,000	23+
Potatoes	40,000	24+
Carrots	30,000	33-
Cabbage	25,000	39-
Cucumbers	39,000	39-
Spinach	16,000	61*
Green beans	10,000	98~
Corn (cobs)	8,000	122~

Code: + good for water problem areas.   - Not as good.   * Not good.   ~ Very bad
Note : 1 Gallon = 4.45 Litres. 1 Pound = 0.454 kilogram

Note: I am sorry that I have not been able to locate the source of the above table and so cannot acknowledge the source. I would be happy to acknowledge the source in future editions. I believe it to be from a very useful book by Larry Wilcox *The Answers* which was lost to Nepal customs.[49]

As a friend has pointed out just yield in pounds may not give a fair comparison. "To make a fair comparison, you would have to compare the food value in , for instance, calories. If you do this you may find that corn may turn out to be the most efficient user of water." Of course it depends what you are looking to achieve. If you are a people whose staple diet is rice and so rice supplies most calories, but you need protein and vitamins for a healthy diet, see which plants should you grow with the limited water available.

In dry conditions remember weeds compete with plants for water.

Improve the soil. Watering soil which is non absorbent, stagnant and resists the deep penetration of plant roots, is a waste of time labour and land. Good soil is porous, well aerated, it absorbs and stores water, and has good drainage. Organic matter in the soil helps the water absorption and retention and makes it easier for the plant roots to penetrate deeper and so be able to reach moisture deeper down in the soil.

Over watering when water is available is wasteful and it results in shallow rooting, then later when water is not as plentiful the crops wither because the shallow roots cannot reach the moisture still available lower in the soil.

7.14 Sources of information on water conservation

For more information on this subject the author recommends.[90]

The author has been recommended but not seen

1. *Rainwater Harvesting: The Collection of Rainfall and Runoff in Rural Areas,* by Arnold Pacey with Adrian Cullis. (1986) London. Intermediate Technology Publications, 216 pages.

2. *Looking After Our Land: Soil and Water Conservation in Dryland Africa.* By Will Critchley, available in English or French, there is also a video to go with the case studies described. Available from Drylands Program. IIEd, 3 Endsleigh Street. London, England, UK.

Also the following list of sources supplied in[99]

Dr R P Singh, Director, Central Research Institute for Dryland Agriculture. Hyderabad-500 659, INDIA.

Dr R C Mondal, Director, Main Institute. Central Soil Salinity Research Institute. Karnal-132001, INDIA.

A B Damania, The International Center for Agricultural Research in Dry Areas. P O Box 5466, Aleppo, SYRIA.

The Director, The Jacob Blaustein Institute for Desert Research. Ben Gurion University of the Negev. P O Box 1025, Beer Sheva 84110 ISRAEL.

Professor R G Wyn Jones, Director, Centre for Arid Zone Studies, School of Agriculture, Forestry and Allied Sciences, University College of North Wales, Bangor, Gwynedd ll57 2UW, Wales, UK.

International Crops Research Institute for the Semi-Arid Tropics, Patencheru P.O. Andhra Pradesh 502 324, INDIA.

Kathleen McCullough, Administrator, Arid lands Information Network, 274 Banbury Road, Oxford OX2 7DZ. England, UK.

Vore Seck, Entre Nous, B.P. A237, Thies, Senegal.

Haramata: Bulletin of the Drylands: People, Policies, Programmes. This newsletter covers issues of drylands agriculture.

Contact Camilla Toulmin, Robin Sharp, International Institute for Environment and Development, 3 Endsleigh Street, London, WC1H 0DD, England, UK.

The Semi- Arid Tropical Crops Information Service. SATCRIS, ICRISAT, Patencheru. Andhra Pradesh 502 324, INDIA.

The Drought Defeaters Project. Contact Dr Phil Harris, Drought Defeaters Project. Henry Doubleday Research Association. Ryton-on-Dunsmore, Coventry CV8 3LG, England, UK.

Sahel Information Network. Contact Josephine Mazza or Ingrid Anderson, United Nations Sudano-Sahelian Office, E 45th St; New York, NY 10017. USA.

Network News. International Drought information Center. The director is Donald A Wilhite. For subscriptions write to: Jan Schinstock, Dept of Agricultural Meteorology. 236 L.W. Chase Hall. University of Nebraska-Lincoln. Lincoln. NE 685833-0728, USA. c:\.2.10waterconfinal 17.3.01

PART 2

CHAPTERS IN COMMON WITH VOLUME 2

LIVESTOCK

B1.1 The positive effect of farm livestock
B1.2 The negative effect of farm livestock
B1.3 How to manage livestock without causing erosion

B1.1 The positive effect of farm livestock

- Livestock have an important effect on both soil erosion and soil improvement.
- The manure of animals is an important way to replace nutrients taken from the soil.
- The nutrients supplied are more beneficial to the soil, and the crops grown in the soil, than non organic nutrients of the same theoretical nutrient value.
- Farm livestock are important as a way of adding value to the crops which are grown, and to bought in crops.
- The manure and urine of livestock stimulates beneficial soil microorganisms, and adds to the organic matter of the soil.
- In most situations mixed livestock and crop farming is preferable to either single enterprise. There is generally a synogystic (added value) affect.

In some circumstances animals such as sheep can help improve the soil and reduce erosion. This has been called the "Golden Hoof" effect. It occurs on light, stony or fluffy land which dries out quickly and is subject to wind erosion. The sheep manure helps to bind the soil while the sheep's small feet firm up the loose soil. So the sheep directly help to reduce erosion; they also help indirectly by their feet pressing the soil to better contact with the roots of the plants. As the manure helps the soil retain moisture and enriches the soil the plants grow better. Stronger denser plants give more protection to the soil and the roots hold the soil against erosion.

However to achieve this good effect there should not be heavy stocking or grazing in the wet season or the good effect will become a bad effect.

B1.2 The negative effect of farm livestock

Much soil erosion is caused by overgrazing or by grazing when the soil is particularly susceptible to erosion either by water or by wind.

> Livestock is one of the main factors causing soil erosion in tropical countries. Uncontrolled grazing during the dry season removes the vegetative cover and leaves the ground open to erosion by rain during the wet season.[89 p11.]

Soil erosion by overgrazing is caused either by, or by a combination of, the following effects.

Fig.1 An overgrazed range in New Mexico. From soils and Men. 1938 yearbook of Agriculture. US Department of Agriculture. Washington. U.S.A. Page 157. Ref 192

a) The plants being grazed too hard, by the animals grazing right down to ground level, or even in extreme cases digging up the roots and eating them. Or, by continual grazing so that the plants do not have time to recover from the grazing, and so are weakened; or, both may occur. The result is that the plants begin to die out, the ground becomes bare, and under the influence of strong sunlight, heat, drying winds, rainfall and the animals feet, the soil deteriorates and is eroded.

b) Heavy treading which wears away the plants, with no resting time to allow the plants to recover, such as in gateways or trails. The soil becomes packed hard and so when the rains come the water runs off rapidly, causing erosion, instead of soaking into the soil and helping the growth of plants.

c) Animals nibble at the shoots of young trees, they also tread on or break them; goats and some other animals also eat the bark of the trees and so weaken or kill them. The result is that no new trees can grow up to replace those cut down or destroyed by the weather, by disease, or by age. In the 1950's goats were allowed to graze on the hills of the Hong Kong New Territories. The hills were bare and ugly, the slopes were eroding and most of the rain which fell ran off and was unavailable for use. In the early 1960's goats were banned and since then the hills have become transformed; forested green and beautiful.

Some animals eg. Elephants, may uproot trees, push them over or break them.

d) When animals are allowed to graze on land which is wet and soft the soil becomes puddled and spoilt. Subsurface sealing blocks the absorption of water, while the surface becomes sodden. If the land is sloping the surface is loosened and tends to move down the slope. When heavy rain comes this mud may easily develop into a surface slide which in turn may develop into a landslide.

B1.3 How to manage livestock without causing erosion

More and more, the importance of zero grazing is being recognised. This means that the livestock do not graze the land but are enclosed or tethered and their fodder is brought to them. There is less waste as grazing animals dung on, urinate on, and trample on much of the grass. Also because the manure is all together it is all available for manuring crops and pastures as needed. By contrast when the manure falls on sloping pastures much of it is lost into the air or washed away during the rainy season.

Zero grazing or stall feeding enables forage to be utilised more efficiently, both because less is trodden, dunged or urinated on, and because a high proportion of the fodder consumed is needed for energy by the grazing animals. The result: enabling greater numbers of animals to be kept with zero grazing than in a free grazing system, and so more manure for land fertilisation.

However, the fodder has to be carried in, and later the resulting manure carried back to the fields, to reduce this work the animals should be moved round the fields as fodder is available, and stall fed in simple shelters. See illustration.

A simple movable shelter in the fields.

Fig.2

Even when fodder is in short supply and has to be carried long distances, on being brought back to the homestead it is frequently simply dropped on the floor in front of the animals, and 50% or even more may be trampled and dunged upon and virtually wasted. Thus, several of the advantages of stall feeding are nullified, and the only gain is that the quantity of manure is increased by valuable fodder being trodden into it.

A number of feeding racks or other devices have been designed in attempts to minimise wastage. One which is worthy of wider

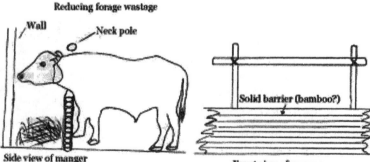

Reducing forage wastage

Wall

Neck pole

Solid barrier (bamboo?)

Side view of manger

Front view of manger

Fig. 3

recognition, and which is more effective in wastage reduction than the more widely - used overhead rack, has been adopted by some livestock owners in the eastern hills of Nepal. It consists simply of a wooden barrier firmly erected in front of the stall fed animal, extending to the height at the base of the neck, and leaving a space 0.75 to 1m. wide to the wall of the building. Above this barrier, about the height of the animals' backs, a pole is firmly fixed parallel to the barrier, so that to reach behind the barrier stock have to extend their heads forward through the gap created. Cut and carried fodder is placed in the space between the wall and the barrier, and the overhead pole prevents the animals from picking it up and throwing it back, as is the habit in particular, of buffaloes. Farmers have to use their discretion as to the width of the gap between the overhead pole and the solid barrier, according to the animals which are being kept - clearly, buffaloes will need a wider space than cattle, and if, as in many buffaloes on the Indian sub-continent, the horns are widely swept back, up or down, allowance has to be made for this. A similar, but obviously smaller device can be erected for goats and sheep, in preference to the overhead racks which are commonly used for these species.[197]

A community the author visited in Nepal had hardly any woodland or forest near them as it had become so degraded. They had to travel long distances to fetch fodder, firewood, building wood, and water. They decided to keep livestock out of one area of degraded forest and only remove dead wood for two years; then only dead wood and carefully cut grass. They were so pleased with their regenerated forest that they decided to convert most of their common land to that system, they reported that they could keep more animals as more feed was obtained than before. Later they found that they had most of the firewood they needed without having to travel far to get it. Soon they would be able to harvest building material. The community would allocate it according to need and the user would pay a fair price. They had their own policing system.

At the same time they had noticed that the soil was no longer being eroded away but instead was improving. ALSO THE AMOUNT OF WATER SUPPLIED FROM THE SPRING BELOW THE FOREST WAS INCREASING, BEFORE THE CHANGE IT HAD ALMOST DRIED UP.

The normal answer to the suggestion that the livestock should not graze but be kept enclosed and the fodder brought to them, is that it takes too much labour; and that when animals are concentrated in one place disease is likely to build up.

The remedy for those problems is to do as farmers in East Nepal. They move the animals about the farm. They may be tied up or enclosed in a field or in the centre of a group of fields with perhaps a simple shelter for shade. Figure 2. Fodder in the area is brought to them and they are moved over the land when the crops are not growing, or kept together and the manure taken out to the nearby area, instead of having to be carried a much longer distance from the farmyard. This is a very important factor when the manure is manually carried out just before sowing time, a very busy season.

Although fodder has to be carried it will normally be for shorter distances, and manure carrying is greatly reduced.

People report that when the animals are not treading on the fields more fodder is produced, when the animals do not have to spend a lot of energy walking about to get their food more meat or milk is produced.

The destructive effect of grazing has been clearly demonstrated in the Kondoa area in Tanzania. This semi arid area had in the early 1970's been converted to a moon-like landscape through intensive grazing. A conventional soil conservation program started in 1973, but building terraces and planting ridges could not change the destructive process.

A chance observation that the vegetation had re-established by itself within an enclosed area led to the drastic decision in 1979 to evict all cattle from the whole Kondoa district. Ten years later the land was again covered with grass and bushes. The formerly dry rivers keep water year round and the area under maize has increased dramatically.

Biologically, the eviction of cattle and goats has been a great success, but the cattle rearing farmers have suffered greatly. What remains now is to find a livestock production system that is environmentally and socially acceptable to this fragile ecosystem.---.[89 p11.]

In one African country I think Burkino Faso, there had been constant trouble between pastoralists (livestock herders) and the crop growing farmers. They have now come to an agreement, the crop farmers have fenced paddocks into which the livestock of the pastoralists is driven at night. They are protected by the farmers in exchange for the value of the manure. The livestock are safe enclosed at night so the pastoralists are benefited by the arrangement, a good relationship now exists between the two groups.

A further means of maximising the benefits of stall-feeding is through the efficient storage of manure. Where dung is removed daily, it is often pushed just outside the stall, where it remains, exposed to sun and rain until such time as it is needed to spread on the land. By that time, many of the nutrients have been lost, either by oxidation or leaching, or both, and the manure is reduced to a fraction of its original fertiliser value. It is surprising how few farmers or extension workers appreciate this problem, but a simple roof to protect the manure from the elements would contribute significantly towards it maintaining its fertiliser value, and towards increased crop yields.[197]

A quiet revolution, not widely observed outside the ranks of researchers in animal nutrition, has taken place in tropical livestock production. The basis for this is the pioneering work of Dr Thomas R Preston in ruminant (animals which chew their cud, = animals with four stomachs,PJS) nutrition- this opened the way to cattle diets based on unconventional energy rich ingredients such as sugar cane and its by-products, supplemented with urea and protein rich browse from trees.

Following successful work in Cuba he was convinced of the need to base livestock productions systems on available local resources and to train researchers mainly in their own country to develop sustainable livestock production technologies suitable to the conditions of their own country.

After his successful work in Cuba, Preston pioneered other systems based on the same concept. A system based on reject green banana was introduced in the Philippines and an intensive feeding system for cattle based on chopped whole sugar-cane was introduced in Mexico and the Dominican Republic. A feeding system using the foliage from the tree Leucaena leucocephala (Ipil) as the protein and fibre supplements in diets based on molasses was introduced in Mauritius and a similar system based on banana pseudo-stems, was introduced in the Seychelles.

Thomas Preston's latest farming system, developed in the Cuaca Valley in Columbia, supports extremely high levels of livestock production, up to 3,000 kg of meat a hectare a year. The system, which can be modified to meet climatic and economic conditions of a large part of the developing world, is based on soil protective perennial crops (sugar cane and nitrogen fixing trees).

The sugar-cane stalk, after removal of the tops, is turned into juice and bagasse using a simple animal powered 3 Roll mill: the tree foliage is separated into leaves and wooden stems. The cane juice is a complete replacement for cereal grains and is the basis (75%) of a high quality diet for pigs.

The cane tops are fed to cattle and sheep; the tree leaves provide protein for both the pigs and the cattle.

The bagasse (the sugar-cane pulp after the sugar has been extracted,PJS) and tree stems are used for fuel.(They could also be used to make fibreboard for use in building, or used for mulching. This author.PJS) In Columbia the system has been extended with recycling of animal excreta through plastic bag biogas digesters, ponds for aquatic plants and fish and finally for feeding earthworms (as poultry feed). These elements complement the system, providing additional fuel, protein for the livestock, and organic fertilizer for humus for the crops. All animals are confined (kept enclosed, PJS) and soil erosion is avoided.

The system is sustainable, building on the concepts of eco- development and self reliance. Almost all needs are grown on the farm with minimum fossil fuel-derived inputs and there is a surplus of biomass energy. Soil fertility is improved rather than eroded. The system is targeted, and had greatest impact with resource poor farmers, providing employment for all members of the family.---.

It is clear from Kondoa in Tanzania that an intensive and sustainable animal production system for small scale farmers best can be achieved through the integration of livestock with agroforestry and crops with high biogas production. As animals in this area in future will have to be confined, dry season fodder will have to come from sources such as legume trees and an energy source that either can wait on the ground (e.g. sugar-cane) or stored (e.g. maize stalks).

The introduction of such systems is now underway in Kondoa on an experimental basis. This may well mark the first stage of transition from traditional, extensive livestock production to more intensive forms of livestock rearing in the eroded semi-arid area of Africa. Dr Preston was awarded the 1990 laurette in farming and food production from Sweden's Innovations for development Association. He can be contacted at CIPAV. Convenio Interinstitutional para la Production Agropecuaria en el Valle del Rio Cauca, AA 7482 Cali Columbia.[89 pp11-12]

Bamboo leaves can be used for good quality fodder. As can the leaves of trees used in alley farming and other forms of agroforestry. Fodder trees can be grown to supply green fodder in the dry season. Hay and silage can be made when extra fresh fodder is available.

It may be worth investigating the treatment of cut dead grass or straw with either Urea or Ammonia to make it more digestible and so of increased nutritive value. In the UK bullocks can be slowly fattened only using straw treated with 1% ammonia.

FORESTRY, AGROFORESTRY, AND BAMBOO

Part 1 Forestry

B2.1 Introduction

Much has been said and written about the need to plant and conserve trees. Films such as "The Fragile Mountain" which looks at an area of Nepal, have had a good effect in highlighting the sort of situations which occur. Unfortunately however, they have often given the impression that all that is needed is to plant trees and the problems of soil erosion will disappear. Though in some areas forestry is the best solution for that particular location, forestry is only one important component of the correction of erosion of soil and soil quality; and responsible land care.

In addition such films have attracted ridicule such as the article "Fragile mountain or fragile theory." An article which should itself be ridiculed for assertions it makes.

Something written in 1944 was very perceptive, if more heed had been taken of what was said and appropriate action taken, North Africa would be in a much better situation that it is in now.

> In many parts of Africa forests are of primary importance in checking the destructive forces of nature, such as violent winds which cause the soil and sand and even stones to shift and drift. If created with knowledge and skill, the forest may not only conserve moisture and regulate stream flow, but actually increase the amount of precipitation in the vicinity. The established and correct practice of forestry in sparsely populated areas may be the means of reconditioning the land and saving whole populations from racial suicide.---.
>
> in the Southern Provinces of Nigeria they are fortunate, for there is still forest cover and in consequence an ample rainfall. But away to the north, even on the southern borders of the Northern Provinces, in many places the orchard bush is giving way to the savannah type, and still further north the rainfall is insufficient to maintain tree growth apart from stunted thorns and a few inferior species.[163 p108.]

It is now widely known that because that warning was not heeded the desert is expanding south from the Sahara and there are at last measures being taken to stop the spread, sadly it may be too little too late.

It has been pointed out correctly that planting forests can in fact create landslides. This is true when forests are planted unwisely as is unfortunately too often the case.

If we study the natural cover in areas unaffected by human activity, we notice that landslides are few, and are caused when excess water comes from another area, or, when due to movement of the earth, the area is tilted to such a degree that the soil inevitably falls off the underlying rock.

In natural forests species are mixed, there is an undercover of grasses, herbs, shrubs, creepers and smaller species of trees. Among and above this lower growth are the larger trees, the largest having wide-spreading roots, which penetrate deeply into the soil and the rock below, anchoring the soil and bringing up nutrients which enrich the topsoil, resulting in strong healthy growth of all the plants. The area is covered by a range of species ranging in ages and maturity.

With most planted forests single species are planted in solid blocks at one time. When the trees mature there is a great weight of tall trees crowded in an area. If the ground is steeply sloping, the whole slope may slip when a threshold weight of trees and weight of water is exceeded. In undisturbed natural forest the timber weight stays relatively constant with trees at different stages, and undergrowth protecting, improving and holding the soil. The soil mantle is secure and will not fall, until, as a result of movement of the underlying rock a steeper angle of tilt than that for planted forest is reached.

If lessons are learned from nature and a range of species planted, with planting and harvesting over a period of years, rather than planting a single species, using clear felling and clear replanting, soil loss would be minimal. For practical purposes this mixed planting could be with alternating strips across the slope, *not* up and down the slope, a practice which increases water runoff and erosion.

Research in Taiwan in the 1960's and 90's showed that planting across the slope also resulted in better growth of the trees, so a shorter time between harvest and a better return on the investment.

I should point out that in the few circumstances where the underlying soil is loose and soft as in shaly soil, and so liable to slump when too wet, it may be best to encourage run-off and so planting up and down the slope would be appropriate.

There is an interesting quote in the book 'I Planted Trees' by Richard St Barbe Baker.

> Way back in the history of their country (Germany), famine had tested their endurance. they had been short of water, too---. They began to study the weather ad the movement of the clouds. They saw the rain clouds driving past the barren wastes, but being sucked down towards evening over wooded areas. They began to realise the connection between trees (or bamboo PJS) and the indispensable rains, and once it had dawned upon them that the famine and drought was their own fault and had been brought upon

them by their own improvidence and thoughtlessness, they began to remedy the situation and applied themselves most earnestly to planting and forest conservation. it was not the shortage of timber that drove them to a progressive forest policy, but the realization of the necessity for soil conservation, to conserve the water supply and control the run-off from the hills, to maintain irrigation and the flow of springs, to ensure a higher level in the rivers in the dry season and to prevent floods from depositing barren soil in the valleys.[163 pp80-81.]

B2.2 Points to be remembered when planning forestry projects:

B2.2a Monoculture (growing single species) is unwise, especially if it is an introduced (not native) species. Governments and organisations tend to become suddenly enthusiastic about something which works or grows well. They then extend it rapidly to have convenient statistics of hectares of forest or a particular crop; it is the "IN" fashion and good figures will look good to supporters. The result is that too many or too much of a certain crop or species is grown, with the likelihood that pests and diseases build up, the system then fails and the system is condemned. A potentially good system or species is rejected because of impatience and poor planning. A range of tree species should be used so that on the one hand there are fewer of each species and so a serious build up of pests is much less likely; and on the other hand if one species is destroyed or severely damaged the other trees survive, and can be used to replace affected species.

As an example of the problem, when this author was in Taiwan there was a sudden extension of the growing of Paulownia trees introduced from Brazil. These trees are fast growing with white straight grained wood ideal for furniture making, they are very easily propagated by cuttings. I was urged to encourage the tribal farmers I worked for to grow them, but I was alarmed at the rate they were being planted everywhere, and so I rather discouraged the planting. Soon a mycoplasma like disease appeared which was spread by insects. The timber became stained and unsuitable for furniture and the trees died. The farmers lost money as there was a glut of trees which had little value.

Another example is the Psyllid insect which has seriously affected leucaena growing.

B2.2b Care needs to be taken when planning forest planting. Care needs to be taken to ensure that land planned for forest planting is not land traditionally used as common grazing, common harvesting of fodder, firewood etc. or is being bought from people in an unfair way.

If there is antagonism the forest will probably be a failure, human death may even result. Promises by the rich and powerful that those who lose out will be compensated are rarely believed or kept, it is usually the rich /powerful who gain most from the forest.

B2.2c Forest projects usually fail unless they are community based. The sort of thing that happens is that:

1) fences are put up to keep people out. These are removed by people wanting the posts and/or the wire, or access to the forest area. Guards have to be employed, often related to the forest staff. They are often bribed to be away when wrong activities are planned. They may even arrange for posts, wire or even timber to be removed, this always happens when the guards are of duty, on holiday, ill or at another part of the forest.

2) Trees planted have a poor survival rate as they do not belong to the people and so are not cared for. Animals are allowed to graze in the forest on foggy days when the guards cannot see them; and when the guards are off duty, such as on government holidays or parades and meetings, which may be quite frequent.

3) Tree seedlings are handled carelessly or even deliberately damaged. I have seen seedlings in plastic root bags/tubes being thrown like hand grenades down steep hillsides to their planting positions. The root bags shatter, the roots and twigs are broken and so the trees have a low survival rate. That means more planting work next year, it may also mean that as the project is given a budget according to how many trees planted each year, if trees can be replanted in the same place without the need to prepare new areas there will be money left over, which is always very acceptable.

4) The poor people know that the main beneficiaries of the forest will not be themselves but rather the powerful and rich. They may leave the trees at first till there is a worthwhile picking of branches and dead trees, then these will disappear. Next trees will be killed and later removed, the forest will downgrade unless much force and expense is used to control damage. Small local and social/community forests are more likely to be successful; so integrated land use is best.

B2.2d Social forestry is best; this may be where government forest adjoining a community is handed over to the people, or where all the local people agree that a forest is needed. It may be degraded forest or land that is eroding badly, and may even be landsliding or threatening to start sliding if not planted and protected. If trees are to be planted the people themselves decide on the trees to be planted; how many fodder trees, how many for fuel, timber or for sale. The community decide together that no grazing be allowed and how this will be policed.

An example of this was at Humin de ghera in Palpa, West Nepal. The government had taken, fenced and planted an area of land for a government forest. It was a failure! fence posts were stolen and the wire was cut; when the guard was off duty goats and cattle would be driven in, and they would trample or chew the young trees, grass would be cut along with the tree seedlings. When in an adjoining government forest the guard arrested people cutting wood there was a public outcry and the guard's life was threatened.

In about 1988 it was decided to use a social approach. The government would transfer control to the local community phasing it in stages over three years. First a general meeting was held, resolutions were made, and a committee was formed to implement the resolutions. It was decided to prevent all grazing in the area, no fencing would be required as all the community agreed to care for the forest. However the government watchman would continue to be payed by the government and he would report any offenders to the committee. He would also keep non locals out of the area.

Each autumn community members would be able to buy the right to cut and remove grass from designated areas of the orchard. The person with that right was responsible for seeing that no damage was caused to the trees and only grass and weed could be removed. A day was set when all the grass would be cut and the committee would make sure that the procedure was fair and people followed the rules. The money received for the grass was used for building and equipping a school for the village.

After the government completed the handover the people would pay for the watchman if they felt he was needed. As the forest developed, people wanting to buy timber would negotiate with the committee and pay a fair price. The collection of dead wood for firewood would also be controlled by the committee but the forest belonged to the village.

When I left the area in 1990 this project was going well and the community was pleased with it. Instead of being a government failure it was a community success. When I saw it again nine years later I could hardly recognize the area; instead of short sparse grass the land between the trees was covered in tall thick grasses and some legumes. From the point of view of fodder alone it was a huge success, and more fodder meant more manure for the fields. The trees had also grown stronger, particularly natural regrowth and naturally seeded growth, and the bamboo we had planted. The pine trees planted by the government were still disappointingly slow except in very fertile areas.

Unfortunately I heard a rumour that the government might take the forest back, I sincerely hope that they do not. It is important that real trust is established.

In Taiwan a similar arrangement established community forests between the villages and the large government forests. This protected the government forests and pleased the local people once they realised that there was no trick, that the community forest did belong to them. Sale of products from the forest was used to repair a dyke to protect the village from flooding; to repair the village school, and build a recreation centre. Roofing materials were bought for old and poor people, to replace thatch which was a fire hazard and difficult to replace as there was less thatching grass grown in the area. The experience of working together and helping themselves improved the local moral, and enabled other cooperative work to develop.

B2.2e It is often unnecessary to plant a forest. In another area of Nepal a returned Gurkha soldier was appalled at the state of the forest in the area. It was so degraded that he realised there soon would be nothing left at all. Fodder, firewood and building material was very short; the water springs were also greatly diminished. He persuaded the people that something must be done and they agreed to leave one area untouched, no one could cut anything, only fallen wood could be removed and no grazing allowed, fines were set for possible offenders. Two years later the trees were looking much stronger and some seedlings were growing, the people were amazed, they could see hope for the future. Permits were given for grass cutting under supervision, and as it was realised how much more there was than before the community decided to set another area of degraded forest aside. Later they agreed that all animals would be stall fed. See B1.3.

At the same time they had noticed that the soil was no longer being eroded away but instead was improving, also the spring below the forest which had almost dried up, was now increasing in flow.

This approach could be modified to a rotation system whereby one section of the local degraded forest area could be rested and allowed to regrow into a healthy forest. Then it could be harvested in a sustainable way and another block similarly treated, until all the forest had been brought back into strong productive sustainable growth; and then careful local management keep it that way.

B2.2f A point to be remembered is that local people do not always feel that forests are good.

a) If there is a forest nearby it could encourage monkeys which will then rob their crops. If as in many Hindu societies the monkeys are regarded as sacred and so cannot be killed, the coming of monkeys to an area can result in serious damage to crops and so malnutrition or even starvation among the human population. Forests also encourage other animals such as jackals and tigers.

b) People often realise that forests harbour pests and diseases of their crops. For example the Rice leaf beetle comes from the forest when the rice plants are young and so the fields nearest to the forest are the first affected. What is seldom realised is that natural predators of crop pests also shelter in the forest and if there is no home for such predators they will not be present and so more and more dangerous chemicals will have to be used to control the pests which over time will become resistant to the pesticides.

B2.2g Grazing should not be allowed in forest areas where there are young trees or where it is hoped that tree seedlings will grow to replace harvested trees. It is probably best to ban goats from ever grazing in forests.

B2.3 Planting across the slope is much better than up and down the slope

As mentioned, research in Taiwan showed that planting across the slope was best for soil conservation, reducing the runoff of water and nutrients, the trees grew better and quicker, so it was also better commercially to plant across the slope. As the people who plant trees seldom sell the grown trees they are not concerned with growth but only convenience; however it is often easier for planting and weeding to walk across than up and down the slope.

B2.4 Clear felling should not be practised if it can be avoided

Clear felling should not be practised if it can be avoided, but rather strip felling with sufficient time between felling for the previously felled area to grow over before the next strip is felled and the soil exposed. When in Taiwan I saw the effects of clear felling of hillsides. After four or five years, when the roots had weakened typhoon rain beating the earth and flowing down the slopes gauged out the soil, the tree roots started to give way and to start to move with the heavy wet earth and large slides formed. In two and a half years river beds and bridges had been filled up and turned into wide tracts of surging water and weirs. Paddy fields which had been producing food for hundreds of years were washed away and when the rivers receded what had been fertile fields were desolate strips of rocks gravel and sand. The soil which had taken many thousands of years to develop stripped away for ever. At the same time houses were destroyed and lives lost all to satisfy a timber merchants greed, yet it would probably be put down to 'an act of God'.

An important booklet to read is *The Man who planted hope and grew happiness*. by Jean Giorno Published by Friends of Nature, Brooksville, Maine, USA 1967

See also part 3 of this chapter.

Part 2 Agroforestry

B2.5 Introduction

In warmer countries organic matter builds up under forest, savannah or prairie conditions; where the soil is undisturbed and is covered with vegetation which protects it from heat and the strong sunlight, and from drying winds. When the same land is brought into cultivation and is exposed, rather than protected, organic matter decreases rapidly. As explained elsewhere when organic matter decreases the number of soil microorganisms reduces, and the structure of the soil deteriorates. The soil becomes progressively less fertile and more susceptible to the forces of erosion.

This can be seen by comparing the rivers and streams in parts of Africa.. Where there is traditional agriculture, cultivating clearings in the forest or between the trees is carried out, the water in the streams is still clear during heavy rains. Where plantation agriculture is practised the rivers turn brown in the rainy season as soil is carried from the land; clearly an unsustainable system even with imported fertilisers.

Such observation lead to research into combining agriculture and forestry - hence the term AGRO FORESTRY.

Agroforestry is a collective name for land use practices which involve a close association of trees and shrubs with crops and/or animals in some form of spatial arrangement or temporal sequence (eg rotations). The interactions between trees and crops (or livestock) provides both ecological and economic advantages to the farmer. Agroforestry is not a fixed arrangement or combination of species, but includes many different types of practices. In fact one of the essential features of agroforestry is that it can be sensitively adapted to various environmental, cultural and economic conditions. A.Daw, writing in.[85]

Over the past decade agroforestry has become an extremely popular concept among agricultural and rural development workers, and is often considered the panacea to many land use problems found throughout the developing world. Problems such as soil erosion; declining soil fertility, (and thus reduced crop yields); deforestation and declining fuel-wood supplies; reduced fodder for animals and water shortages, have become acute in many areas, forcing many farmers to suffer a reduction in their living standard, and to clear more and more marginal lands unsuitable for agriculture.

The causes of this ecological and economic plight are complex and controversial, and include the effects of climatic change, population growth, economic and political policies and inappropriate land use practices. **It has become clear that alternatives are required to the high-input, simplified cropping systems which have been promoted in the last few decades.** Modern farming technologies adapted from temperate zones have often had poor results on fragile tropical soils. The rapid decline in the productivity of tropical soils under continuous cultivation even with supplementary fertilisers is well documented, (Kang B T. Van der Kruijs A and Couper D C 1986. Alley farming for food crop production in humid and sub humid tropics. Alley Farming Network for Tropical Africa. IITA and ILCA, Ibadan, Nigeria.) Moreover, in many countries agricultural inputs are unavailable, infrastructures are poor, and farmers are usually financially unable to modernise farming methods. Many so called 'modern' farming techniques are simply inappropriate given the degree of environmental and economic risk under which farmers till the soil.

Agroforestry is seen as a sustainable way of tackling multiple land use problems, and of encouraging product diversification on farms. It is a system which builds on the positive associations between agriculture, trees and livestock for soil and water conservation. Moreover it is a land use practice which is valued for the role it plays in building self-reliance among farmers rather than dependence on outside resources. Sally Jeanrenaud 1990 writing in.[85]

Much research has been carried out in Africa; other people hearing of it have also tried to develop systems appropriate for their own situations.

There has also been criticism of Agroforestry. As is too often the case people tend to jump on the latest band wagon or fashion without careful consideration, then when things go wrong there is a tendency for a pendulum swing against the concept. As with all new suggestions, careful consideration and local trial should be made before larger scale adoption. Some of the trials which were carried out with alley cropping in:

small research plots where the advantage of alley cropping is overestimated, particularly in semi-arid states or on acid soil where tree roots tend to spread many metres laterally. In Hyderabad, India, it has been observed that roots from nearby leucaena plots reduced the yield of sole crop treatment by 34% compared with no root interference.---.

In Lampung, Indonesia, measurements of roots of one-year-old trees indicate that most trees have a saproot reaching into the subsoil, as well as many horizontal roots. Regular pruning in alley cropping significantly changed the rooting profile in otherwise deep rooting species, such as albizia. A local tree that does not fix nitrogen, *Peltophorum pterocarpum*, appears to be the most suitable for alley cropping because of its compact canopy and deep rooting profile. In contrast, the fastest growing nitrogen species, *Calliandra calothyrsus*, was too competitive.

The earlier synthesis of ICRAF's alley cropping experience indicates that the current strategy of using fast growing species may be counter-productive; they may be too aggressively competitive with alley crops. It would therefore be more worthwhile to select trees with non aggressive rooting habits from climax vegetation, especially if these are also adapted to very low nutrient soils. Promising tree species are *Grevillia robusta* and *Markhamia lutea*, which farmers in East African highlands plant on boundaries.[203 p19] An adapted version of an article in ICRAF's July -September 1994 issue of "Agroforestry Today".

B2.6 Alley farming/Alley cropping

In Nepal I visited a farmer Mr K. B. Gurung of Damauli who has been working on this, using one of the agro forestry techniques known as alley farming; he finds that using rows of the legume tree Leucaena leucocephala (Ipil Ipil), two metres apart has given excellent results. The area planted in this alley cropping system gives more maize and soyabeans than the same area without the trees. Each year the yield improves and the soil certainly looks and smells better, with many worms casts visible while there were none visible on the treeless areas. We looked at the soil comparing the area of alley farming with the other half of the field planted with the same crops as the alley farming half, it was clear that the soil was much better in the alley farming half.

Fig. 1. Tree root ripping plough.

Fig. 2. Tree root ripping plough.

In addition to these benefits he has the direct value of the Leucaena trees. He cuts them short in late winter and spring using some of the branches for his own fuel wood and selling the rest. He feeds the twigs and leaves to his cattle, buffalos and goats, though much of the leaf has fallen onto the soil between the rows. About every 10 metres he leaves straight trees to grow large for use as building material or for sale.

However he has found it necessary to root prune the Leucaena to stop them competing with the alley crops. He does this by using a special very narrow version of the local plough which he ran along about 1 foot (30cm's) from the side of the alley trees(Figure 1). He found it best to fit a blacksmith made metal sock (Figure 2) instead of a traditional share peg (Figure 1).

The hedgerows should preferable comprise legume species that can fix nitrogen and be cut back a few weeks before an arable crop is planted. Typically two prunings are carried out during the growing season of most crops to minimize shading and reduce competition with the arable crop. When the crops are harvested, the hedgerows are allowed to grow freely as part of a fallow management practice. The choice of a hedgerow species depends on the location and the farmer's preference. Weed control in this system consists of weed suppression by the hedgerow canopy during the fallow period, smothering of weeds by the prunings left as mulch on the soil surface, possible effects of leachates (chemicals released and washed into the soil) from the decaying mulch on the growth of weed seedlings.

This system not only controls weeds, but reduces soil degradation through its beneficial effects on soil physical properties, organic matter maintenance, erosion control and nutrient recycling." An adapted version from a paper presented to the First International Weed control Congress in Melbourne, 1992. Written by I Okezie Akobundu, a scientist with IITA Ibadan Nigeria.[77 p9]

Because of the obvious benefits this author includes alley cropping in the good news step by step terracing system.

Another report from Zambia describes the use of Sesbania sesban.

Experiments in Zambia with an indigenous tree Sesbania sesban, locally called ger-geri, have proved that it can double maize yields and replace chemical fertilisers ---.

The experiments which began in 1987, show that one hectare plots fertilised by ploughing back Sesbania sesban saplings into the soil produced an average 78 to 90 90 kg. bags of maize compared with 30-40 bags from plots to which chemical fertiliser were applied.

Mr Phillip Chikasa, a scientist with the International Centre for Research in Agroforestry (ICRAF), which carried out the experiments in conjunction with the Southern African Development Community (SADC), said the plant holds great promise for food security in southern Africa. "We are convinced that Sesbania sesban can replace fertilisers" said Mr Chikasa, who attributes high yields to the presence of essential nutrients such as nitrogen, potassium and phosphates at optimum levels.---.

Another advantage over chemical fertiliser is that the dead plant material not only improves soil structure but greatly minimises loss of plant nutrients through leaching. It is also estimated that the tree can fix between 150 and 200 kg. per hectare pre year, and that when the trees are harvested after one, two or three years of growth they continue to contribute respectively 30, 70, and 100 kg. of nitrogen to the maize crop.

Apart from improving maize production, the tree is also a source of energy; it is deep rooting and has 80 per cent of its roots in the 0-50 cm depth of the soil. The plant reaches the water table in under 18 months and is able to capture rich nutrients in the deeper horizons of the soil.[105.]

This author dislikes the vagueness of the above report eg. it does not state how much fertiliser was used in the comparison, it does not state how deep is the water table; however it does point to the potential of agroforestry to replace chemical fertiliser.

More recently there has been a reaction against alley cropping. An article in 'Agroforestry today' January/ February 1995 - the Journal of ICRAF the International Centre for Research in Agroforestry uncovered negative results worth considering.

The encouraging results obtained in demonstration plots were not so easy to obtain in farmers' fields, probably because the trial and demonstration plots were small and so the effect of the tree roots robbing the crops of nutrients were not as great. That is why in the good news terracing system, the alley trees are planted in steps in the terrace walls, the roots do not compete with the crops or interfere with ploughing, but they hold and stabilise the terrace.

> Based on ecological theory and limited data on tree-root profiles, there is growing agreement among researchers that competition between trees and crops is likely to outweigh the positive benefit of mulching, especially on high acidic and low moisture soils, Selection for nitrogen fixing and fast growing trees may not be the solution, rather there is an urgent need to look for complementarity in rooting of trees and crops.---.
>
> The evidence from the long term trials indicates that because of experimental and interpretational problems, researchers have probably overestimated the capacity of the technology to increase crop yield.[140]

This statement is a useful reminder of the folly of taking at face value the data coming from research institutions. Once a technique seems to be promising it is surprising how much apparently clear evidence is produced to back it up. Now the system will go into reverse gear as it is fashionable to prove that alley farming is not so good. For all the so called statistical analysis of research institutions local trials carried out in an appropriate manner are the best way forward. The following chapter suggests better ways of carrying out trials.

I believe that it could also be that farmers had not been well enough instructed about controlling the size of the hedging shrubs. The trees must not be allowed to grow too big between lopping. There should also be root pruning by ploughing deeper at the sides of the strips. *The Nepali farmer mentioned above had made a chisel type plough which he used to run along the sides of the hedge strip to cut off roots which were spreading into the cropping area.* See Figure 2.

In areas where rainfall is light it is obvious that competition for the water will occur, so the technique is either inappropriate or tree/shrub species which need much water should not be used. Likewise in areas where the growing season is characterised by overcast skies shady plants should not be used for the hedges. See B2.7.

A survey by IITA the International Institute for Tropical Agriculture indicates that "alley cropping is most suitable for food production on Alfisols and other high base soils in the humid and subhumid tropics". [115 p19.]

Another factor would be alley cropping in areas where the weather makes it inappropriate. If rainfall is light and is hardly sufficient to sustain the ground crops, planting water hungry hedges would obviously be inappropriate. However in wetter areas or monsoon areas where the main crops are grown during and following very heavy rain; the alley hedge plants would not be competing with the ground crops, and could enable the absorption of more of the rain water; particularly over time as the mulch they produced increased the organic matter of the soil. Shading can be a problem in monsoon areas, but in such situations the monsoon deciduous (leaves fall off in the monsoon season) leguminous tree Acacia albido mentioned in the following section B2.7 could be the answer.

I have heard that work in Tanzania has shown that farmers did not think the extra work involved in alley cropping was economically justified. I have heard from my son who has worked in Tanzania that there is not the pressure on the land in that country that is found in many other countries, the government is encouraging outsiders to come and farm empty land. I wonder therefore if the reasons for the lack of appreciation of alley cropping in that country could be for one or more of the following reasons.

1. Land farmed has often been cleared from jungle fairly recently and so is not yet so deficient in soil organic matter and so soil micro organisms, so that the alley farming effect of improving the soil was not so noticeable.

2. Because of the better state of the soil the addition of relatively small amounts of chemical fertiliser makes an economical difference in crop yield.

3. As there was still jungle nearby to supply organic matter to add to the soil there was no need for the manure from alley farming.

4. Because of the availability of firewood locally there was no economic benefit from the firewood from the alley farming.

5. The alley farming was not carried out long enough to realise the longer term benefits it could give by improving the quality of the soil.

B2.7 Types of trees to use

As I mentioned in B2.2a. I would be wary of using only one species of tree for agro forestry, especially if it is an introduced (not native) species.

Acacias are one of the alternative legume trees.
According to Masanobu Fukuoka's book, 'The one straw revolution',

> Acacia roots increase soil aeration and drainage by breaking it up. Acacia roots also "pump" other nutrients essential for plant growth from the subsoil to the leaves. The nutrient rich leaves of the acacia add humus to the system, increase soil fertility, and make nutrients available in the topsoil for other plants. Like most other legumes, acacias form a symbiotic relationship with soil bacteria (rhizobium) which fix otherwise unusable nitrogen. The taproot of the acacia, in a similar relationship with mycorrhizal fungi, transforms unavailable forms of phosphorous into forms available for plant uptake.. Mycorrhizal fungi exploit other minerals in the subsoil as well. Acacia baileyana is being used by Masanou Fukuoka, a well known Japanese organic farmer, as an inter-crop to rejuvenate the bare nutrient -poor, clay soil.---. Acacia baileyana is native to Australia---. In Western countries it is grown commercially for the production of cut flowers. Other qualities make it valuable for agroforestry systems. It grows rapidly; it re-sprouts quickly after cutting; it has hard wood; the leaves make good livestock fodder; and its flowers attract bees. Quoted in [84]

One of the obstacles to farmers taking up agroforestry is that in many countries the main growing seasons are/ is during the monsoons. At this time the weather is either raining or overcast, the result is that crops growing in the shade of trees do not grow as well as those out in the open. Farmers observing this are rightly wary about planting more trees.

When I pointed that out at a conference I was told about the legume tree 'Acacia albida' also known as Faidherbia albido (Del), Apple ring acacia, Winter thorn, and Camel thorn. I have since seen various articles about it, pointing out that it grows over a wide range of conditions and is particularly well suited to agroforestry. The following is an example:

> Acacia albida is a leguminous tree native to much of Africa which sheds its leaves at the beginning of the wet season, fertilising the crop beneath it, and grows them again for the dry season. This means the crop beneath is not shaded during the critical growing season. The 40 foot deep tap root draws up nutrients which have been washed deep into the soil and makes them available for the growing crops.
> Nitrogen fixed by the roots is returned to the surface soil by leaf fall, which also helps to build up the surface organic mater. Twenty Acacia albida trees per hectare have been able to contribute about 40 kilograms of Nitrogen per hectare and keep yield up at double the subsistence unfertilised level.
> Farmers have known about the benefits of this tree for generations and have protected them in their fields and not cut them down. But generally they have not practised germinating the seeds and planting the seedlings, and so the tree is not very common.
> Acacia albida seeds have evolved so that they germinate only after passing through a ruminants (four stomached animals) gut. Now germination is achieved through the slightly less exotic method of boiling them for a short period and seedlings are distributed to farmers at nominal cost.

From an article by Martin Whiteside of Christian Aid, PO Box 100 London SE1 7RT [82 p22].

Acacia albido grows under a wide range of conditions in Africa, from semi desert shrub bush or grassland to semi arid woodlands and short grass savanna type to sub-humid forests woodlands or wooded grasslands. It may be the dominant species in the Altitude range between 1400m - 1900m ASL. This ideal farm tree has been the subject of several studies and scientific publication [88]

---Its foliage is a valuable fodder for all types of stock. The young leaves and shoots are usually avidly browsed. Livestock that eat Acacia albido remain in good condition during the dry season when it is often the only green growth available. The tree is often lopped and branches carried to the camels, cattle, sheep, and goats. The pods, too, are eaten by livestock, especially cattle, and also by elephants, antelope and baboons. The nutritional value does not deteriorate on drying (which is the case with many other acacias), so Acacia albido is fed dry in many parts of Africa. In the Sudan, trees produce an average of 135 kg of pods per tree. The yield from a stand of 12 trees in the Sudan has been calculated to be 200 kg of protein from the pods alone. This compares favourably with 180 kg of crude protein from a crop of unshelled groundnuts. The two crops can be, and are, grown together, since the groundnuts (and other crops) are grown during the wet season when the Acacia albido is leafless.---.
The seeds contain up to 27 percent crude protein and are eaten by people in Rhodesia (Zambia and Zimbabwe) during times of famine. The seeds are boiled to loosen the skin and then reboiled to separate the kernels.---.
The sapwood is dirty white. The soft, yellowish -white heartwood is subject to attack by borers and termites. Although it is easy to work, it springs and twists after sawing, even when the wood is seasoned. The old bark is rich in tannin (28 percent), the root and pods are not (5 percent). In northern Nigeria, the pounded bark makes a packing material for pack saddles for oxen and donkeys. In West Africa and Tanzania the tree yields a gum of good quality. From [141 pp111-113.] There is more information on this tree and others in the book.

An example of another legume species suitable for agroforestry is Ethyrina which the author was told is good for fodder, for soil improvement, shade, and for live fencing, it is easily propagated by hardwood cuttings. It grows rapidly and coppices (grows up again after being cut) well, it grows in both the tropics and subtropics in a wide range of conditions.

B2.8 pH, salinity, and inoculants

As legumes often need inoculating with rhizobium this may seem to be a limiting factor if rhizobium is not available, however in the authors experience this may not be a problem.

When conducting trials of Leucaena I did not have sufficient inoculant for all seeds to be planted. I went about the area collecting soil from under legume shrubs and trees. I mixed the soil and added it to the soil the seeds were planted into. We recorded the plants with inoculum and those with the legume soil mixture; there was no observable difference in the effect.

Soil pH can also be an important factor, as most tropical soils are acid we carried out comparisons of seedlings planted into acid soil. When about 5.0 grams of lime (2 level coke or beer bottle caps with the lining removed) was worked into the soil into which each seedling was planted, the seedlings grew well. When identical seedlings were planted without the lime they grew very poorly, in two years they were only a quarter the size of those with lime. It seemed that lime was only required in the initial stage, after that the trees grew vigorously without further addition. It may be that less lime would also have been effective, readers will need to find what quantity suits the soil they work with. The soil we planted into was poor, hard red laterite soil. This method is obviously much better than considering liming all the soil.
In many third world conditions lime is not available or is too expensive for normal use as it will be leached out of the soil in the rainy season; particularly if the soil is short of organic matter.

Practically speaking, if we keep up the organic matter of the soil it has a neutralising affect; recent tests have confirmed this; green manure crops were added to soil and the effect monitored:

The soil had a pH of 4 and soluble aluminium saturation of 50%---. Sesbania cochinchensis, an aluminium sensitive leguminous tree was grown as a test crop for 4 weeks, plant dry matter weight and chemical composition were used to determine the liming effectiveness of the manures. Soil chemical analysis was done prior to (before) planting. Since all treatments had adequate (enough) fertiliser before planting, plant growth was used as an indicator of aluminium toxicity (poisoning effect). The manures increased pH and reduced soluble aluminium at varying levels.

Using plant growth as a measure of Aluminium toxification (poisoning effect) cowpea and leucaena were quite effective as 'liming' sources. Leucaena was the most effective overall; however, cowpea was more effective at the lowest dose.---.

The results of this study show that green manures can replace lime in detoxifying soils at least on a short time basis, which is enough time to allow young seedlings to establish (get started). Freshly added organic materials can reduce soluble Aluminium and increase crop yields.

The results also showed that cowpea dramatically increased the uptake of potash by the plants grown with the cowpea added to the soil.[37.]

Salt tolerance in Leucaena. James Brewbaker, University of Hawaii.

Leucaena's salt tolerance is effective only along shore lines where calcium levels are high. Inland saline areas are often not calcareous and so Leucaena cannot tolerate them. Quoted in [86 pp36-37.]

B2.9 Rationalizing Agroforestry

After years of attempting to eliminate shifting cultivation in what is now Zaire, Belgian agronomists developed a scheme for rationalizing shifting cultivation in thinly populated areas. This story is told in great detail in a book by Jurion and Henry (1969). Large forested areas were divided into strips or corridors approximately 100 m wide and oriented in the east west direction to maximise sunlight penetration. Every other corridor was cleared every year. (till the system was extended over the planned area) In this fashion every cultivated corridor was flanked by a forest fallow on each side. The sum of the number of years in crops plus the number of years in fallow determined the number of corridors in a management unit.

The corridors were as long as the topography permitted.

Individual or communal boundaries were drawn at right angles. In Yamgambi, Zaire, the cycle consisted of 3 years of cropping followed by 12 years of fallow. Thus the available land was divided into 15 corridors, laid out in contours along the slopes in rolling terrain.[74 p381.]

The rotation was :
Year 1 Clearing in November-December.
Year 2 March; maize: July; bananas, rice: September; manioc (cassava).
Year 3 Harvest of manioc and bananas.
Year 4 May; maize: September; Abroma (fibre), ground (pea) nuts.
Year 5 Fibre ready for cutting.
Year 6 Regrowth of forest
From a chart by Dumont (1957) in[76 p50.]

The system was set up supposedly in agreement with the population. Each family received a strip of forest of 8-12 hectares.

The strips of forest provided shade and seeds for the germination of the cultivated area when they were in their fallow or resting period. Each corridor though farmed by different families in different sections was planted to the same crops.

The regulation of land by the (paysannat) system has a number of advantages:

1) The ratio of cultivated land to fallow land, and thus the preservation of soil fertility, is under control.

2) The concentrated cultivation of particular crops in one belt, where each family has a section, facilitates the introduction of improvements like rotations, cultivation of cash crops such as cotton, better varieties, row cultivation, planting at optimum times, denser planting, weed control, burning off old cotton stalks, plant protection etc.

3) The concentration of cultivation reduces the expenditure of labour on transport. Building tracks becomes an attractive proposition. Marketing and processing can be organised cooperatively.

Difficulties resulting from varying soil fertility and the varying work capacity of families can be largely overcome in more flexible paysannats.[76 p51]

This author would feel that the basic idea is good but questions some of the components. Programs like this which state that the plan was drawn up with the approval of the people rarely are. It may be with the approval of the local vocals who have their own reasons for encouraging the program. It may also be with the agreement of the people involved not because they agree with it, but because under the program they will be given land and perhaps other help. Perhaps because it is better to go along with it even though it does not seem the best way, because if they turn it down they could be worse off than if they accept it. The aid or the government help may go elsewhere, the government may regard them as uncooperative people and so establish health posts, schools, water systems etc. elsewhere. It may be they agree out of fear of reprisals if they do not, or out of fear or regard for their tribal or group leader; these things are often complex. I would criticise lack of flexibility and diversity.

Looking at the way the program is set up I would feel that the hidden agenda may be to enable mechanisation, which would put people out of work. Long strips all growing the same crop could pave the way for plantations to take over after the peasants had done the hard work of clearing the land and breaking it in. The peasants may have realised that and so after independence the system was discontinued.

There are cases where the previous governments policy was good but was automatically discontinued after independence. I have heard Africans admit that colonial policy for soil conservation eg, not allowing clear felling of areas, but only strip felling; not allowing vertical cultivation of slopes, etc was in fact wise and good. That since independence all such colonial rules had been rejected and the land had deteriorated as a result. The national governments were now starting to bring back those policies. A similar situation was seen in Taiwan where the soil conserving policies enforced under Japanese colonial rule were rejected when the Japanese rule ended. Later after much damage had been done to slopeland and watersheds, it was realised that policies regarding management of hillsides were necessary.

B2.10 Other indigenous practices.11

An article entitled 'Soil conservation by stealth' in the May/June 1999 edition of International Agricultural Development, describes how an indigenous practice called the Quezungual system in Western Honduras, not far from the border with El Salvador works well.

> Generally the area has suffered severe land degradation over the years as most farmers practice slash and burn agriculture, but in the southern part of Lempira Department there was an exception. Farmers have developed their own agroforestry system which precluded burning and appears to sustain cropping over a period---.
>
> The Lempira region of western Honduras is a mountainous area. Annual rainfall, which falls from early May to October, varies from 1400mm to 2200mm. In the dry season there are often strong winds which can dry out the soils.
>
> Small holder farmers cultivate small plots of less than two hectares on the steep slopes. The steepness can vary between ten per cent and twenty five per cent, though in some instances the angle may be as high as fifty per cent.
>
> The distinctive feature of the Quezungual system, which is named after one of the villages practising the system, is the three levels of vegetation. During clearing many of the taller trees such as Cordia alliodora (laurel), Diphisu robinioles (guachipilin) and Swietenia spp. (caoba) are left. Also left are other trees and shrubs which are pollarded (cut off after which the tree sends up shoots from the stumps which turn into new branches and leaves,) at about 1.5 metres, and underneath them the crops of beans and maize are grown.
>
> Many of the trees and shrubs are leguminous, such as Gliricidia, Bahuinea, Leucaena, and Guanacaste. The trees are pollarded in the dry season and the cut material is left on the ground.
>
> There are two cropping seasons. The first one, the primera, is planted in April/May, once the rains start, and harvested in August/September. This is followed by the postrera crop, which is harvested in December.
>
> Maize is usually sown in the primera season, and followed by beans. Some clearing of vegetation may be needed after the dry season and then the seeds are dug underneath the mulch of dead prunings.
>
> Many farmers increase the cropping season by planting fruit trees, such as avocado and maranon.
>
> Farmers manage the system by ensuring that the trees and shrubs do not create too much shade for the food crops. For instance, the shrubs are first pollarded at the beginning of the rainy season in time for the first sowing of maize, a crop that needs plenty of sunlight, (the primera crop). Further pruning is carried out three to four months later in the second half of the rainy season for the sowing of the beans and

sorghum plants (the postrera crop), that mature after the rains stop. Lower branches of the trees may also be pruned should they be restricting light. The pruned material is often separated - the woody pieces are used for firewood, some of the leafy material from the leguminous species may be used as forage, and the rest will be left as a mulch. Some farmers use the pollarded material to make barriers against soil and water run-off.

The tall trees such as Cordia are either harvested after some years and the timber used for roofing, or left for 15 to 20 years when the timber has more general construction uses. The life of the trees is not reduced by pollarding. Gliricidia, for example, survives up to 30 years with annual pollarding. As there is plenty of tree seed in the ground, there is ample regeneration of both the timber and the shrub species occurring all the time---.

Farmers using the quezungual system are getting higher crop yields, and, even more important, the yields are sustainable over time. Average maize yields under the system are 50qqs per hectare compared to 20qqs under slash and burn. "As well as better yields, farmers are also getting firewood and timber, which makes the system more profitable than slash and burn", says Ian Cherrett (working for a FAO project in the area).---. "The better yields can be attributed to nutrient recycling, mulch covering, improved soil structure, increased infiltration of the rain and the deep penetrating roots of trees such as laurel, guanacaste, and aceituno.

Managing the system requires less labour, 30-40 percent less", says Ian Cherrett. Drying of the beans is made easier because they can be draped over the pollarded shrubs for drying immediately after the vine is picked.---.

"Farmers' motivation in adopting the system is not to conserve soil and water per se", says Jon Hellin (also working on the FAO project in the area). "The advantages they mention are improved soil moisture and sustainable production. The quizungual system is, therefore, an excellent example of where soil and water conservation is achieved indirectly or by stealth". [133 pp13-15.]

The article adds that the authors can be contacted on e-mail at cherrett@sdnhn.org.hn and hellin@xelha.u-net.com

B2.11 Other agro forestry methods, and sources of further information

As may be seen in the earlier chapters, the author favours the use of agroforestry techniques in soil conservation work. The terrace risers should have narrow steps along the centre line; in which are planted trees which provide soil improving materials. They also provide fuel and fodder which reduces the pressure on forests. When planted in the middle of the wall their roots stabilise the terrace walls; they do not interfere with the crops on the terraces or with ploughing.

Agro forestry covers many different techniques apart from Alley cropping, such as using a tree fallow with legume trees to protect and improve the soil during the fallow period, used when the fertility levels have become too low for worthwhile cropping. This reduces the time needed before the fertility is restored and wood is also produced.

Agro forestry also includes growing crops in clearings in the forest, wind break hedges, and mixing crops and trees. Some crops can grow happily on the ground beneath the trees. Some plants such as passion fruit can be planted at or near the base of trees and allowed to climb up the trees, as the fruit ripens it falls to the earth and the farmers simply harvest from the ground beneath the trees.

Another practise is to mix pasture and forestry. Inside a forest the climate is less extreme than outside, it is less cold in cold times, and cooler and less dry in hot times. Growth of grasses has a longer season in the forest. The animals are protected from bad weather, fodder trees may also be part of the forest.

Pomiforestry is another technique, where fruit trees are grown in clearings or between forest trees. The fruit trees are sheltered and so there is less or no wind damage, there are less pest and disease problems than in typical monoculture orchards. There may be natural predators in the forest to prevent the levels of pests becoming too serious. In open orchards there is less cover for these predators and the ones which would be present are often killed by neighbours spraying insecticides.

In 1979 ICRAF the International Council for Research into Agroforestry based at Nairobi Kenya was set up. This has seen Agroforestry emerge as an explicitly interdisciplinary applied science [85] *Ref 85.*

The following are some sources of information which the author has not read but have been very favourably reviewed they should give interested readers much more information.

1. The June 1990 edition of the ILEIA Newsletter (Volume 6 No 2) ILEIA c/o ETC Foundation, Kastenjelaan 5, PO Box 64 The Netherlands.

2. The Agroforestry Seed information Clearing House. AFDSICH Project, Department of Agronomy, UP LOS BANOS. College, Laguna, Philippines 4031.

3. Agroforestry Today. ICRAF's Quarterly magazine, is published in English and French. The magazine includes articles on agroforestry research, development, training and outreach activities. Write to ICRAF. PO Box 30677 Nairobi Kenya.

4. Agroforestry Technology Information Kit. Published by the International Rice Research Institute. PO Box 933, Manila, Philippines. 1990. 262 pages $40.

5. Agroforestry Extension Training Sourcebook. By Louise Buck, 1989. Produced by CARE, a two volume set for field workers doing agroforestry extension. To order CARE International, Regional Advisory Office PO Box 43864 Nairobi, Kenya.

Part 3 Bamboo

Bamboo: An Undervalued Crop Plant

B2.12 Introduction

When I went to Nepal after 16 years of multi-disciplinary, agriculturally related work in Taiwan, I was shocked to see the way the soil was degrading and eroding. I discovered that in the past when the fertility of land reduced to an unacceptable degree people cleared more land or moved to another area. That system is no longer possible, due to the greatly increased population, and because in most of Nepal there is now nowhere to move to. The fertility and moisture retaining capacity of the cultivated land is declining; the soil structure is deteriorating and so the soil is more and more susceptible to erosion. The uncultivated land remaining is poor, usually steeply sloping and with shallow soil which soon loses both fertility and soil, when exposed to UV light and monsoon rains.

Initially my work was to be to develop the new agricultural section of Palpa Community Health Program (CHP). I felt that the most important work we should be doing, would be to look for a way to reverse the present trend. To find and demonstrate a practical, integrated, sustainable, and if possible improving system of agriculture for Palpa District and the many similar districts in Nepal. This work would be at least as important as the medical aspects of the community health program.

Taiwan is amazingly like Nepal, not only that the two countries are on similar latitude and comprise about one third plains and two thirds hills and mountains, they have the same combinations of rocks and soils, the same steep hills, crops, and erosion problems.

The results of careful research work in Taiwan by the Chinese/USA JCRR: (Joint Commission for Rural Reconstruc-tion) in the early 1960's had shown that 'ON VERY STEEP HILL SIDES, ERODING AND LANDSLIDE AREAS, AND ON VERY POOR LAND, IT WAS BEST TO GROW BAMBOO'. THE GOVERNMENT THEN DESIGNATED SUCH AREAS AS BAMBOO AREAS AND NOTHING ELSE COULD BE GROWN THERE.

THE BAMBOO PLANTING HAS PROVED TO BE A VERY GOOD POLICY AND TRANSFORMED WHAT WAS BECOMING USELESS LAND, INTO VALUABLE INCOME GENERATING, IMPROVING LAND: CONSERVING WATER, AND CONSERVING AND IMPROVING THE SOIL.
See Fig. 1 and Fig. 2.

In Taiwan bamboo cover occurs principally within the tropical and subtropical climate zones (from sea level to approximately 1,600 metres elevation). The bamboo type consists largely of three native species: Phylostachys makinoi (bambusoides), Dendrocalamus latiflorus, and Bambusa stenostachya. Bamboo has considerable economic importance since its stems are used for a wide variety of purposes and its shoots are used for food. It also appears to have promise for watershed protection since it produces a large amount of litter, grows in dense stands, and its root system has good soil holding features.

B2.13 The value of Bamboo for soil and water conservation

Due to its fast growth, easy propagation, soil binding properties, and short rotation, bamboo is an ideal plant for use in afforestation, soil conservation and social forestry programmes, Various aspects of research on utilization carried out on bamboos in India have been summarized by Vaemah and Bahadur 1980.[30]

Bamboo is an extremely valuable plant for use in soil and water conservation, and land stabilization. At the same time providing many other benefits such as fodder for livestock, which in turn results in more animal manure for the fields. It can be used to generate income from very poor land., it is a quick growing source of building and cottage industry materials, and firewood.

Bamboo is very versatile, some species growing in wet conditions others being amazingly drought tolerant, particularly if grown widely enough to shade the ground, and cover it with the mulch from its own litter. Bamboos will grow at widely different elevations and temperatures, from tropical areas to areas with winter snow. There is a great potential for its wider use.

A key part of my approach was to investigate the possibility of growing bamboo on a much more extensive scale than was the case in Nepal. Bamboo could be planted on very steep areas, obviously eroding and landslide areas, and on very poor land, thereby reversing the deterioration of the areas concerned and conserving the precious soil and water. At the same time, it would transform those bad areas into valuable areas, in a time scale much more appropriate to village people than that taken by conventional forestry. Bamboo harvest can start at 4 years and by 7 years will be giving good income, given a transport possibility which in some cases could be by river rafts, in others by jeep trail, see Figures 12 and 13 later in this chapter. In such a case, the income could be much better than the declining income from any alternative use of that category of land. In addition, the more extensive growing of Bamboo will provide building materials, material for cottage industries, fodder for animals, and food and firewood for humans. Whilst doing this it steadily improves the soil it grows on.

To start this work I planted bamboo propagation trials in January 1981. Though my main work was with agriculture and horticulture, I was able to carry out research into bamboo culture on a small scale.

In the autumn of 1987 I was granted the use of a piece of steep rocky hillside and we were able to start trials on a larger scale. Figures 3, 4, 5, and 6. In 1988 Wolf Andler of the Tinau Watershed Development project saw the work. He gave us encouragement, and asked if we could do some work on bio-engineering and Bamboo propagation, in one of the project forest areas. We agreed to a joint program, and much was learned from that work, see Figures 8 and 9.

Trials were carried out in different situations, damper and drier conditions and at different times. Different types of cuttings, planted upright, diagonally (sloping) and horizontally (level). With single node, double node and multi-node cuttings, also root cuttings, 24 species were tried at that time. Helvetas a Swiss organization gave further support by enabling me to spend extra time in Nepal to write up the work and make recommendations. The result was a book "Bamboo a valuable crop for the Hills"[28.] I have since re written the book for wider use under the title "Bamboo: a valuable soil conserving/improving and income generating crop". A practical guide to the growing of Bamboo.[29.]

I have seen farms which have been protected by a barrier of bamboo. Landslides have swept down the hillsides at each side of the bamboo barrier, but the farmhouses and the families in them were safe. I have also seen bamboo planting used to confine the course of a river, protecting the land behind the bamboo. See Fig.11. Bamboo can be used to convert growing landslides into stable useful areas providing very useful material for building, fodder, firewood. cottage and factory industry. Figures 7, 8, and 9.

We have planted culm (stem) cuttings of bamboo in gullies and to stabilise landslides. In two places houses were threatened by landslides which had developed below them. After the owners asked us to plant bamboo, landslides stabilised and instead of the area being a threat, became a source of income to the householders. On a return visit 4½ years later the owners were delighted with the result. Not only was the land now stable but they had already sold poles for a good price. **I believe that there is great scope for soil conservation using bamboo.**

It is my hope that a wider understanding of the value of bamboo for countries like Nepal will be realized, and arising from that realization, greater implementation of work in bamboo propagation. I believe that there are

Fig.1 A bamboo forest (Phyllastachys bambusoides) in Taiwan. This species grows in a wide range of conditions from northern Japan where the winter temperature may go down to minus eighteen degrees C to South Taiwan which is near the northern Philippines. From sea level to 10,000 ft above sea level in Taiwan.

Fig.2 Bamboo plantation provides excellent economic returns and good soil protection. Bamboo should be considered in planning the rehabilitation of eroded sites. The illustration is of a newly harvested bamboo forest.

Fig.3 Cuttings are planted in the spring.

Note : The method of propagation varies with different types of bamboo.

Fig.4 This trial area was planted mostly by culm (stem) cuttings of many species in the spring of 1988.

Fig.5 A view of the same area in winter 1990

many countries in which the more extensive planting of bamboo would prove very beneficial. Where jungles have been removed exposing land to rapid deterioration, bamboo forest could take the place of the vanished jungle. This in a more appropriate, and socially beneficial way, than the pine forests commonly planted. Bamboo harvesting does not need chain saws or bulldozers. A heavy slasher and simple bogie (see Fig.12) is all that is needed.

We sometimes see publications which read as if planting trees would solve all of the problems of hill people and their land. Others maintain growing more fodder crops or soil engineering would solve the

problems.

These approaches are too simplistic; an integrated approach is needed, flexible enough to fit different social, and physical needs and differing climatic and soil conditions. I do believe that given those considerations, Bamboo culture can make a very significant contribution towards creating a viable future, for people in hill areas, and also indeed in many other areas.

Bamboo is not the solution on its own but it has an important part in the creation of practicable integrated sustainable systems

Fig.6 Five years later, bamboo has been harvested for 4 years already.

Fig.7 If this situation is left, the landslide will spread across the terraces all the area will be wasted, if instead the area is converted to bamboo forest using creeping bamboo species, it will be stabilised and produce a sustainable crop.

without which there is no hope for the future of the land or the people.

B2.14 Other aspects of the value of bamboo

Several years ago I cut down the side of the roots of a bamboo clump during damp misty weather. The leaves were absorbing water from the damp air, and pumping it into the surrounding dry soil, which was moist around the roots. Presumably the plants can later reabsorb part of the moisture and with it plant nutrients from the soil. It is interesting to walk under a clump of bamboo on a very humid day, it seems to be raining. However, on moving away from the canopy, we discover it is not rain. The cold leaves condense water from the air

Fig.8 Preparing to plant bamboo cuttings on a landslide. Such landslides are now stabilised and producing fodder, fuel and income.

which drips and so it seems to be raining. This is corroborated in the following quotation.

Young growing bamboo shoots sometimes exude water on a considerable scale Riviere A&C (1878) Les Bambous Paris. It was noticed in the Hamma garden in Algiers, that when shoots were just about to emerge from the ground, the soil became damp, while in the two or three days after emergence, the bud was found to be wet, especially in the early morning. At first this moisture was mistaken for condensed mist or dew, but it was found that it appeared even when the buds were protected, so it became evident that it was a natural exudation. This wetting effect was seen especially in Bambusa Macroculmis Riv; B. Vulgaris Schrad; Dendrocalamus Hookeri Mun. and Phyllostachys Mitis Riv.
The secretion of water is not confined to the shoots in their bud state. On certain August evenings, a regular rain was observed to fall from the foliage of clumps of bamboo in the Hamma garden.[4p63.]

In a forest situation there will be shade and mulch from the culm sheaths, leaves, and dead branches, under the bamboos. The water which drips will be absorbed for the use of the bamboo roots.

Walking through a bamboo forest it is always cool. In Taiwan it was found that after bamboo forest was planted, the annual number of days of rainfall increased in the areas concerned. It could be that as clouds pass over bamboo forest the cooler air above the forest causes moisture in the clouds to condense and rainfall begins. The days following the rain the area would be even cooler and so encourage more precipitation (rainfall).

After planting a conventional forest it will take perhaps 20 years before there is income. Bamboos brings much more income than maize does after only 4 - 6 years. Therefore it is feasible for small farmers to plant bamboo. This is particularly beneficial since bamboo can be planted when the soil has become too poor for maize, and so normally needs resting from cultivation for a number of years.

B2.15 Bamboo is very versatile

There are between 600-700 species of bamboo, growing in a wide range of habitats, from dry to waterlogged conditions. Some species used on river banks to hold the soil and to stabilise the course of the river.

The range of conditions under which Phyllostachys (makinoi), or bambusoides, (the most common species in Taiwan Japan and Korea) is grown in Taiwan, is really surprising. From Subtropical, up to 42 degrees C(109 degrees F) in the shade in East Taiwan to Temperate conditions at 10,000feet above sea level. It grows in Fukai in Northern Japan where the lowest temperature is -15.1 degrees C (5 degrees F).

This Bamboo is so versatile, it has I believe very great potential for foresting or reforesting devastated areas. It is native to East Asia but it is also growing in India where it has indiginised in Saharan, Upper Bashahr, in Himachal Pradesh. Though it grows best in

Fig.9 Bamboo will grow on infertile landslides like this one where not even weeds will grow. This was one year from a culm (stem) cutting being planted. In such conditions growth is slow for the first 3 years, then it is quite amazing.

Fig.10 This acid soil has become crusted and hard, yet as we have proved it could be planted with bamboo and become an income generating resource, in addition to the bamboo the leaves used for fodder and so adding to manure for the local fields.

the northern half of Taiwan, it also grows, in the south of Taiwan which is on the same latitude as Mandalay in Burma, (Myanmar). Chittagong in Bangladesh. Mecca in Saudi Arabia and the middle of Cuba. In Taiwan we were selling it for export to Japan and Korea. It can stand frost and snow and even grows reasonably well outside in London England. It is an extremely useful bamboo, used for water pipes, for house walls and house rooves. The culms are split, the node divisions are trimmed away and then the halves are laid side by side like UUUUUUUUUUUUUUUs concave (like the gutters in corrugated tin sheets), running up the slope. Then similar sized half culms are laid the other way up, that is convex upwards or like the humps of corrugated sheeting. They fit over the gaps between the bottom half culms; so that when it rains the water falls on the humps which are on top, and then moves down into the gutters of the half culms underneath, just like pan tiles.

Alternative methods for walls are to flatten the culms to make pliable boards, and weave them to make matting or lattice walls. Or split bamboo is woven between uprights and then daubed with mud for walls. This particular species of bamboo is also used for making furniture, chop sticks, hats, and baskets, for scaffolding, roof purlins etc. I have seen it growing along river banks to stop them eroding, but it grows quite well in drought prone areas though not as quickly as in areas with more regular rainfall.

In Nepal, rivers are now only used for tourist rafting, they could also be used for transporting bamboo rafts to suitable collecting points or factory sites next to the river.

Melocanna baccifera, the bamboo used most commonly for making paper in India could also be grown in many other areas adjoining rivers or road.

Soil erosion is a common problem above and below roads, if bamboo was grown the bamboo would stop or greatly reduce the problem. The bamboo can be harvested and lowered directly to truck pick up points along

the road. In Taiwan zig zag trails are made down slopes, for removing bamboo poles, small bogeys are made from scrap car or motorbike wheels. The bamboo poles are placed long ways between the retaining bars, they are tied together and the loaded bogey is lowered down the slope and taken to the truck pick up point. The brake rope is attached to the hand brake levers, on the brake drums, so that as required the brake is used to slow or stop the bogey. See Fig. 12.

Bamboo has rightly been called the poor mans timber, for harvesting neither tractors nor chainsaws are required, the bogeys/carts can take the

Fig.11 Bamboo has an amazing ability to hold soil.

bamboo to a jeep road where it can be taken to main roads, railways or local factories or markets.

B2.16 Other uses of Bamboo

Many people wonder what can be done with bamboo when it is grown. The following article by R.C. Gaur of the Indian Forest Research Institute and College. Dehra Dun, U.P., India. is a good description of the use that can be made of bamboo.

The traditional and other uses in India are summarized below:

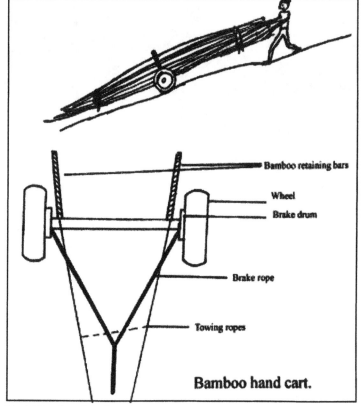

Traditional Uses

The strength of bamboo culms, their straightness, lightness, combined with hardness, range in sizes, hollowness, long fibre and easy working qualities, make them suitable for a variety of purposes. In the humid tropics houses are built entirely of bamboo without using a single iron nail.

Large suspension bridges are made solely of canes/bamboos by the tribal's.

Fig.12 Bamboo hand cart.

Among the sophisticated uses, the manufacture of a variety of writing and other paper, charcoal for electric batteries, liquid diesel fuel obtained by distillation, enzymes and media from shoot extracts, used for culturing pathogenic bacteria are important. The white powder produced on the outer surface of young culm for the isolation of a crystalline compound is medically useful. Tabasheer or Banislochan, is a popular medicine which is a silicious secretion found in the culm of some species. It occurs in either fragments or in masses (2 cm thick) chalky, translucent or transparent and tasteless and is used as a cooling tonic and aphrodisiac and in asthma, cough and other debilitating diseases. (Raizada et al, 1956).[32]

Bamboos are also commonly used as agricultural implements, for afforestation of river banks, anchors, arrows, boats, bows, broom, brushes, chairs, chick containers, cooking utensils, cordages, dustbins, fishing rods, flutes, flower pots, fuel, furniture, fish traps, hedges, hats, kit frames, ladders, lamps, mallets, musical instruments, papers, pens, poles, pulp rafts, rayons, roofing, ropes, scaffolding, tobacco pipes, toys, tool handles, table mats, tubs, umbrella handles, walking sticks, water pipes, and wrappers. Notwithstanding, the status report mentioned that, certain areas like cytology, the physiology of flowering, tissue culture and the revision of their taxonomic position etc. need accelerated research in India.[31]

Water pipes

An article in the journal 'International Agricultural Development' July/August 1989. states that

Water pipes made from easily obtained bamboo are proving stronger than either steel or plastic and are enabling villages in Tanzania to have piped water.--- research results show that wood and bamboo, when properly treated, were superior to any other conventional pipe materials. A preserved wooden/ bamboo pipe lasts an average of 50 years, compared with 40 years for plastic and 10-15 years for a steel one. Cost comparison also shows bamboo and wood to be cheaper.

For animal fodder

In an article on Bamboo in the Kathmandu based 'The Rising Nepal' newspaper B.O.Lamichaney states,

> Bambusa and Dendrocalamus species, (the most common in Nepal) produce fodder of high protein content, high total digestible nutrient (T.D.N.) percentage and good palatability, and are compared quite favourably with leguminous fodder sources. New shoots of bamboos are used as a vegetable in many areas of Nepal. This use has become a commercial enterprise in Japan, China, and Taiwan; and canned (dried) shoots are sold throughout the world. Again bamboo is the principle raw material for paper production in South Asia.

Paper making

More than 90% of the paper made in India is made from bamboo. Melocanna baccifera is a forest bamboo good for covering and holding poor land. It is used for paper making in Chittagong Bangladesh, where it is often transported by floating down the river. This could be done in Nepal and other countries, if grown on the eroding hillsides, at the sides of convenient rivers. In Nepal for instance, bamboo could be floated to the paper factory at Narayanghat, which now uses straw which should be used to maintain the fertility of the land it is grown on. The straw makes very poor quality paper, and as already mentioned the rivers are at present only used for tourism.

> In tests of nearly 100 species of bamboo the culms of Melocanna baccifera were rated among the top six for paper making. This species is also used for house walls. It is stated that white ants (Termites) do not touch it. Tests showed the average yield of culm per acre at 8715 tons = 217875 tons per hectare. From [33]
> pp190-196.

In the USA the value of bamboo versus wood pulp for the production of paper has shown the following.

> Estimates based on carefully documented records of the United States Forest Service indicate that plantations of slash pine managed on a 35 year rotation gave, at a time when most of the trees were under 20 years of age, an average annual yield of 1.13 tons of oven dry, sulphate, kraft pulp per acre. Bambusa vulgaris, on the other hand according to records of the Trinidad Paper Pulp Co Ltd--- has produced more than 4 tons of pure dry cellulose pulp a year on a 3 year cutting cycle---on a 4 year cutting cycle it produced up to 4.5 tons---.
> For the Rayon industry also, certain bamboos have been found to be well suited by virtue of superior technical properties, including a high alpha cellulose content. McClure F.A. in Grass the yearbook of agriculture. 1948. The United States Department of agriculture.[65-p735.]

Bamboo also makes excellent plywood which can be moulded into shapes such as stacking chairs, trays and very durable bowls and plates. Bamboo chipboard is another useful product.

Mulching Material. There is always plenty of self mulching material from the almost leaf like culm sheath's of the bamboo' When culms are harvested and the lighter branches and leaves removed they can be used for mulching the fields.

B2.17 The need for national and local trials of potentially useful species

Bamboo flowers infrequently, and in Nepal the seeds are usually sterile, so in the past, the normal way of propagating bamboo has been by roots (Rhizomes). These are large, heavy and will die if dehydrated.

Due to these factors **there has been little movement of bamboo species from one place to another. So there is great potential for the *careful introduction* of species from one area to another.** This should be done with careful checking that material infected by pests or diseases is not used. We brought

Fig.13 An old jeep or pickup converted to transport bamboo poles.

a bamboo which is very useful and the favourite species in Okhaldunga in Eastern Nepal to Palpa, West Nepal, It is growing very well.

I visited a farmer near Chitwan at about 0200-400 ft ASL on the Terai, the North Indian plain which includes South Nepal. This man had been a Nepali living in Assam, India but had to leave with many others. He had brought two bamboo roots with him Phidim, and Philingi (Lahure) which appears to be Melocanna baccifera. He told me that in Assam, there are three kinds of bamboos used in house construction. All three are found in the same conditions and so if one kind grows well in Palpa, the others will probably also grow. As we have Philingi growing very well in Palpa, at 2,000ft and 4,700ft ASL. the other two should also grow in Palpa. The other kind of the three, which he did not bring to Chitwan is a much bigger bamboo. It is what is called in Assam, Bom Bamboo. It has large culm like 'Dungrei'(Dendrocalamus giganteus ?) up to 9 inches (23 cm's) thick, but the culms are solid or almost solid. This kind is used for the posts and main beams of houses. Philingi (Melocanna baccifera) is used for the walls. Phidim (sp unknown) is used for the rafters of the roof. I was very impressed at the production of poles in just 3 years from planting rhizomes of Phidim. Subsequently we obtained planting material and these species are all now growing well in Palpa on a trial basis at the HASP bamboo demonstration area, below the Mission hospital compound, Tansen . Palpa. Nepal.

Wood for posts and beams is getting scarce and expensive, in Nepal as in many other countries. Bom bamboo could well be the answer. I was told that the species was originally brought to India from Africa to produce building materials for the houses of the tea plantation workers.

Although what I have described occurred in Nepal, it indicates the potential for trying Bamboos in areas different from those where they are grown now.

When the methods of culm cuttings we developed, taught in[29], are used, the successful movement of propagating material is much easier than when using roots or rooted cuttings. Seedlings can be used when seeds are available, however Bamboos have many years, sometimes over 100 years between flowering, seed viability is short and seedling development is slow. In addition there is the problem that seedlings should be planted at the beginning of the rainy season when there is too much other urgent work needing to be done. Another problem is that animals grazing the area will trample on or chew at the seedlings which are easily destroyed, so propagation by seedlings is often not practical.

At the demonstration area mentioned above which we developed on a rocky hillside at Tansen, West Nepal, there are now 27 species of bamboo. We had a route plan explaining the different methods of propagation into damp or dry conditions each of the different species. As there is nothing like it in Nepal, the Royal Botanical Gardens only having about 4 species I hope that it be recognized as a valuable resource for Nepal.

B2.18 Factors which have held back the growing of bamboo in Nepal, and possibly other countries

a) There are traditional beliefs about people dying when bamboos die. It could be that a drought following a poor rainy season triggers flowering and die back or death of bamboo. This would also mean famine conditions, and so also result in the death of people. Therefore people made the association of bamboos dying and people dying.

b) Traditionally people have only been able to propagate bamboo in the monsoon. It may be that the reason is that commonly, when bamboo or any plants are dug up in Nepal, they are treated very roughly. They would thus have little chance of surviving if moved in the dry season. Damaged plants planted in the wet season would survive rather better. This is not a good time both for success rates, (Bamboo planted carefully survives better in dry conditions than damp conditions;) and for available labour. Therefore, bamboo growing was restricted.

c) Lack of control over grazing animals, which find bamboo leaves and shoots palatable. The animals also trample and break new plants. and new shoots.

d) Theft. When there are relatively few bamboos theft can be a problem.

Flowering in Bamboo

This is an important point to discuss as there are many misconceptions about this feature of bamboo growth. There has been the fear that if a bamboo forest is planted, it will all die at a set time. This would be a problem particularly if from any kind of cuttings as when the parent plants died so would all the cuttings. The short answer is that while in a sense that can seem to be the case in fact it is other considerations which trigger the death of bamboo groves. This problem is particularly noticeable with clump bamboo. If for instance 10 cuttings or seedlings are planted under similar conditions of soil, moisture, ventilation and sunlight, then after thirty years the soil in which the clump was growing became exhausted, and the clump was dense, choking itself and under stress, the clumps would produce flowers to reproduce themselves before dying. If a drought year came when the clump was already becoming less vigorous, the stress would increase and trigger flowering. If however not all of the clumps were growing in the same conditions the flowering would occur at different times. If for instance a road was cut through or there was a landslide at the side of one of the clumps, so that the roots were cut and the soil round the roots became exposed and dry, the clump would be stressed and flower. If a cattle shelter was erected near another clump so the soil became more fertile, the clump would thrive and so would not flower. The point is that it is the nutritional status rather than some timing mechanism which decides when flowering takes place. Also there are two kinds of flowering,, light flowering which does not lead to the plants dying , and gregarious flowering when the plants completely exhaust themselves and what remains is almost like a pile of compost. In the pile either seedlings grow, or new shoots come from some life which remains in the pile, and the clump comes back to life. Normally in Nepal the bamboo seeds are sterile so if it were not for the clumps recovering the bamboos would die out.

To quote from the useful book "VENU BHARATI" (Bamboo in India) By Vinoo Kaley:-
- Bamboo is one of the world's best engineering materials.
- Bamboo is the world's fastest grower.
- Bamboo is the world's second best photo converter and at a much less water cost than the first.
- Bamboo plantation enriches the soil, arrests soil erosion and tames flash floods.
- Bamboo offers stakes to trees, fodder to animals and food to humans.
- Bamboo is recurringly harvestible with first mature stems in 5-6 years of plantation.[208p.167]

I THINK THAT THESE FACTORS ILLUSTRATE THAT BAMBOO IS AN EXTREMELY VALUABLE BUT UNDER UTILIZED CROP IN NEPAL AND INDEED IN MANY PARTS OF THE WORLD. IT ALSO HAS GREAT POTENTIAL FOR STABILISING STEEP OR ERODING LAND AND CONVERTING BARREN LAND INTO USEFUL AND IMPROVING LAND.
WHEN SOIL CONSERVATION IS BEING PLANNED BAMBOO PLANTING SHOULD BE AN IMPORTANT CONSIDERATION.

For information on working with Bamboo to make a wide range of products see 208 and 212.

For a practical guide to bamboo culture see my book

B A M B O O

A VALUABLE MULTI-PURPOSE AND SOIL CONSERVING CROP
A PRACTICAL GUIDE TO THE GROWING OF BAMBOO

P. Storey

Available from 38 Bellingham Road. Kendal.Cumbria. LA9 5JW England.UK.
Tel: 01539 727 841 email: pj@rosebeck.freeserve.co.uk

APPROPRIATE FIELD TRIALS

B3.1 Introduction
B3.2 Comparing trial layouts
B3.3 Practical field trials

B3.1 Introduction

Why should this chapter be included?
As pointed out in "Change" Philosophy 4.4. in my book 'Quality Rural Development' [175] we need trials and demonstrations, in order to find and demonstrate the best methods of soil and water conservation and soil improvement. In a particular area, which grass and legume cover is the most suitable for grassed waterways? which green manure crop has most beneficial effect? It is important to try alternatives and then demonstrate the suitable ones. The trials must be accurate, and yet practical and appropriate, sadly too often this is not the case, credibility is lost, and so instead of encouraging a change to better land management, change is discouraged.

Farmers look at a new thing as:
- Can he do it ?
- Can I do it?
- If I do it, will it be better for me?
- Could it cause problems with my neighbours?
- Is it worth me trying it?

Unless those questions are answered satisfactorily, most and particularly the poorest farmers, will feel reluctant to try it.

The confidence to try different techniques follows careful observation and questioning.

1. It may be that there are other places where the alternative method/crop etc is being used successfully by local people, if so, we should take some people from the project area to see the method and talk with the people about it. If the project people are convinced enough and try it, perhaps adding their own modifications; and they get it to work well, the new method is likely to take off much more rapidly than if the method was introduced by the project staff. It could be that problems develop which the project staff could offer advice about, so a combination results in success.

In some cases while the people feel more open to the new idea after the visit, the project may need to have a trial locally before it is acceptable enough for people to try it themselves.

If there are no such examples it is possibly best for the project to try out the method under local conditions, the reason should not be that we have no confidence in local farmers but that the project should take the first risk.

Granted if you have been in a place for some time and the people have developed confidence in you, some people will be happy to try your suggestions straight away. Or, as is more common, wealthy people who can afford to take a gamble may try the new method.

It can be risky for projects to try new ways of doing things, but failures in farmers fields are even worse, unless failures are comparatively rare, and the farmers know that the new method has been well tried first.

Change is often threatening, particularly for the poorest people, so it is important to avoid things which are needlessly threatening, or make them less so.

When people feel uneasy they tend to reject that which causes the feeling. That is why a slower approach, showing respect, being ready to be humbled and risk failure and persevere until the best way is found, is usually more effective. Failure can also bring sympathy, and when perseverance pays off- respect.

2. Another way for relevant research is that used in Taiwan by the "Farmers Associations". Farmers pay a levy on their crops which goes to their own association. They control it and employ their own research and extension staff. The farmers decide what they want research on and how it should be carried out. This has proved much more effective than the government system although the FA's members also make visits to, and give requests for, research or advice from the government improvement stations and extension workers.

3. Some points which should be remembered when planning or engaging in trials.

Trials vary greatly, some are needed, some are not.

Some are practical and take into account all the factors which may effect the subject under practical conditions, too many do not.

It should first be ascertained if the trials are needed, it could be that they may have been carried out in the past, or there may be someone locally who has developed a local strain of the crop or animal which is as good as or better than that which was to be tried, it is good to check for such things. There is too much 're inventing the wheel' in development work.

It is best if the trials are performed under difficult conditions as well as or instead of average conditions.

For instance, in Nepal improved maize does mature faster and give a higher yield in a good year, so it may appear to be superior, but in a dry spring will die. In a dry spring the local maize will go into a semi dormant stage, then grow away again when rain does come. It is usually better for rural people to have a lesser crop every year than more some years but none in others, particularly as the maize will cost more if people need to buy in a bad year, while spare maize will fetch a better price. Similarly the local maize is more resistant to attack by pests and diseases.

B3.2 Comparing trial layouts

Is normal scientific method suitable for field trials in the third world?
When scientists carry out trials and comparisons they use formulae which state that the position of treatments within a trial must be selected at random. This is in order to avoid bias which may be made unconsciously for or against one of the treatments. However this author and others with experience in field trials do not believe such criteria are suitable for agricultural field trials; particularly those in hilly conditions.

Publications such as "How to perform an agricultural experiment" by G Stuart Pettygrove and published by VITA insist on randomization as the following quotation shows.

> Randomization simply means that treatments are assigned to plots in a random fashion or are randomized within a block. The reason for randomization is the elimination of any bias which might occur if we assigned treatments to plots in either a haphazard or a regular fashion. The randomization procedure should be completely objective and may be accomplished by flipping a coin, drawing cards from a deck, or by using a specially prepared table of random numbers---.

I hope to show why I and others disagree with this opinion; and I will suggest a more suitable alternative. Unless soil, pest, weather and other conditions are unusually regular, or the trials carried out on a very large scale, the alternative will be more likely to give a true result than one obtained by randomized design of where treatments will be placed within a trial.

In typical third world farmers fields it is rare to get perfectly even conditions, particularly in a hilly situation. One side of the field or strips down the middle may be different to the rest. Perhaps more or less fertile than the rest, more or less dry, more or less liable to wind damage, disease, insect, bird or rat attack; even people stealing from the crop. For this reason it is important to equalise the effect of such factors for the different treatments.

Conditions vary within fields and within years; if the difference in performance in a trial is so small that it takes scientific analysis to determine whether one treatment is better than another, it is probably of little if any value in normal third world farming situations. It would only be of value if that difference was consistent over at least three years.

This author was shown a trial being conducted for a PhD thesis, which was comparing different insect traps.

The layout had been made by correct statistical method using randomized placement of the treatments. The result of randomization was that there were more of trap B towards one corner, which happened to be nearest to a piece of woodland. Obviously as the insects came into the field from the woodland they would be more likely to be trapped in the nearest traps; so it would appear from the trial results that trap B was most effective. The results of the trial would be worse than useless, as they favoured one trap which might in fact be the worst of those tested. When I pointed this out and suggested a rearrangement, the person whose trial it was replied that it did not matter he had carried out the trial according to scientific method, it would not affect his PhD. I feel practical method is more appropriate. Perhaps there is a place for statistical analysis of the likelihood of a trial being an even reflection of possible conditions in a trial before the trial is actually planted or otherwise commenced. Normally statistical analysis is only carried out at the end of the trial.

I believe that we should first design a suitable and balanced layout according to the number of treatments which will be involved. For instance if we are using three treatments we should design a trial with A, B and C plots within a balanced trial layout. Having designed as even as possible a distribution of the plots we then use a random way of selection such as names picked out of a hat to decide which treatment goes in the A plots, which in B, and which in C plots. In that way there will be no bias in the selection of positions within the trial, but each treatment gets as fair a trial as possible.

0N	80N	160N
160N	0N	80N
80N	160N	0N

1

A good rule for agricultural experiments should be that there should be at least three replications (repeats) of each treatment, method or variety, with as equal a distribution of each treatment, method or variety as possible.

The trial 1 was set up by a person working for the British Open University. It was to compare the growth of trees given Nitrogen fertiliser at 0, 80, or 160 kilograms per hectare.

A farmer looking at 1 layout will notice a basic fault; if there was a diagonal strip from the top left to the bottom right which for some reason was different to the rest, the results of the trial would be distorted. If the strip was a depression it would be more moist and it would perhaps receive some nutrients from the other plots. The difference between the 0 nitrogen and the added nitrogen would be less. If that strip was more or less fertile it would affect the accuracy of the results. If the layout had been as 2 it would have been more obvious, but fertility drifts and similar factors in agricultural situations are not confined to up and down or sideways orientation.

160N	80N	160N
0N	0N	0N
80N	160N	80N

2

In 2 it can be seen that the top strip of three plots has a total of 400 kilograms of nitrogen fertiliser; the middle strip has none, and the bottom strip has 320 kilograms of nitrogen.

If the ground was sloping slightly from the top, and so when it rained some of the nitrogen from the heavy nitrogen strip at the top would soak down into the middle strip; also, the bottom strip would have some of the nutrients removed by erosion of water from above, there would be less difference in yield than there should normally be. If there was a depression across the middle of the trial the extra moisture and nitrogen may make the plants softer and more susceptible to pest or disease attack and so the affected plants produce less. If there was a general moisture shortage the plot in the depression would grow better.

Pests of some sort may move into the bottom strip. The results of the trial would perhaps indicate that both the application of 160 kilograms or 0 kilograms gave good results while the application of 80 kilograms gave poor results. Therefore the results would give an inaccurate and meaningless result. A wise farmer would be very sceptical of such trials and in any case the trial would be a waste of effort.

For these reasons it is important to have the treatments distributed as evenly as possible.

It is impossible to avoid one or other treatments being in such a strip in a 3 x 3 layout with the treatments balanced within the trial.

We could try a 4x3 layout but we still find the same problem, if we want to balance each part of the trial we would still have diagonal strips of one of the treatments when using only three treatments.

Another factor to be considered in trial layouts is that as the plants on the outside plots have better air flow they may be less prone to disease but more prone to pest attack than the inside plots, as in the illustration with the insect trap trial. Even with guard rows*, one side or part of the area may be more affected by wind, another may be more prone to weeds, insects, vermin or birds coming from neighbouring land.

For a balanced 3 treatment trial we could use a 4x4 layout. <u>3</u> We will use letters for simplicity.

A	B	C	B
C	A	B	C
B	A	C	A
C	B	A	B

<u>3</u>

If we count the numbers of each treatment
In the top half we have   Ax2   Bx3   Cx3
The left half   Ax3   Bx3   Cx2
The bottom half   Ax3   Bx3   Cx2
The right half   Ax2   Bx3   Cx3   This layout is as well distributed as possible for three treatments.

If not possible to have so many replications it would be better to have a 9 x 1 layout or a 5 x 2 with one plot discounted. Guard rows would be essential.

   *Note. Guard rows are rows of the same crop as that of the used in the trial. They are sown or planted surrounding the trial area to avoid the edges of the trial plots being affected by factors such as more than normal exposure to wind damage, drying out, trampling etc. The guard rows are not counted in the trial.

With a 9 x 1 layout chosen by random selection it could be as in <u>4</u> :

1	1	2	3	2	3	3	2	1

<u>4</u>

If the soil is more productive in the centre of the block, or rabbits or birds are more damaging at the ends of the block, or there is more of a drying or other effect at the ends of the block, the above trial will give inaccurate results. The 1 plots are towards the ends and the 3 plots towards the centre.
Or, it could be <u>5</u>:

1	1	2	3	1	3	3	2	2

<u>5</u>

If the right side of the field was better than the left the 2's and 3's would do better than the 1's.
   We should try to correct this factor. I would simply arrange the three replications as follows in <u>6</u>:

1	2	3	1	2	3	1	2	3

<u>6</u>

If using four replications I would arrange them in the same way.

Alternatively a 5 x 2 layout could be used with one of the plots unused in the trial but planted the same as the guard rows which should always surround the blocks to avoid edge effect on the trial areas.7

1	2	3	1	2
3	1	2	3	

7

It is best to call the treatments 1, 2, and 3, as above, instead of 0, 80, and 160, or whatever were the descriptions of the treatments; we can then use this layout for any three replications (repeats) of three treatments.

It is best to use larger multiples of the number of treatments, particularly if we need a square design, for example

3 x 4 = 12 mini plots, or 15, 18 or 21, plots if we are using three treatments.

Similar equally distributed layouts can be used for different numbers of treatments and replications. PROVIDED WE PLAN THE LAYOUT BEFORE WE DECIDE, PERHAPS BY DRAWING LOTS OR DRAWING LETTERS FROM A HAT, TO DETERMINE WHICH TREATMENT WILL BE CALLED 1, WHICH 2, AND WHICH 3, OR SUCCESSIVE NUMBERS IF MORE THAN 3 REPLICATIONS ARE BEING USED, WE CANNOT BRING BIAS INTO THE TRIAL.

If we know or suspect it could be, that conditions are better at one end of the field we are using, it will be better if the plots are running from one condition to the other. 8 is wrong 9 is right

8

Good end   Not so good end

WRONG LAYOUT

One treatment has the good conditions, one the not so good, and the other the intermediate conditions.

Good end   9   Not so good end

CORRECT LAYOUT

Each treatment covers the good and the not so good conditions equally.

However if working on terraces or in other situations which are not ideal for trials but where trials are needed, we can and should arrange our trials to get the most meaningful results.

It can be useful to ask the farmer what he thinks about the layout; he knows the conditions and possible problems, and so these can be taken into account in planning a successful trial.

When farmers notice something wrong about the conditions of a trial, the trial and the trials organisers can become the butt of many a joke, though the organisers may not hear of it, and credibility is ruined. In any case the aim should be to get the most accurate unbiased results possible. *It is wrong to stick to a book method designed to avoid bias under laboratory conditions and ignore possible biases brought about in field conditions.*

The table 10 was selected in such a way and would be regarded as correct by experimental methodology. It is given as an example in the publication "How to perform an agricultural experiment" previously mentioned; I do not think it suitable for a field experiment for the reasons already given.

3	4	2	5	1
1	2	5	3	4
4	5	3	1	2
2	3	1	4	5
5	1	4	2	3

10

If the block is split diagonally the bottom right half has three 1's + one half 1 in a row, with only one 1 in the other half.

There are three 5's + one half 5 in a row at the top left half and only one 5 in the other half.

There is a diagonal row of three 3's in the centre.

If there is a top left to bottom right fertility drift, or a more or less favourable strip running diagonally from bottom left across to top right, the trial will give distorted results.

If we rearrange we can improve the distribution. If we start by 1 2 3 4 5 on the first line, then move 2 places for each subsequent line we get the following arrangement in 11.

1	2	3	4	5
3	4	5	1	2
5	1	2	3	4
2	3	4	5	1
4	5	1	2	3

11

IN EACH DIRECTION DIAGONALLY, HORIZONTALLY, AND VERTICALLY THE DISTRIBUTION IS EQUAL. A total of 15. THERE ARE NO GROUPS OF THE SAME TREATMENT IN A ROW.

To try another example:- Three treatments four replications 12.

1	2	3	1
2	3	1	2
3	1	2	3

12

This is unsuitable as there are diagonal rows of 3's and 1's and the 2 treatments are at diagonal ends.

If we move the numbers about we can put them in the following layout 13 which gives a much better distribution.

2	1	2	3
3	2	1	1
1	3	3	2

13

With four treatments and three replications-14.

1	3	2	4
2	4	1	3
3	1	4	2

14

Four treatments and four replications 15.

1	2	3	4
3	4	1	2
4	3	2	1
2	1	4	3

15

Or 16.

1	2	3	4	1	2	3	4
4	3	2	1	4	3	2	1

16

Five treatments with three replications 17.

1	5	4	3	2
3	2	1	5	4
5	4	3	2	1

17

Five treatments with four replications 18.

1	2	3	4	5
3	4	5	1	2
5	1	2	3	4
2	3	4	5	1

18

We should try out proposed layouts before using them; perhaps using statistical analyses on the layout before conducting the trial. When we feel that we have the most equal distribution, we should draw lots to determine which treatment is used in place of which figure.

There are too many inappropriate trials; they waste valuable time and resources and spoil credibility because the results are not replicable in farmers fields.

Other points which are important differences between laboratory tests and farm field tests are the differences in the conditions from one years trial to the next. Because of this it is much more appropriate to have six replications of a particular trial one year and six the next rather than twelve replications in one year; even more appropriate would be to have four replications for three years. Better still three replications for four years. It is not good to have less than three replications unless the conditions are very regular.

As an example of what may happen, a peanut trial, a new crop to the area. The first year showed extremely promising results, compared with the local crops. The second year the peanuts were infested with insects and gave a low yield. The third year the insects were controlled, but due to a wet season disease was serious. As it was still under trial, farmers had not yet been advised to grow the new crop and no harm was done.

In another case a trial was carried out to compare the growing of sorghum instead of the local millet. For two years the sorghum out-yielded the millet, and the farmers were told that it was clear that they should not bother with millet but only grow sorghum, as it gave more food and was of nicer taste than the millet. So the third year the whole village planted sorghum instead of their traditional millet. Unfortunately for them the sorghum midge appeared and attacked the flowers of the sorghum. The result was that there was very little grain. Much of what there was, was eaten by birds, which did not bother their millet very much. Then what was harvested was eaten by insects and rodents which did not eat the millet, the people were starving. That was the last time they took notice of advice from outsiders, development was ended because the trial was not continued long enough. The third year the trial should have been carried out on a bigger scale but not to replace the traditional crop.

Another example of a new idea being promoted too quickly, (unfortunately there are too many such mistakes;) is the introduction of modern steel mouldboard ploughs to an area of Africa. The organisation concerned had very good intentions; they felt that the local simple ploughs only stirred and loosened the soil, while mouldboard ploughs would be able to do a better job and be able to bury straw, farmyard manure, and green manures much more effectively.

They used the new ploughs on their own trial farm and offered to plough the fields of the local farmers. They were disappointed when the local farmers did not want the new ploughs, but expected that they would be convinced after two or three years.

The yields were better where the mouldboard ploughs were used. However the local farmers continued with the simple ploughs as they had for hundreds of years. After two or three years the land of the foreign organisation started to yield less, while the outside fields continued as before. The crops in the trial ground looked poorer each year and when the situation was investigated it was realised that the strong UV light and weathering spoilt the top 1 or 2 inches (2.5-5.0 cm's) of soil. The mouldboard ploughs turned that dead soil under and brought up fresh soil which was also spoilt and then ploughed in. The process was sterilising the soil. The simple ploughs loosened the soil but the dead soil mostly stayed on the top, while the roots of the plants went into the rich soil underneath which was protected from the weather by the soil which remained on top.

Unfortunately although that lesson was learned about fifty years ago many foreign organisations and extension people are still introducing mouldboard ploughs into areas without long enough testing to find out the longer term situation. Proper trials are very important, too often it is a case of more haste less speed or no speed at all, development being actually set back because it is assumed that the existing situation can quickly be improved. An ignorance, or an arrogance towards traditional ways.

Another example of a trial which I feel should have been planned more carefully was prepared in conjunction with the British Open University. It is the layout for a walnut tree trial in the diagram 19.

```
C \     C      C      B      B      D      D      E      E      E
C      C\     C      B      B      D      D      E      E      E
C      C      C\     B      B      D      D      E      E      E
C      C      C      B\     B      D      D      E      E      E
C      C      C      B      B\     D      D      E      E      E
C      C      C      B      B      D\     D      E      E      E
B      B      B      E      E      C      C\     D      D      D
B      B      B      E      E      C      C      D\     D      D
B      B      B      E      E      C      C      D      D\     D
B      B      B      E      E      C      C      D      D      D\
D\     D      D      C      C      E      E      B      B      B
D      D\     D      C      C      E      E      B      B      B
D      D      D\     C      C      E      E      B      B      B
D      D      D      C\     C      E      E      B      B      B
E      E      E      D      D\     D      D      C      C      C
E      E      E      D      D      D\     D      C      C      C
E      E      E      D      D      D      D\     C      C      C
E      E      E      D      D      D      D      C\     C      C
E      E      E      D      D      D      D      C      C\     C
E      E      E      D      D      D      D      C      C      C\
```

19

In the event of a \ diagonal fertility or \ infertility drift as indicated by the diagonal dashes, the mid section would contain:-

The C treatments 8 half squares = 4 plus 38 full squares = 42
The B treatments 4 half squares = 2 plus 29 full squares = 31
The D treatments 8 half squares = 4 plus 6 full squares = 10
The E treatments 16 full squares = 16

There would be a similar unequal distribution if the diagonal drift was in the opposite direction. If bottom or top strips were better or worse the E's and C's would again have a much greater representation.

This kind of layout would be suitable under controlled laboratory conditions but seldom under third world field conditions.

The total number of replications of the treatments is not equal.

C and E have 52 replications, and B and D have only 48 replications.

I wondered if a better chart could be planned and found it was not very difficult. See 20. The chart below I feel sure gives a much better layout.

B	B	B	C	C	C	C	D	D	D
B	B	B	C	C	C	C	D	D	D
B	B	B	C	C	C	C	D	D	D
B	B	B	C	C	C	C	D	D	D
E	E	E	D	D	D	D	B	B	B
E	E	E	D	D	D	D	B	B	B
E	E	E	D	D	D	D	B	B	B
E	E	E	D	D	D	D	B	B	B
C	C	C	B	B	B	B	E	E	E
C	C	C	B	B	B	B	E	E	E
C	C	C	B	B	B	B	E	E	E
C	C	C	B	B	B	B	E	E	E
D	D	D	E	E	E	E	C	C	C
D	D	D	E	E	E	E	C	C	C
D	D	D	E	E	E	E	C	C	C
D	D	D	E	E	E	E	C	C	C
C	C	C	C	C	C	C	C	C	C
D	D	D	D	D	D	D	D	D	D
B	B	B	B	B	B	B	B	B	B
E	E	E	E	E	E	E	E	E	E

20

With this layout there are no more than 2 direct line treatment linkups.

Each part of the area has an equal number of replications (repeats) of each treatment.

Each treatment has 50 replications.

B3.4 practical field trials

SOME POINTS WHICH SHOULD BE REMEMBERED WHEN PLANNING OR ENGAGED IN TRIALS.

Using a grid for improving the accuracy in assessing probability.
For accurate conclusions of trials when we have limited resources or under variable soil conditions we should have many replications. However this will mean that the percentage of edge area which cannot be included in the results, will be higher than normal. Instead of many separate small plots it may be better to have fewer larger plots but overlay them with a grid. This does not need to be a physical grid but may be simply say five clumps or rows wide by five clumps or rows long, recorded as one small block **21** within the main block.

So instead of having one figure for this area we have 30 . We can discard the outer cells which are edging, and therefore unreliable; and have 12 useful figures or results for comparison. If we have 4 big plots we will be able to record 48 useful figures, rather than having 48 small plots having a high proportion of edging (unreliable effect), so reducing the value of the figures / results obtained.

Of course the 48 smaller plots would cover a greater possible degree of variation, than that of fewer larger plots. That has to be borne in mind when designing the trial. On a variable hillside results obtained from say one large plot containing many cells, would not give a fair assessment of the results which should be obtained if the whole hill field was treated in the way indicated to be the best in that one large plot. I would say at least 3 or 4 of the larger plots should be used. The more widely variable the conditions in the particular field, the more of the bigger plots should be planned.

Care needs to be taken if rows or clumps of the crop are grown so that all plots are treated in the same way. Eg if recording rice clumps, measurements should be taken from the same position say at the right of a clump to the other plot border also being at the right of a clump. If area only is calculated it could be that in one plot a higher number of clumps is included, in the next a higher proportion of inter-clump spaces is included and so the grain and straw yield would have a lower potential than that of the first plot .

Of course care should be taken to choose rows or clumps which are the average distance apart. It may be appropriate to count five clumps of rice in one direction, and ten in the other direction. If rows are planted, five rows 2 metres long.

If we use a measuring frame which encloses one metre squared it is important to have a standard procedure. Eg The frame must always be placed so that one side bar presses on the outside base of one row of plants and the bottom bar on the inside base. The clumps inside are counted <u>22</u> is right <u>23</u> is wrong.

X	X	X	X	X	X
X	X	X	X	X	X
X	X	X	X	X	X
X	X	X	X	X	X

<u>22</u>

CORRECT, 24 CLUMPS INCLUDED.

X	X	X	X	X	X	X
X	X	X	X	X	X	X
X	X	X	X	X		X
X	X	X	X	X	X	X
X	X	X	X	X	X	X

<u>23</u>

INCORRECT, 35 CLUMPS INCLUDED

For this purpose a frame <u>24</u> can be made to enclose 1 metre squared eg with internal measurements of 161.8 x 61.8= 9,999.24 which is near enough to 10,000 square centimetres or one square metre.

Add on twice the thickness of the bars to be used. Eg if the bars are 1.5 cm's thick prepare two bars 164.8 cm's long, and two bars 64.8 cm's long. Drill holes 0.75 from each end of each bar. Fit a rivet or a screw or bolt and nut at each joint. *Do not make them too tight. Then it can be folded for carrying and adjusted slightly if needed to fit the rows of plants or clumps.*

B.Another point to bear in mind when planning trials is the time of stages such as ripening of the neighbouring crops.

We may need to run a dummy mini trial the season before the main trial in order to know when the stages will occur with the different treatments in the particular environment of the proposed trial.

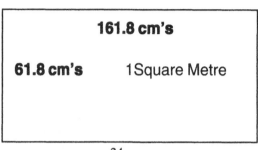

To give an example:- In Taiwan, if rice seed is sown direct into a paddy-field in Taiwan on the same day as seedlings are transplanted, the direct seeded rice will catch up with the transplanted rice and may even be ready for harvest a few days earlier. There are no problems in having a plot of direct sown rice in the middle of a large area of transplanted rice. By contrast when we have the same trial in Nepal the rice is much older at transplanting and so will head much earlier and harvest earlier. Due to this there cannot be a simple trial of direct sown rice, sown on the day the other rice is transplanted. The reason is that pests and disease of the flower heads of rice accumulate during the flowering of the transplanted rice. As the flowering stage passes, the by now high populations of pest and disease move to the slightly later direct sown rice. As the rice approaches harvest birds, and rats, move into the transplanted rice and start to feed. At harvest time as the rice is cut, the pests from the large area of transplanted rice move into the smaller area of trial plots of direct sown rice, so the effect of the pests is much greater on the trial plots. When the timing is adjusted so that flowering occurs at the same time, this problem will not occur and fair comparisons can be made between the two methods. If on an equal basis the yield of the direct sown treatment is better, that method can be used on a bigger scale; perhaps with minor control plots of transplanted rice, also adjusted so that flowering and harvest coincides as closely as possible. In this way the main crop will not dominate the trial plots and distort the results.

Some trials should be carried out as basic trials.
That is to say, trials of untested methods which may or may not be suitable or may need refining before being appropriate for promotion. These should be carried out under as average conditions as possible while not in a normal farm situation so that failure will not be a serious matter. Also the conditions must be carefully controlled and so are best under proper trial conditions which cannot be influenced by personal preference or the covering up of poor performance. After it seems evident that the method or locally modified method is reasonably reliable the methods should be moved to the farmers field trials. When proven under average farm conditions the trial becomes a demonstration.

It is best to have the trial in a public place, so that there can be no accusation that extra fertiliser or special treatment was given to the newer methods or treatments in the trial. Failures will also be seen but they give credibility to the successes and farmers respect the trials as being similar to their own conditions. Farmers know that failures do come, when the reason for them is overcome respect grows. Trials without some failures are not trusted.

Farm trials: should be trials which are fairly certain to be successful or at least where results will not be worse than the normal practice, or trials farmers themselves are interested in so accept the risk themselves; these are carried out directly with farmers, in farmers conditions.

In some cases outsiders may give suggestions or advise when asked but it is the farmers who carry them out. A farmer may be impressed by a different way of doing/growing something and ask an extension person to advise him how to use it, or make a comparison with the traditional on the farmers land.

In some cases an extension service may ask a farmer to let them conduct a trial on the farm. With or without the farmer doing all the work.

In other cases the farmers are shown how to conduct balanced trials and the farmers do it themselves. This can be the most effective way of introducing or encouraging alternative methods. As an example:

A Letter from Rwanda.
Louise's eyes shone as she told me about that day's meeting.

We didn't tell them what to say, just to explain to the new women what they did. And do you know how they started? Turi abashakashatsi'-'we are researchers'. My assistant and I were over the moon---. Why don't farmers jump at the chance to improve their yields?

There are several historical reasons. Rwanda is a very organised country, so the farmers are used to being told what to do by the local agents of the Agriculture Ministry: often young people who know nothing about farming. The advice they relay from the Ministry-plant in rows, plant and harvest at certain dates, dig anti-erosion ditches etc- is sometimes useful, but often of little benefit, or even positively harmful. However, farmers are fined if they don't follow orders, so they obey when necessary, but treat all such "advice" with suspicion. After all, it is their bellies that are empty when a "new improved" variety is a flop. They may have very good reasons for ignoring advice. Fungicide treatment of bean seeds was found to be useful against certain seedling diseases becoming more and more common. But in certain areas the farmer puts the beans in her mouth and spits a seed into each hole, leaving both hands free for the digging. Seeds enriched with fungicide will not improve her   health---

Examples of unsuitable "improved" techniques and varieties abound- like the wonderful bean variety that nobody wanted because it was black. Researchers are not farmers, and cannot be aware of all the intricacies involved in farmers' decisions about what is best for them---.

All the time the emphasis is on the fact that the farmers' opinions matter, that they know a lot about beans, that they are "experts" collaborating with the researchers. Not easy when farming is a low-status activity done in large part by women, who are considered about on a par with cows (who actually enjoy quite a high status) hence the jubilation when these women, explaining what they do to a new group, call themselves "researchers".

These women, with their new-found confidence, could help spread new ideas to their communities. And the information they can bring to the researchers will be invaluable for keeping the research on the right lines. If the researchers will listen, Louise is American. How can Rwandan researchers respect farmers when their life's aim has been to get a good education, so as to have a better status than farmers? How far do we have to be cut off from our roots before we start to yearn for them?

From Food Matters Worldwide No 9 Feb 1991 Pg 5.

Local investigative trials

These are when local people are asked who is the best at producing a particular crop or rearing a particular animal. The researcher then visits the local expert and *listens* to what he or she says about how the best results are obtained. Then the advice given is incorporated into comparative work to find out if the recommendations do appear to be better under careful trials. While very often such local experts do know or sense techniques which are most suitable for their area (and may even be for general use in improvement of the particular enterprise) sometimes their conditions are not practical or replicable (repeatable) for general use. It could be special skill or ingredients are required or what appears to be exceptional may not be so when not specially treated. For example large animals may have poor reproductive efficiency and so be less economical under average livestock management. Of course it is important that the local expert is given credit for his expertise whether it is replicable or not.

Adaptive trials

It is important to interest the local people in what you are doing and learn as much as possible about their own experiences in the matter being considered.

When we had just met, Virgilio stood up and said; 'Lucas, I understand you want to know. You are a scientist and you want to know. But there is only one way to know what I know about cassava. Speak with me; don't speak to me like the others did. ask me about my life and I will tell you about cassava.'---.
-Farmers get tired of being talked to when they should be talked with and here it happened: the man who

sweeps the street in the small provincial town of Moncion, (Dominican Republic) telling me the same things social scientists repeat time and again. I had been sent to consult Vergilio by a well respected cassava cultivator in the neighbourhood who told me 'that man knows more than anybody else; he is a wise man.' I spoke at length with him and it appeared that although he had benefited from only three years of formal education, he could write down much of his information. Although he was the poorest of the village, he was duly respected for his knowledge about cassava. Although his social position was marginal in many respects, he was part of an invisible network of those who experiment with cassava.---.

We took his advice to heart: asked about lives and heard about cassava. We reconstructed the history of cassava cultivation and so got an idea of why people had changed from one variety to another, or had opted for another cultivation practice. In so doing we designed trials together; we redesigned the experiments the farmers had been doing themselves to make them more reliable to the scientists we worked with and we redesigned the scientific experiments so they could respond to local conditions. We called this mutual adaptation and the experiments, adaptive trials." [8] p61.

When setting out to arrange a trial it is best if the decision about the trial is made in public. *This is to protect against distortions and damaging comments being made later.*

There may be reasons for farmers or others to work against a fair trial.

1. It is quite common for farmers to resent those with more education and privilege, they often have good reason for those feelings. This may mean that they enjoy the prospect of the extension officer or other advisor being made to look a fool. It could be that they start in good faith, but the extension person or trial organiser starts to mess the farmer about, or appear condescending or bossy; the result is resentment.

2. It could be that the farmer agrees to the trial thinking that he is going to receive special favours, when he does not he feels cheated and resentful.

3. It could be that his neighbours or his landlord or the local priest/holy-man/witch-doctor or a combination do not want the trial to be a success. They may not want to lose influence and power to someone else.

The following illustrations should help to show the sort of problems which occur.

❏ Just after a trial comparing two methods of growing rice was started the author without thinking asked the farmer what the result would be ? The farmer stated quite categorically that the traditional way would be best, there was no way the direct seeded rice would grow as well as the transplanted rice. I immediately realised my mistake, the farmer would now lose face if the direct seeded rice performed better, that is what happened. Neighbours started to comment that the direct seeded rice was catching up with the transplanted rice and that it looked better. Soon afterwards I heard by the grapevine that the farmer had put fertiliser on the transplanted plot and the surrounding plots but not on the direct seeded plot. I asked him if he would be putting fertiliser on his rice and pointed out that if he did he should also put it on the direct seeded rice. He said that he did not use fertiliser, however it was clear to see that the transplanted rice suddenly went greener and lush; even so the direct seeded rice gave as good a yield as the transplanted rice.

As he had denied using the fertiliser he had to say that there was no point in using the new method since it was no better than the old method; so though the new way was better it was not taken up.

❏ In a similar trial, when we were choosing the field for a rice trial in which one half would be used for one method and the other half of the field for the second method, the farmer told me that the field was uniform. When at harvest the new method yielded most he said in public "Oh well that side of the field always does do better." I believe the change was due to pressure from the local priest.

❏ Another trial was being carried out on a the local landowner/moneylenders land, he was very keen to have this new method on his land when I told him I expected the method may give about 25% more yield with the same inputs. He was a very intelligent man yet he consistently did the opposite of what we told him to do. Then we said he did not need to do anything we would do it all.

The new method using direct seeding of rice needed less irrigation water but at different times to the main transplanted crop; however that was no problem as the trial field had its own supply and did not effect the other fields. When we irrigated the trial area in the morning, we would find the field drained at night. When we drained the field to encourage tillers (side shoots) we would later find it flooded. He denied knowledge of this but one of our staff hid and saw him change the water regime. The new method still proved better but not in as spectacular way as it should have. We had to spend a lot of effort for nothing; we were convinced that the method was better and could make a dramatic effect on the rice growing of the area. The poor farmers who seldom had enough water would have sufficient, the yield would be increased by 24%, and the rice be of better quality. The crisis time of transplanting when there is not enough labour would be eased, and though an extra weeding was needed that would be after the peak labour period had passed; the farmers would be much better off. From subsequent experience I realised that the money lenders did not want the farmers better off. They would be able to start to repay their debts which were paid back with 80% interest. As the landlord had a relatively small amount of land it was better to sacrifice his own better yield in order to keep his stranglehold on the village he controlled.

Trials carried out on good farms

The idea is supposed to be that there will be a trickle down effect. In fact it seldom works, the poor farmers reason that they are not clever enough or their land is not good enough, for the new techniques. That if problems arise the rich farmer will be able to get prompt advice and help. He will be able to get needed inputs when he wants. He can make the new method work and will because of the perks he gets from the local agricultural office. He is clever and he has good land, he is wealthy enough to take risks. The ordinary farmers are not in his position.

The best effect is gained if trials are successful on a poor farmers land, everybody knows that if he or she can do it they can all do it. The whole village is likely to take up the better method, if not in the first season soon after.

It is important to advise farmers not to be too quick at taking on the new method. Only part of the land or crop should be used at first.

As an example of what may happen, a trial of peanuts, a new crop to that area showed extremely promising results, compared with the local crops, the first year peanuts were grown. The second year the peanuts were infested with insects and gave a low yield. The third year the insects were controlled, but due to a wet season disease was serious. As it was still under trial, farmers had not yet been advised to grow the new crop and no harm was done.

In another case a trial was carried out to compare the growing of sorghum instead of the local millet. For two years the sorghum out-yielded the millet, and the farmers were told that it was clear that they should not bother with millet but only grow sorghum, as it gave more food and was of nicer taste than the millet. So the third year the whole village planted sorghum instead of their traditional millet. Unfortunately for them the sorghum midge appeared and attacked the flowers of the sorghum. The result was that there was very little grain. Much of what there was, was eaten by birds, which did not bother their millet very much. Then what was harvested was eaten by insects and rodents which did not eat the millet. The people were starving. That was the last time they took notice of advice from outsiders, development was ended because the trial was not continued long enough. The third year the trial should have been carried out on a bigger scale but not to replace the traditional crop.

In order to initiate trials by poor farmers it may be necessary to insure the trial, agreeing that if the trial yield is less than the control yield compensation will be made. It is important that safeguards are made in order to avoid cheating, for instance the farmer may put less manure on the trial area, and more on the control, knowing that he will be compensated for the difference in yield. When working with the poorer farmers care needs to be taken in choosing the farmer. It may be that he may run out of food or money during the growing season and so go away to earn money. The field is not weeded, and other management practices not carried out when they should be, and so the trial wasted.

If the difference in performance in a trial is so small that it takes scientific analysis to determine whether one treatment is better than another it is probably of little if any value in normal third world farming situations. With such a small difference the reason could be due to variations in soil or microclimate etc. It would only be of value if that difference was consistent over at least three years.

WEED CONTROL

B4.1 Weeds should not be automatically regarded as bad for the land

Under tropical conditions it is often better to allow weeds to grow and so protecting the soil, rather than having bare soil, exposed to heat and to harmful ultra violet rays, which burn up humus and sterilise the top of the soil. If their seeds could be a serious threat to adjacent land, they could be cut before flowers are mature.

   The weeds also protect the soil surface from heavy rain and reduce or prevent wind and water erosion.

   It has been noted that a metre of rainfall (typhoons can yield as much or more in 24 hours) falling on one hectare of exposed tropical soil in the humid tropics, would have a mass of 10 million kg; and fall with an impact energy of 200-300 mega joules as trillions of raindrops with an average diameter of 1-3 mm, and an impact velocity of 9m a second. These figures give an idea of the risk that exposed arable fields face when complete weed control is achieved, especially in crops that do not develop full canopy (cover of the ground) for most of the growing season.

   Weed free crops give their best economic yield if also fertilised and protected from insects and diseases. However, if the soil in which crops grow is not protected, the top soil may be washed away over time, become degraded and thus yield sustainability will become uncertain (impossible PJS). Weed scientists should advocate that weed control practitioners adopt weed management strategies that embrace crop protection and production, such as integrated weed management, to minimize soil degradation---.

   Soil degradation can be reduced by integrating techniques that minimize erosion, add organic matter to the soil, provide a favourable environment for microbial activity, and suppress weeds. [77.]

From an article by I Okezie Akobundu a scientist with IITA Ibadan Nigeria. This was presented to the First International Weed Control Congress in Melbourne, 1992.

I believe we should suppress rather than destroy or eliminate weeds, except in the case of weeds which prevent the growing of crops economically. However the presence of such troublesome weeds may indicate a need to change the land management. Perhaps constant use of fertiliser has destroyed the soil micro organisms so that the plant foods are bound up and unavailable to the crop plants, as was the case in vol 2, 4.13a. When the particular fault is eliminated the weed problem may reduce considerably, so it may be better to first improve the soil before rushing in with herbicides or other controls.

The giant grass weed Imperata cylindrica is found in many countries of the third world. In Taiwan it choked out the crops and took over land degraded by fertiliser use which lead to overcropping,. Although crops such as maize would no longer produce more than weak plants 3 or 4 ft high, this grass soon grew to 4-5 ft and in ten years time from the field being abandoned it was growing thickly and over 6 feet high.

To thrive so well on land spoilt and degraded it must have the ability to extract and unlock plant foods in order to produce such a large biomass. It is natures way to protect and eventually restore fertility.

Many weeds can be regarded as a ground cover and green manure, as long as they are not allowed to spread their seeds. They are often better at extracting plant foods and trace elements from the soil than the crop plants, after their decay these plant nutrients can be used by the crops.

In considering weeds, it is important to realise that weeds can often be an indicator of soil condition. A Chinese agriculturalist told this author that wild ageratum (which grows in Taiwan, China and Nepal) is regarded as an indicator plant by the Chinese, the deeper the colour the better the soil. It is certainly the case that plants growing on poor soil have pale washed out looking flowers. The richness of the colour may be an indication of potash availability which soils in the countries mentioned are usually short of. If potash was in good supply the other important nutrients would probably also be sufficient. Similarly other indicator plants could be used as simple nutrient indicators.

Ecologically speaking,, the plants which are attracted to a soil are often the kind of plants the soil needs to correct a particular situation. As they change the situation, other plants move in, and eventually suppress the earlier species. It is good to discover what soils the weeds growing on a piece of land prefer, they could be acid land weeds, indicating it may be worth adding lime or ashes to the soil.

They could be deep rooted plants which will bring up plant foods deficient in the surface soil, or perhaps the soil is becoming less absorbent and so dry and more suited to deep rooted plants. If we loosen up the soil and add organic matter the weeds may be less competitive and much less of a problem.

When the soil needs organic matter, weeds which generally produce the maximum sustainable amount of organic matter under the conditions, may thrive. The previously mentioned Imperata species are among these, they will colonize degraded land, and after a few years time there will be a dense stand of grass 6-7 feet high, producing a large amount of organic material each year. It is resistant to burning, in areas where burning occurs for various reasons, so it is able to protect and hold land which would otherwise be very susceptible to erosion and loss of fertility, due to the effects of sun and weather. It will tolerate poorly aerated soil such as that which has been compacted by machinery or heavy grazing, and it can also grow on sandy soil.

In some situations if it were not for the Imperata grass the land would be cropped until more severely degraded, and then left bare as occurs in West Nepal. The grass which then grows is very sparse and poor, it supports few livestock and so does not receive much manure. It takes much longer for it to recover, and in fact erosion continues during the fallow period.

> In Taiwan and several other countries Imperata species of grass can be a problem as it is very vigorous and has tough rhizomes which are hard to kill. If small pieces are left in the ground the grass quickly recovers and chokes out crops planted on the land. Particularly when the fertility of the land has declined and fertiliser is being used. There are several varieties of Imperata; in Asia and in other hot moist areas I. cylindrica is most important. I. Major and I. latifolia grow in Asia. Var. africana throughout Africa; var. europaea mainly in the Mediterranean region; var. condensata in South America---.[59 p76.]

B4.2 Control of Imperata cylindrica

It would seem that in some situations it could be good to encourage this grass to grow and then incorporate the mass of biomass, and the organic matter and plant foods it contains, into the soil. This is difficult manually as the roots are extensive and parts of the roots not removed, quickly regrow, it takes a lot of digging to kill it.. We used herbicides successfully (see below) but they may be unavailable or too costly, and they can be very dangerous, particularly where the equipment it not as good as it should be. An alternative may be to cut it and spread it on the soil in a thick mulch which would smother new growth. What does grow through should be cut and added to the mulch, this would continue until no more shoots appear following rain. The soil micro organisms would be encouraged, the soil protected from the harmful sunlight and the soil texture improved. It could be that the mulch would allow conversion to zero tillage once the grass was controlled.

The following describes what was done by the author to control Imperata in an area of East Taiwan.

First the grass was cut and put into piles to make compost. I learned that turning Imperata into compost is a slow process, the stems and the rhizomes are very woody and a little piece of rhizome (which can live for a long time,) will grow into another plant. I realised why the tribal farmers burn this weed. The farmers use extra long mattocks to dig out the rhizomes and roots and after first leaving for a time to dry they burn them with the top growth. The labour involved in this had made the land clearing uneconomical as labour costs had soared with Taiwan's increasing prosperity.

We then tried various herbicide sprays and combinations of sprays as a controlled trial. We found that the best results were obtained with Dowpon (Dalapon) sprayed when the grass is growing, followed a week to ten days later by Gramoxone (Paraquat). The reason being that Dowpon works slowly moving down to the roots but it takes a while before the top dies off. The top growth is still feeding the roots and helping them to survive the herbicide. We gave it time to move through the plant and then sprayed with Gramoxone to destroy the foliage. In four - seven weeks if green shoots emerge, we sprayed them again with Gramoxone. This worked, we had no more trouble and I felt that we had found an economic way to control the Imperata over larger areas.

Since our work with Imperata, the herbicide Glyphosate has appeared on the market and may do better.

However we need to think carefully before using any of these herbicides. Chemical companies have been known to be economical with the truth. The authors knowledge of herbicides is not up to date but I have learned from experience in the past that herbicide use anywhere needs great care, in the developing countries even more so.

See paraquat poisoning below.

An organic method of control of Imperata Cylindrica.

The following is extracted from an article 'Cover crops critical to sustainable agriculture', in the International Agricultural Development journal July August 1999.

"In southwestern Benin, Mucuna pruriens, or velvet bean, was introduced as a green manure crop", said Mark Versteeg a former IITA scientist now working in Uganda. "First attempts to get farmers to plant it as a cover crop didn't meet with much success. Then farmers found that it was particularly good at smothering Imperata cylindrica, a pernicious grass weed.---.

After the mature palms are cut down, maize is grown between the young palms, but Imperata grass soon re infests the land, and farmers may have to hand hoe their maize crop five times to keep the grass under control. Even then, yields of maize average only 500 kg per hectare, which is insufficient to support the farming family.---.

Farmers who participated in the trials were more impressed by mucuna's ability to suppress Imperata, and thereby reduce hand hoeing of maize crops---.

Researchers reported farmers getting yield increases of approximately 500kg/ha for a local maize variety and 800 kg/ha for an improved variety following a one year fallow with mucuna. They got these higher yields with less effort, because there was far less Imperata grass competing with maize. Many of the early adapters also earned additional revenue by selling mucuna seed to various buyers as the idea spread.[138p5.]

Farmers embrace a creeper

Farmers in southern Benin have found the best way to deal with their biggest problem, a grass weed, is to smother it with another plant, a ground creeper. And in doing so, they have taught researchers once again the valuable lesson that working with farmers rather than for them is the key to increasing production.

The problem arose in the mid-1980s. The traditional farming system with a long fallow period had collapsed because of pressure to produce more food. The ultimate result was a drop in production because soil fertility took a nosedive. As soils degraded, fields became infested with the grass weed *Imperata cylindrica*, known as spear grass, and were then abandoned.

Researchers at the Institut national des recherches agricoles du Benin joined forces with the International Institute of Tropical Agriculture in looking for a leguminous cover crop to improve soil fertility. They selected Mucuna pruriens, the velvet bean, brought in from Latin America.

Mucuna, an annual leguminous, ground-creeping plant, produces a lot of growth, and when it dies down during the dry season, it leaves behind large amounts of organic matter. Initial trials in 1988 and 1989, showed that if maize is planted into this thick mulch at the beginning of the next rainy season, grain yields more than double. The maize crop benefits because the mucuna debris provides nitrogen and helps the soil retain more rainfall.

Demonstration plots with farmers produced some sensational increases in yields-as much as tenfold. But that didn't convince most farmers to plant mucuna. They weren't interested in a crop that yielded no food.

Some farmers did persist with mucuna, however, because they saw its potential in another more important direction. They found it could eliminate imperata grass from badly infested fields. If they cut down the grass just before the rains and then planted mucuna, the creeper had the chance to outgrow the imperata and smother it. In its search for light through the thick carpet of mucuna, the imperata uses up its root reserves, and by the end of the season there is very little left in the field. Next season maize can be planted into the mucuna mulch.

That finding, spread from farmer to farmer by word of mouth, was enough to get more people to plant mucuna.

Because imperata does creep back within three to four years, farmers will have to re-introduce mucuna periodically to suppress the imperata once again, and this also ensures soil fertility is maintained. So in a roundabout way, the researchers have achieved their objective.

Experience has shown when maize follows mucuna, yields are increased.

Some farmers have got yields of 2 000 kilograms per hectare, others have seen yields treble. But mucuna only supplies organic matter and nitrogen, so it may be necessary to apply phosphorous and potassium if these nutrients are deficient.

David Dixon *is a writer and broadcaster specializing in agriculture.*[206.]

The above method is much preferable to using chemical control which can be very dangerous.

B4.3 Paraquat poisoning

In 1975 I took an eight weeks Overseas Pest Management course. On that course we went to research and testing centres including the famous Rothamstead Research Station. We were assured that Gramoxone (Paraquat) was perfectly safe to use. When it came in contact with the soil it was broken down and inactivated so it could not harm the soil or organisms in the soil. If it was swallowed; as soon as it came in contact with the contents of the stomach it was inactivated, and so it was safe. There was therefore no need for marking the herbicide with a skull and crossbones, the usual mark of dangerous poisons.

The author accepted those repeated assurances, I felt that Paraquat was a very useful and safe product; and so encouraged its use and invited sales representatives to show their film advocating the product at farmers conferences, I also used it myself. I was horrified to discover later that it was not safe, I have wondered if anyone who died of it would not have died if I had not recommended it as being safe.

In (1994) it was reported that there have been many deaths from Paraquat in Malaysia; and that the herbicide did not carry the usual skull and crossbones symbol for a warning.

It is interesting to note in the book Weed control in the tropics, 1971[Ref 59] available at the Centre where the course was held has an advertisement for Gramoxone in the back of the book. This states "Progressive growers want five properties in their weedkillers: 1. Safety, 2. Quick action, 3. Economy, 4. Effectiveness, 5. Action in all weathers." Yet in the book it states that "Both compounds (Paraquat and Diquat) can cause skin irritation and eye injury, and the concentrate can cause cracking and loss of nails. Inhalation can cause nose bleeding and damage to the lungs; lung damage following oral ingestion of Paraquat is usually fatal." Page 20.

The book also states that "decomposition in the soil appears to be very limited and on pure organic and very sandy soils, where adsorption may not occur, toxic residues may accumulate." I understand that in the soil, while the product itself may not be harmful, on decomposition its residues are harmful to micro organisms.

I still feel that Paraquat is very useful but it is certainly not safe. In some cases it would be better to leave weeds and to use Paraquat to kill off foliage; as it only destroys chlorophyll; leaving the roots of plants alive and holding the soil; while hoeing or cultivating the soil increases erosion. However reports have shown that its regular use does have a harmful effect on the soil organisms. The makers deny that Paraquat itself is harmful, I believe the problem is that it is the chemical that results from the breakdown of Paraquat which harms soil organisms.

It is now admitted by the makers that it is dangerous, but that was not until many people had died a horrible usually slow death, sometimes taking up to 6 weeks. The makers now admit that livestock should be kept out of treated areas for 24 hours after spraying and that it is harmful to fish.

Gramoxone looks like Coke and so is sometimes drunk by children. As I had difficulty in discovering the signs and symptoms of poisoning. I include them here.

The immediate effects when 'Gramoxone' concentrate has been swallowed are usually vomiting, abdominal discomfort, soreness and inflammation of the mouth, throat and oesophagus, difficulty in swallowing and later diarrhoea. Vomiting may be pronounced with marked nausea, sweating and involvement of the central nervous system, perhaps with tremors and convulsions. Signs of poisoning may be mild, even though a lethal dose may have been swallowed.

Signs of kidney and liver damage may appear within one to three days of ingestion, the severity of this damage depending on the amount of Paraquat absorbed.

Signs of pulmonary (lungs) dysfunction may develop gradually a few days after ingestion, but can begin as much as seven days later.

In the first instance, pulmonary function tests may give abnormal results in the absence of any clinical signs of lung damage. The pulmonary effects subsequently include dyspnoea (difficulty in breathing) with pulmonary (lung) oedema (watery swelling) progressing to a marked pulmonary fibrosis (a morbid growth of fibrous tissue) and death from respiratory insufficiency.

There is no antidote, so hospital treatment is required.

Hospital treatment includes

1. Give stomach washout---.

2. It is important to purge the gastro- intestinal tract immediately, within 4 hours if possible. (The author has been told that four hours is normally too late). Give up to one litre of 15% Fullers Earth (Surrey finest grade), including 200 ml 20% mannitol in water. Alternatively, sodium or magnesium sulphate (Epsom salts) can be used as the purgative. Administration should normally be orally but, if this is not tolerated, stomach or duodenal intubation can be used. Continue purgation until the stools are seen to contain adsorbent.

3. CONTACT NEAREST POISONS INFORMATION CENTRE FOR FURTHER ADVICE ON TREATMENT.

4. Maintain and monitor fluid and electrolyte status on a daily basis.

5. Carry out haemodialysis or haemoperfusion (using a charcoal column) to remove paraquat from the plasma. This will only be of use if carried out within 48 hours of ingestion. In some cases renal failure may necessitate the use of haemodialysis at a later stage.

6. In the event of respiratory difficulties, delay the use of oxygen as long as possible as it enhances (strengthens) the toxicity (poisoning effect) of Paraquat.

7. In severe cases, particularly where shock has supervened, consider additional supportive therapy such as the use of steroids.

ADVICE MUST BE SOUGHT FROM A POISONS INFORMATION CENTRE SINCE SERIOUS SYMPTOMS (SIGNS) ARE FREQUENTLY DELAYED AND DEATH CAN OCCUR WEEKS AFTER INGESTION IF CORRECT TREATMENT IS NOT APPLIED EARLY.

After I had been telling people in Taiwan that it was safe I began to hear of people dying after being poisoned by Paraquat. Sometimes it takes up to six weeks before they finally die. Other workers who have used it without proper masks develop lung problems, many tiny dead areas, which make them unable to do heavy manual work.

Unlike pesticides it did not have a skull and crossbones on the label and so it was handled less carefully. The contents of containers were even sub divided and distributed in coke or similar bottles. As it looks like coke it was, and probably still is, swallowed accidentally by children or adults.

That should not have mattered as we were told that it would be harmless, it would be inactivated by the contents of the stomach.

Point one what if the person does not have much, or nothing, in his or her stomach ?

Point two what if instead of going to the stomach it gets into the lungs, either by the person coughing and spluttering when they discover it is poison; by it being absorbed into the blood stream and being carried to the lungs; or by the spray mist being breathed in. The latter could be from a spray plane passing over, or spray drift blowing over a person. It could be a spray operator having it drift on him if he was spraying up a bank side, instead of only spraying below him which is sometimes more difficult, or a sudden gust of wind blowing the spray.

I understand that the chemical stops the lungs from emptying the fluid which naturally builds up inside them. The person slowly dies of drowning.

(In developed countries a blue dye is now added so that it is no longer mistaken for coke or similar brown soft drinks. The formulation in third world countries may still be brown, as will old stocks of the product).

According to the Pesticide Action Network

> In Malaysia, a 1986 government report found that Paraquat caused more poisoning and more deaths than any other pesticide---. in 1981 and 1982 it was estimated there were 1,200 deaths a year in that country, most of them suicides---. While Paraquat manufacturers claim that they are not responsible for unintended uses of their products, the WHO and many others have argued that Paraquat's easy availability increases the rate of suicide deaths. Efforts to stop suicidal use of Paraquat via education, by adding blue dye, an odour, and an emetic to the pesticide have had uneven results.
>
> It has at least 40 other names as well as Gramoxone so this further confuses users---.
>
> Paraquat is extremely toxic to mammals by all rates of exposure. Inhalation toxicity is dependant on particle size- within limits, smaller particles are more toxic---.
>
> Swallowing as little as one teaspoonful of Paraquat or absorption across broken skin can be fatal. Death results from suffocation due to fibrosis (scarring) of the lungs where Paraquat is concentrated. There is no effective antidote or treatment for severe Paraquat poisoning.
>
> One of Paraquat's degradation products, Qina chloride, does not bind well to soil and therefore is a potential groundwater contaminant. *United States Environmental Protection Agency report 1987.*

In many places in the third world workers cannot read easily; even when they can, the instructions are often in the wrong language. Pesticides in Nepal are rarely sold with instructions in the Nepali language, only the main Indian languages and English. Even if they can read, the instructions as to application rates are not understood, as they describe pints and gallons per acre; measurements not understood in Nepal.

I have several times asked farmers coming from pesticide shops if they knew about the safety recommendations for the pesticides they have purchased ? they have not. Often the sales men or women themselves do not know. If the person has Paraquat inside him or her, the only thing to save their life is to make them swallow Fullers earth, this must be done very quickly. If that is not available, or, they have the chemical in their lungs in more than a very small amount of dilute spray mist, they will die.

Other weedkiller's can affect the operators children due to birth defects. A common weedkiller 2.4-D is related to 245-T which caused so many abortions and birth defects in Vietnam.

B4.4 Herbicide (Weedkiller) usage

The points I am making are:

(1) Chemical weedkiller's can be very useful but they should be used very carefully and in moderation. Not as a regular part of management. We do not know whether there may be accumulative affect, unrecognized side effects, or affect animals fish or people as the chemicals they break down into, get into the plants and the ground and streams.

For example Glyphosate (trade name Roundup), is being used more and more to kill weeds. How many developing countries can monitor such things as local drinking water for Glyphosate. Many people in the third world particularly the poor, depend on fish for their livelihood or as protein food for themselves, but increasingly, more wealthy people have fish farms, or buy fish for pet food or farm livestock feed. If drinking water becomes polluted and dangerous in the west people can have piped water or even buy drinking water, third world people cannot. When chemicals enter the soil it has been estimated that it may take up to 30 years before the chemicals or their derivatives (the chemicals they change into) reach the water bearing layers, and seriously contaminate the water. That means that it will take at least as long from the chemical being banned until it stops polluting the water, as it is still moving through the ground.

Glyphosate is:

1. harmful to fish. (Unless used in accordance with the guidelines for the use of herbicides on weeds in

watercourses and lakes.)

2. Irritating to eyes and skin.

3. Minimum interval to be observed between application before harvest and a/ harvesting oilseed rape. or mustard, 12 days. B/harvesting barley, oats, wheat, or peas, 7 days.

4. Application to water courses may lead to illegal pollution. Before using in or near water courses read the guidelines mentioned in 1.

In addition it is advised "do not use straw from cereals treated pre-harvest as a horticultural growth medium (for example compost) or as a mulch."[60 p168.]

The best use of chemical controls is probably to help deal with a severe, or persistent weed problem; but then to use good management to keep the problem from getting to such a serious stage again.

> (1) **We cannot simply accept scientific reports about the safety of agricultural chemicals.**
>
> (2) **Chemical weedkiller's can be very bad for our health if not used in a very careful controlled way.**
>
> (3) **Under third world conditions it is often very difficult to be sure that they are used in a controlled way.**

B4.5 Alternative methods of weed control

The above being so we should try to find alternative methods of weed control.

These may be to use lightweight push cultivators, including rotary weeders. By making a small simple hand push combination planter, cultivator, hoe and earthing machine, weeding in Taiwan was made much easier and quicker.

They may involve biological methods such as cultivating land before it is needed, to encourage weeds to germinate, and then again cultivating at the sowing of the crop, to kill the germinated weeds.

Another way is to use a mulch, I have been told that I should have used the cut razor grass straw, I mentioned above, as a mulch to smother the after- growth of the shoots of the roots of the grass, perhaps by concentrating the grass on half the area. After two or three months lift the grass straw and place it on the other half. Where there are persistent green shoots in the first area, indicating the remaining plants, these are dug up and burnt.

Another way is to grow crops such as sweet potatoes which allow weeding in the early stages, and later the sweet potatoes smother the weeds. Some opinion suggests that the sweet potatoes have other qualities which inhibit (reduce) the growth of weeds.

Crops such as Sunnhemp grown as a straight green manure crop choke out and kill weeds. Or they can be grown between rows of other crops such as maize or cassava, smothering the weeds and balancing the soil exhausting effect of the main crop.

> Imperata can be eradicated by flooding, by regular cultivations and by plants which give a deep shade; competitive crops like cassava and sweet potatoes " (Note the author disagrees with the effectiveness of cassava having farmed land on which the reverse had occurred Imperata had smothered a cassava crop. Imperata in Taiwan grows quickly after being cut down; up to 8 feet high.) Competitive ground covers include Centrosema pubescens, Pueraria spp; Crotelaria juncea, (authors note: In Africa it has been found that Crotelaria ochroleuca is very effective in controlling Imperata.) Leucaena glauca, and the tree Vitex pubescens---. (Proceedings of the 8th California weed conference 1956.) Sowing cover crops such as Tephrosia candida after ploughing is said to virtually eradicate (destroy) Imperata (R.S. Baker Abstr. meet. Weed Sci. Soc. Am. 1967, 5-6) A dipterous gall Pachidiplosis (G.W. Ivens PANS(C), 1957,3,209-23) and the parasitic plant Sopubia ramosa (Keeley P.E; Thullen R.J; and Miller J.H, Weed Sci. 1970. 18 393-5) both affect Imperata.[59 p76.]

> Van den Hove. 1965 found that competitive plants, especially leguminous cover crops, eg -Pueraria phaseoloides (Puero) plus controlled grazing can eradicate Imperata cylindrica.

The author had useful experience with folding turkeys over Imperata stubble in Taiwan. The turkeys ate off the shoots as they appeared, they also scratched around the roots and picked up quite a lot of feed for themselves. This was supplemented with a small amount of chopped sweet potatoes and rice bran. The turkeys grew very large and when given a little extra food for the last two weeks were very heavy with excellent tasting meat. At the same time the Imperata was killed and was no trouble to subsequent crops planted on the area. As pigs dig up and eat the rhizomes they could be used in a similar way.

The Benedictine Fathers at Peramiho in Tanzania have done very good work which have showed the many values of Sunnhemp, (Crotelaria ochroleuca).
   The first reason for introducing it was for weed control.

   The area round the farm house was overgrown with couch grass. (Cynodon dactylon, a particularly serious weed especially in irrigated plantation crops.) It disappeared after planting Sunnhemp for one season. Since then no herbicides have been used except 2,4-D against Micandra physalodes Scop. A missionary at Urambo (Tabora) used last year one ton of Sunnhemp seed (for 100 acres(40 hectares) to fight the "Devils Weed"(a red flowering plant killing maize completely) and he has ordered another ton for next season.[66 p75.] (Authors Note : Why not keep own grown seed?).

   Striga is a semi-parasitic weed pest, it grows in fields where the soil is becoming exhausted. It has been reported by W.L.Watt in East Africa Journal 1936,1, 320-2, that farmyard manure and by S.D.Timson in Rhodesia Agric. J; 1939, 36, 531-3 that compost reduce the incidence of Striga, possibly by increasing the moisture content or microbial activity of the soil.

   I do not have experience of Striga, but I have read about a biological control which seems promising. In issue 12 September 1992 of Footsteps news letter there is a letter from Father Gerald Rupper of St Benedicts Abbey. P O Peramiho Tanzania.
   He writes that

   a missionary priest near Tabora, where Striga is very common, discovered that sunnhemp (Crotelaria ochroleuca) will kill Striga completely.
   As Sunnhemp whether Crotelaria ochroleuca or C. Juncea is an excellent green manure crop this method seems very appropriate.

   Sunnhemp may smother other weeds also, as it grows tall and thick about two metres high, if left to go to the flowering stage. It will grow on very poor soil making a large amount of rich biomass and in my experience it had nitrogen nodules the size of peas on the roots.

   Orobanche species (broomrape) are important parasitic weeds in hot, dry areas. Crops which will induce a parasite to germinate but are not themselves parasitised are known as trap crops. Trap crops for Orobanche are O. ramosa-lucerne, maize, clover, rape, mustard, capsicums, castor-bean sesame and pennisetum and setaria millets.

   O. cernua- capsicums. O. crenata- linseed. Host plant remains may act similarly---.

   Seeds of orabanche are very susceptible to flooding; inundation for one month is said to result in complete loss of viability (ability to germinate)---.

   Livestock can help in controlling Orabanche. In Australia O. minor is grazed by sheep and the plants do not grow again.[59.]

In the (former) USSR, Orobanche in sunflower can be destroyed (q.v.) by the use of Phytomyza orobanchia.[59 p10.] Good farmers should always try to find and try out non chemical methods to avoid chemicals if possible.

B4.6 Are GM crops the answer?

The present answer for farming problems is we are told, genetically modified crops. For instance maize which is not killed by glyphosphate weedkiller, so that it can be sprayed and only the weeds die. Isn't this ignoring ecology? Ignoring the reason for weeds being a problem, and so mining the land further. What about possible pollution of ground and river water by the weed killer or its derivatives as B4.4.

B4.7 Research organisations should be encouraged to investigate biological or management methods of control.

The prickly pear (Opuntia spp) was an introduced ornamental cactus which became an extremely serious weed pest in Australia. More and more farmers were unable to grow crops, the land was being taken over by the cactus until government action resulted in the CSIRO research which found the cactoblastis insect, which feeds on that particular plant. It was introduced throughout the worst affected areas and rapidly destroyed the prickly pear; other similar work has also brought dramatic results some less dramatic but still important. Some work is being carried out on fungal organisms which only attack the target species.

B4.8 It is important for workers in the field to be on the look out for non chemical control methods. However it should not be forgotten that methods of weed control involving cultivation may increase soil erosion and so do more to destroy the future than herbicides.

These may be modified sometimes so that they chop the weeds and leave them on the surface or partly incorporated into the soil and so reducing erosion. Improved design of hoes can make weeding easier while disturbing the soil less.

Hand winnowing machines for cleaning grain, and also cleaning seed of weed seeds, can be very beneficial and reduce the labour of grain cleaning, freeing women for other work. These could be bought by village cooperatives, or, as in some areas in the UK, could be a small business with the winnower mounted on a vehicle, which could simply be a rickshaw and taken round an area, cleaning grain and seed for a reasonable fee.

Crop rotation also reduces weed problems, as does the growing of smother crops and inter-cropping.

Stubble cleaning is a useful technique to reduce weeds such as wild cereals. As soon as possible after harvest the land is ploughed and may also be cultivated. The weed seeds germinate and are killed by being ploughed in, before they can produce seeds, normally in preparation for the next crop. The side effect can be that the soil had plant cover instead of being naked to the harmful effects of the sun and weather, it also adds organic matter to the soil.

Growing green manure crops in between the main crops, instead of leaving the land free for weeds to spread and set their seeds may also be a feasible proposition. The green manure will pay for itself in improving the soil, at the same time reducing weeds.

Techniques such as the following may be suitable:

Molasses grass is most valuable in initial land development, as it helps smother the initial weeds and carries the subsequent cleaning up fire. It is sensitive to fire, but produces a very hot fire which carries well. Of course this technique is only used when there are severe weed problems, it should not be used at a time of year when erosion is likely to occur, and quick growing crops should be sown as soon as is practicable. If the land is to be used for pasture it should be sown with more perennial, hardy grasses such as guinea grass, which will form the eventual basis of the pasture. Guinea grass is very palatable and has a high feeding value, especially if it is not allowed to become too coarse.

An alternative is to grow guinea grass on its own. G.grass is tolerant of shading from other plants and this enables it to grow under such weeds as Lantana a serious problem in Australia. When a sufficient bulk of grass has become established it is possible to set fire to it and so destroy the weed. The grass then recovers quite rapidly and replaces the lantana.

Cutting weeds on waste land before they produce their seeds can be a community action to help keep weeds in check. When living among very weedy rice fields in one part of Nepal I was horrified to see that rice weeds were often left until they had already developed seeds. Then they were thrown onto the bunds and edges of the fields, some were fed to livestock or used for bedding, so ensuring that there was a continuing weed problem.

Land management is another important factor. Draining land will reduce or destroy certain kinds of weeds as well as possibly improving the yield of the crops grown.

Research in Taiwan showed the importance of getting paddy fields level. The research was aimed at reducing the water needed. With level fields the water level does not have to be kept so high. This has a side effect of encouraging tillering of the rice plants; improving quality and yield, it also has the effect of reducing weeding as there are no high patches standing clear of the water where weeds thrive and spread. Those weeds set seed which keep the weed problem troublesome.

The high patches also have poorer yields of rice and straw. Farmers in Nepal are much less careful to get the fields level and so have more weed problems. One reason is that at least in the areas I have been in, the simple but effective equipment which is used in Taiwan for levelling the fields is not available. In Taiwan when bullocks or buffaloes are used for cultivation a sloping spike harrow is used. It has spikes about 15" (38 cm's) long. The angle of slope is altered for different work. It can be set flatter for running through the soil to lift out the vines of crops such as sweet potato. When the vines have accumulated on the spikes the farmer lifts the harrow and the vines are dumped in rows for removal. At other settings the harrow is used for breaking up clods, preparing a seedbed or mixing the mud when puddling rice fields. When used as a bulldozer for levelling rice fields a board is tied over the spikes with the tips of the spikes just showing below the board. As the draft animal moves across the field the farmer can control the harrow, pressing into the mud on the high areas and easing up when he wants to dump the mud being pushed in the low areas. It is a simple but effective use of a very useful locally made harrow. I understand that similar harrows are made and used in the Philippines and Thailand.

B4.9 Cover Crops

These are crops such as cucumber, pumpkin, melons, sweet potatoes, Irish potatoes, soyabeans, peanuts, lentils grown with a maize crop. In the dry season crops grown such as buckwheat, chickpeas, field peas. They may not be worth harvesting but give cover to the soil, encourage micro organisms and when ploughed in improve the soil quality and structure. They also smother or reduce the weed population which may otherwise grow strongly in the empty field. In some countries it may be possible to grow a permanent cover crop such as a legume and sow the food or other crops through the legumes. Or perhaps using a method such as ploughing a narrow single furrow through the legume to make the rows in which the crop is sown.

B4.10 Mulching

This is mentioned in other chapters in volume 2 but also helps suppress weeds. With alley cropping the leaves and light branches can be used as a mulch to protect and improve the soil and smother the weeds.

B4.11 Trap and catch cropping

This can be used in some cases, for instance for the control of Striga mentioned earlier.

> Trap and catch cropping are probably the best ways of ridding the land of *Striga*. Trap crops stimulate the parasite to germinate but are not themselves parasitised; catch crops stimulate the parasite to germinate and are parasitised and must therefore be destroyed before the Striga sets seed. Catch crops tend to cause a higher percentage of Striga seed to germinate than trap crops, and it may be worth while to sow catch and trap crops in alternate closely spaced rows and weed out the catch crop a month or so after sowing. Trap crops for S. asiatica include soybean, field pea, sunflower, cowpeas, linseed, castor bean, and cotton; S.hermontheca parasitises a number of legumes, but soybean, lucerne and sunn hemp and cotton are suitable trap crops; however differences in varietal reaction should not be forgotten. Cassava may have some effect in this way as growing the crop for three years has found almost to free the land of the parasite. Sudan grass (Andropogon sorghum) is a very effective catch crop and growing it for 5 weeks before cutting and sowing sorghum in the stubble greatly reduced the infestation of S. hermontheca in E. Africa. In Rhodesia catch crops suitable for S. orobanchoides are cowpea, dhal and velvet bean. Quoted in[59 p101.]

IT IS ALSO EXTREMELY IMPORTANT TO STRENGTHEN THE CASE FOR RESEARCH OF THIS KIND AGAINST THE POWERFUL AGRO CHEMICAL LOBBIES.

RESEARCH AND DEVELOPMENT NEEDS
What are the needs for Research and Development for Genuine Sustainable Soil Conservation and Improvement; Particularly that Appropriate for Normal Third World Situations?

B5.1 The situation
B5.2 What have soil scientists been concentrating on for the last 50 years?
B5.3 The role of soil organisms in sustainable agriculture
B5.4 Sustainable Development
B5.5 Where is the real emphasis too often?
B5.6 What should research be involved in?
B5.7 A key requirement for effective development
B5.8 Local research and demonstration is vital

B5.1 The situation

Appropriate and effective research and development work is of value not only to the countries directly concerned but for the world wide economy and peace.

Is most R and D Agri-culture or is it Agri-mining ?

Agriculturally related work must not only be sustainable but also regenerative and improving. PJS. Sustainable agriculture is sustaining agriculture as it is – that is not good enough.

We are now using up soil capital which was accumulated over tens of thousands of years. Put in another way we are mining the soil.

As an illustration of what is happening we can compare with Oil wells.

When first tapped an oil well will gush oil. The oil is easy to extract, we simply put in a pipe and the oil flows up it, and out. After a time the oil pressure falls until there is a need for pumps to suck the oil out of the ground. The oil level falls and so the oil becomes difficult to extract. Now water is poured into the well. The oil floats up and so more can be pumped out until the time comes when all the oil has been exploited and the well is of no further use.

We start with soil which has developed to the stage of supporting a large biomass, the primeval forest or jungle, we clear the land and start the exploitation. At first the soil gives freely of its reserves of nutrients, then comes the time when it's fertility starts to fall. We then may bring in improved seeds, these are sometimes more compact plants, or plants with very strong root systems, so that they work more efficiently. Or, they may be hybrids with hybrid vigour and so grow more strongly. They are able to draw more nutrients in a given time, than the local seeds. Again production declines and so we supply some external inputs, synthetic plant foods, which enable crops to be grown, for a further period, when that is having less effect we are told the answer is Genetic engineering. More mining till the soil is depleted of its accumulated nutrients and its structure. We can never in a time span meaningful to human life, return the soil to what it was.

I have only seen pictures of how the American prairies became dust bowls. But I have seen the Darling downs in Queensland Australia; where deep dark soils built up under forest were cleared and cropped with wheat year after year. There was no need for fertilizer, that was the way to farm or so they thought. Some thoughtful people noticed that the land was shrinking, the soil was not as deep as it used to be. But 'not to worry' 'no sweat, that was until things suddenly started to change. Poor rocky sterile soil started to be turned up by the ploughs, fertility fell off, and the cheap wheat became not so cheap, fertiliser had to be poured on at an increasing rate. Farm houses were standing above the fields on the rich soil which had also covered the now worn out fields. The land had been mined, not farmed. Instead of Agriculture, Agrimining.

In temperate regions the deterioration of the soil is slower than in the tropics but without good management it does occur. If we cannot develop sustainable land use there will be more poverty in rural areas and more migration to over crowded slums. As less food and other rural products are produced the cost of the products will increase, which will increase the cost of living in the urban areas.

Politicians and others who want social stability, should realise that in the developing countries most of the population live in rural areas; dis-satisfaction in those areas leads to social instability, which easily leads to political instability. This can be seen in many countries, the present situation in Nepal is an example. The unrest started in the poor and neglected Far West region, and has now spread East. In July 2001 it was said that 40 of the 75 district councils of Nepal were under the influence of the revolutionary movement and their movement was steadily increasing, in area and in power.

B5.2 What have soil scientists been concentrating on for the last 50 years?

Soil scientists claim a competence in studies of soil erosion and the practice of soil conservation. Yet Soil science has manifestly failed to relate this competence to the challenge of developing sustainable land management systems in the tropics.

At best science is the passive provider of data sets; at worst it is by-passed directly.

The potential demand on soil science by land managers, soil conservationists, agriculturalists, rural development planners and other clients is huge. For tropical and developing countries, the key issues in land degradation and rehabilitation centre around the development of land use systems that: 1) provide the goods and services needed by land users; 2) are practical; 3) are within the resources of land, labour and capital of land users; 4) produce an acceptable economic return with the minimum of risk; and 5) ensure the maintenance of the resource base and long term productivity. To varying degrees, the knowledge of soil scientists is applicable to all five, but sadly the provision of information has been lumpy, inadequately presented, indigestible to the user, hidebound by specialised terminology and uninterpreted.

M. Stocking School of development studies University of East Anglia. Norwich, UK. Abstract of paper presented at the British Society of Soil Science Easter meeting, 31st March 1992.

'Sustainable Land Management in The Tropics: What role for soil science ?'

Someone has pointed out that:

Agricultural Research in many countries, frequently relates more to the career needs and convenience of the research scientists; or, to the interests of influential people and organisations, rather than the problems faced by farmers.

Not many research workers are keen to abandon their own special interests and the techniques they have been trained in; they like neat formulae and standard equations to work with. Consequently, as few if any are trained in minimum chemical input or non exotic livestock they are seldom whole hearted in doing research on these more appropriate techniques.

There should be more practical trials of the potential of plants and not only legumes, to make plant foods available and to improve soil fertility, with or without micro-organisms.

B5.3 The role of soil organisms in sustainable agriculture. Much more research is needed.

As pointed out in Volume 2:

Soil organisms may be defined as playing a variety of functional roles in natural and agricultural ecosystems: increasing the amount and efficiency of nutrient acquisition by the vegetation; regulating the retention and flow of nutrients in the system; syntheses and breakdown of organic matter; modification of soil physical structure and water regimes; parasitism, pathogenesis and predation.

The subject will be discussed within the general hypothesis that conditions which promote diversity in the soil community also favour sustainability=. In this context sustainability may be demonstrated by a non - declining and stable annual trend in yield and essential resources. A secondary hypothesis is that >agricultural practices such as continuous mono cropping, mechanised tillage, fertilisation and pesticide use diminish soil community diversity and destabilise plant soil relationships=. Examples will be given in relation to tropical cropping systems of differing complexity including cereal monocrops, inter-crops and agroforestry systems. From an abstract of a paper by M.J.Swift, N.Sanginga & K. Mulongoy of the International Institute of Tropical Agriculture, Ibadan, Nigeria, presented in [12]

I believe such serious consideration of the effects of conventional agriculture is not sufficiently included either in the teaching of agriculture, particularly in the third world or in the activities of governments, research and aid organisations.

3.4 Non chemical means of improving the availability of plant nutrients

Nitrogen

It is widely known that legume plants have the ability to nourish bacteria on their roots which convert nitrogen from the air into nitrogen plant food. What is not as widely known is that other plant forms can do the same. Several sources in this book have information on this subject of which some points will be quoted in this chapter. See also microorganisms 2.3 3.13 and chapter 5.

Recent work has shown that more significant nitrogen fixation may result from the activity of bacteria occurring in the rhizosphere's (roots and the surrounding soil) of many tropical grasses and cereals, where they are able to derive energy from substances present in the root exudate's. For example, an unimproved cultivar (variety) of the grass Paspalum notatum, (This authors note: Bahia grass) occurring extensively on sandy soils in Brazil and locally known as 'Batatais', specifically stimulates the nitrogen fixing bacterium Azotobacter paspali which becomes permanently established on the root surface in association with a thick layer of mucigen (a gum-like secretion PJS) developed on parts of the root system---. It was estimated that up to 100 kg N per hectare per year could be fixed. On the other hand soil separated from the roots had only one hundredth of the activity of the latter. (Dobereiner and Campello 1971; Dobereimer et al 1972a). A similar association, estimated to fix about 67 kg N Ha/year, has been demonstrated in the rhizosphere of sugar cane, where the organism responsible appears to be Beirjerinckia, but in this case enhanced activity was also found in soil samples taken from midway between the cane rows, which was attributed to the concentration in this zone of run-off from the leaves providing sugars and stimulating an increase in the number of bacteria (Dobereiner et al; 1972b.)
Most of the important tropical grasses and cereals have now been shown to possess considerable nitrogenese activity in their rhizosphere's, including sorghum, maize, rice, sugar cane and grass species of importance in improved pastures.[18 p70.]

In a 3 year experiment in Nigeria, Jaiyebo and Moore (1963) recorded increases in soil nitrogen of 595 kg ha/year under regenerating forest containing no legumes, and 90 kg's per year under the grass Cynodon plectostachyus. The undoubted importance of blue green algae as nitrogen fixers in paddy fields is dealt with (elsewhere in[18.]), but there is a lack of information on the significance of these algae under other conditions. They require light moisture and substantial quantities of nutrients other than nitrogen in order to effect agriculturally significant gains in nitrogen---. On the other hand, pronounced growth of blue green algae is often observed on the soil surface in the humid tropics and Singh (1961) has estimated that an algal crust in a maize field added 90 kg/hectare in 75 days.
Free living nitrogen fixing bacteria have been found in many tropical soils, those most commonly reported belonging to the aerobic genera Azotobacter and Beijerinckia and the anaerobic Clostridium genus. Meiklejohn (1962) examined the occurrence of these bacteria in Ghana and found that both forest and savanna soils were plentifully supplied with nitrogen fixers. As found elsewhere, the numbers of Azotobacter decreased with decreasing pH and this genus is probably not significantly active in fixing nitrogen in soils more acid than the pH 6.0. Beijerinckia predominated in acid soils and is probably of particular importance in the tropics as it has no need of calcium and is very tolerant of acid conditions. Clostridia occur in all soils and are independent of pH. As they are anaerobes, Clostridia may often be in a resting condition in many soils, but they may fix nitrogen in the anaerobic pockets which often exist in soils normally considered to be well aerated. (Author; where would the nitrogen come from in such

cases?) Latterly, it has been shown that a variety of other bacteria can fix nitrogen, among which those commonly isolated from tropical soils have included species of Klebsiella, Enterobacter, Escherichia, Derzia, Spirillum and Bacillus (Jurgensen and Davey, 1970). However, as the simple, readily decomposable organic substances which all these bacteria require if they are to fix nitrogen are commonly present only in limited amounts in the body of the soil, it is doubtful if they make a significant contribution to the nitrogen available to crops. [74 p372.]

(Author: this may be another pointer to additional values of keeping up the organic matter of such soils.)

Phosphate

Work in India to inoculate seeds with phosphate liberating bacteria has given results with increase in yield of wheat up to 37% [22 p133.]

In **maize,** uninoculated plants   compared with inoculated plants
averaged 2.4   3.37 and 6.1 grams dry weight   3.0   13.3   14.4
Tobacco 1.65   2.72
Strawberry 6.69   16.04[98 p15.]

Pigeon Pea and Chickpea release Phosphates. The following is based on an article in International Agricultural Development April 1992.

We all know that legumes such as these two plants add nitrogen to the soil. Now scientists at ICRISAT in India have shown that they make available more phosphates. They do not add phosphate to the soil, but rather break up phosphate compounds in such a manner that phosphate that was already present but unusable by plants is now available. If you work where phosphate is one of the most limiting nutrients (a common situation in tropical soils), you might want to work these crops into your rotation.

How do they work? Studies show that the roots of pigeon pea exude acids (piscidic acid) which release phosphorous when it is bound up with iron. Chick peas release another acid (Mallic acid) from both roots and shoots. In calcareous soils (alkaline soils with a high calcium content), this acid breaks up insoluble calcium phosphate. Normally this release would only occur if the pH of the soil were lowered.

Both plants are deep rooted so their ability to release more phosphates means that valuable nutrients are being brought up from the deeper soil levels. Residues from both crops are adding extra phosphates which will benefit the crops which follow in the rotation. It is possible that some varieties exude more acid than others. So this trait could be another characteristic for selection by plant breeders. [87 p38]

Potash

Though I cannot now locate the sources of information I have read that microorganisms have also been found which increase the availability of potash. A possible indication could be in the following quote from the classic >Soil Conditions and Plant Growth= by Russell 1950.

A further effect of mulches is that they increase the amount of exchangeable potassium in the soil, although the reason for this has not been established. Thus , I.W. Wander and J. H. Gourley found that it was increased considerably to a depth of 8 inches (21cms) in two years under a heavy straw mulch, and to a depth of 24 inches (60cms) in twenty to thirty years---. The extra potassium may have come from the straw, but potassium is the only mineral element showing this strong downward percolation. Further a grass mulch, made by putting land down to grass and mowing it with a gang mower, such as is used on golf greens and leaving the grass cuttings in place, also increases the available potassium in the soil. This method is used in many apple orchards, both in this country and overseas, and it causes the apple trees to take up much more potassium, and also more phosphate, from the soil than if the orchard is kept either clean weeded or sown to annual cover crops which are ploughed in. Hence, this extra uptake of potassium, and possibly also of phosphate , appears to be a specific effect of the mulch due to some cause yet unrecognised.

These effects of mulches show the reasons why they are so valuable with shallow rooting crops in hot weather, for all these effects help to maintain the surface layers of the soil as a suitable environment for root growth. Indeed, for some of the very shallow rooting crops the root system tends to develop from the soil into the base of the mulch. [25 pp640-641.]

It could be that the use of mulches provides a welcome environment for soil micro organisms, which in turn make more potash and other minerals available for plant use.

B5.4 Sustainable Development

Sustainable development has now been one of the current buzz words among development policy makers for about 15 years. It is vitally important as was pointed out in a 'Keynote Address of the British Society of Soil Science Easter meeting in 1992. Sustainable Land Management in the Tropics.

> For any system of land management to be sustainable, it is essential that the soil not be degraded. Degradation may be chemical, physical or biological'. Greenland D.J. CAB International, Wallingford, Oxon. UK. Quoted in [12.]

While lip service is paid to the importance of sustainable agriculture, to an experienced observer the actual practice differs from the stated policy. An example of this is the British Overseas Development Administration. (The forerunner of the present Department for International Development - DFID). It produced in March 1989 a useful *'Manual of Environmental Appraisal'*. On page 9 we read "ODA is interested in projects which demonstrate the wise use of natural resources to provide sustainable livelihoods for local communities." On page 18 we read:

> Among the main categories of project and policy proposals with a significant impact on the environment are the following:---. major changes in the use of land and renewable natural resources: Forestry: farming; changes in farming practices; introduction or intensification of use of pesticides and fertilisers.

Page 45 states

> Fertilisers, though generally safer to handle than pesticides, also represent a hazard to the environment if they are misused. The addition of small amounts of fertilizer may alter natural ecosystems. Excessive use of fertilizers may lead to nitrate (or other) contamination of water resources. with adverse effects on aquatic organisms. (No mention of the possible effect on people.)[113]

Page 88 points out the dangers and problems inherent in the use of pesticides and suggests the use of alternative options eg. use of resistant varieties of crops, biological control.

It is a pity that there is no mention of the effect of fertilisers and pesticides on the soil; or the decreased resistance to erosion. Or, concern for soil conservation and improvement or at least the maintenance of fertility in a sustainable way; nevertheless it is a good start. We need to consider if the appraisals recommended are carried out.

Another publication of the ODA is *'The environment and the British Aid Programme'*. On page 10 it states **Land degradation**- A United Nations Environment Programme (UNEP) study has estimated that 20 million hectares a year globally are reduced to zero productivity. A further six million hectares are reduced to deserts. Much of this degradation is concentrated in developing countries. In one state in Nigeria 10 per cent of the land has been wasted by gully erosion.

It illustrates the effects of soil erosion on pages 10 12 and 26.

A third ODA publication is 'A strategy for research on renewable natural resources' it:

> sets out the priorities for ODA support for scientific and economic research on renewable natural resources with a view to meeting the needs of developing countries and outlines how the ODA proposes to organise and manage this support.---. The overall aim of research supported by the ODA is to promote the conservation and sustainable development of Renewable Natural Resources (RNRs) in developing countries.---.

There are 6 objectives listed the first two of which are:-

The strategy therefore:

1. Sets long medium and short term objectives and priorities for ODA research in the areas of agriculture, livestock, forestry, fisheries and the environment.

2. Develops an integrated research framework between the various discipline areas in the interests of sustainability.

That seems good, surely the sustainable methods of conservation and improvement in the productivity of the basic resource, the soil, would be high on the list; after all it was highlighted in the policy manual. On page 15 is a photograph showing a man looking at trees which have their roots exposed due to erosion. The caption reads 'Inspecting soil erosion in a citrus plantation, Belize.'

B5.5 Where is the real emphasis?

When we look at the research allocations in percentage terms we see that Integrated Pest Management has the highest allocation with 33.5%. When we note what they are involved in, we read:

> Integrated pest management component technologies:
> the aim is to develop improved systems of pest control which are neither commodity specific nor region specific.

(Authors comment: a bit vague, and questionable as to its being sustainable. I think most people seriously involved in pest management realise that the need is to be more specific rather than a blanket approach; which increases the development of resistance to pesticides.)

> - A major element is the development of safer ways of using agrochemicals and the avoidance of the emergence of new pests and pesticide resistant species.

This has been going on for years and mainly comes down to educating the users into safe ways of handling the pesticides and wearing protective clothing. When a growing percentage of the users are illiterate due to the debt burden reducing investment on education, it is increasingly difficult to inform the users of the correct use of the increasingly dangerous pesticides in use; particularly in third world countries, due to dumping of products banned in the first world.

When we come to the wearing of protective clothing

Point 1. It is not possible to wear full protective clothing for long periods when using a knapsack sprayer in the tropics, particularly if as is often the case, climbing up and down a hillside even more so. This author speaks from experience.

Point 2. Just to buy a face mask costs a Nepali peasant more than one months pay, and most will already be in debt.

Such so called research work should be carried out by the pesticide companies and not charged from aid to help the poor countries.

The only mention of prevention of soil erosion or fertility replenishment in the whole plan is in the sub section agroforestry where it is stated:

> *Agroforestry systems*: this component will focus on the role trees and shrubs play in agricultural systems, with particular attention to the benefits that trees bring in soil fertility replenishment, prevention of soil erosion, provision of fruits and livestock browse, and more generally in the context of social development.

Why state as policy that 'The overall aim of research supported by the ODA is to promote the conservation and sustainable development of renewable natural resources in developing countries.' When so little of it is concerned with such promotion?

Perhaps it is not realised that soil is a non renewable natural resource, once it has gone it cannot be returned or renewed. But in that case why the statements about the seriousness of the loss of soil, the development of deserts and the importance of sustainable systems ? How can agricultural systems be sustainable without suitable soil ? If we are not taking seriously the need to at least maintain the fertility of the soil we should not be talking about concern for sustainable development.

B5.6 What should research be involved in?

A keynote address in[12.] was given by A. J. Bennett of the ODA in 1992. The abstract of which states:

> Meeting the needs of the present, without prejudging the ability of future generations to meet their needs for economic and social development, presents us with the challenge to devise and adopt better land use practices. Growing world populations, increasing land degradation, loss of biological diversity and possible climate change increase the urgency to have sustainable resource management systems in, and for, the tropics.

> A basis for better land use must be a thorough knowledge of the physical, chemical and microbial characteristics and processes of tropical soils.

> The soil scientist must not only provide the knowledge of these processes, but also play an active role in

developing improved land use systems, by working as part of an interdisciplinary effort which ensures that better systems are not only technically sound, but socially, economically and politically acceptable.[12]

That abstract points out correctly the urgency to address soil degradation, and also the need for all disciplines to cooperate in this urgent task.

A letter from a very experienced tropical agriculturalist in the Tropical Agriculture Association Newsletter points out in September 1999 that:

There is a characteristic lack of interest by urban -based tropical governments in the remote and often mountainous regions from which their rivers flow. There are few votes in guarding steep watersheds and few staff to administer more rational regimes of land use. Posting to the hills is seen as banishment by government staff and posts are poorly paid. Lack of schools, medical services and shops further reduce incentives---. As population pressure reduces farm sizes in the valleys, the less successful survive by clearing ever steeper forest land. The resulting hydrological degradation is threatening the future on a formidable scale. Salinity continues to destroy prime irrigated cropland in the commands of the Indus/ Ganges Euphrates/Tigris and Nile. As the green revolution is reaching its limits there are no foreseeable research avenues to replace it.[205 p23.]

To this author the above indicates two things, one is a reminder that farmers are the key people in soil management but they may need help from government or other agencies. Help with transportation and marketing, help in matters such as tenant right so that tenant farmers can benefit from improvements they make.

The other point is that we should not expect government managed extension programs to achieve much. Practical research and demonstration in appropriate and strategic locations is needed with the actual work carried out by farmers, then taking representative farmers to see and discuss with the participating farmers how the successful methods work and how they could be modified to fit local conditions. See B3.12/. for a very effective extension program which could be used.

As has been pointed out by Professor Anthony Young 'If a country has no major source of non agricultural income, and runs out of land it will hit the 'Policy options : Zero' wall. It is time the governments in such countries realised this.

B5.7 A key requirement for effective development

I believe a key requirement is to train multi-disciplinary people who can better understand the broad situation in which third world farmers have to live and work; people who are better able to work in local situations to fit together the various components in a way that farmers can use.

There is too much research and development work which is inappropriate and is failing because the experts have too narrow a training. In this fight against time we cannot afford for such waste to continue.

B5.8 Local research and demonstration is vital

The contents of this book give some guidelines for protecting and maximising the value of the most important natural resources, soil, and water, in third world situations.

Local research will be needed to find out if and how the concepts suggested will need modification to suit particular situations. Unless the techniques have local credibility it is unlikely that they will be seriously taken up by the people who matter - the farmers.

Too often the R and D is carried out without careful consideration of the circumstances in which the farmers who will have to use the results of the R and D are situated.

Where the scales are most heavily tipped against the farmer, it is fruitless to invest in improving research and extension, without first ameliorating or removing these overriding constraints---.
The main reason for the Training and Visit (T and V) system of extension being successful when it was first introduced into the irrigated areas of India in the late 70's and early 80's was because it had become the key constraint in places where there were unutilized improved technologies available, good infrastructure, marketing and input supply facilities and a cadre of well educated staff that could be trained to implement the system relatively easily. Attempts to introduce it in rainfed areas of Africa in the mid 70's were faced with a very different situation; relevant research technologies were often non

existent and the majority of farmers were just emerging from the subsistence sector, roads were poor and marketing infrastructure was rudimentary if it existed at all---.

Any development program has to be set in it's policy and infrastructural environment, and key constraints addressed if it's potential benefits are to be realised---.

In the more extreme situations, there is no meaningful link between research and extension, let alone with the farmer---.

For instance most crop recommendations emanating from research are single commodity ones to give an optimum production increase for a crop grown in pure stand; and they are derived on the research station without any attempt to address the reasons why farmers are not able to improve yields. This one way flow of information has usually meant that only the richer or more progressive farmers, who can come closer to conditions on the research station and have access to all the factors of production can benefit. Compared to the resource poor farmer the wealthier farmer can more readily add labour to complete planting or weeding in a more timely manner, invest in additional inputs, or risk changing his cropping pattern---.

There is thus often a mismatch between what extension is delivering and what the resource poor farmer needs.[102.]

To give an example from this writers experience and related to soil conservation. A research project into the propagation of Bamboo for the dry hill areas of Nepal was using a system which firstly required the bamboo cuttings to be kept in a forestry nursery, with daily watering during the time of year when there is often little enough water for drinking. The cuttings were heavily manured and fertilised, they were shaded from hot sunshine. Secondly after the bamboo had been accustomed to such moist conditions the rooted cuttings were planted out into dry hillside soil, naturally most died. The result was to turn farmers against the use of bamboo. In fact under practical conditions bamboo propagation using cuttings planted directly where they will grow is very suitable in the same area.

EXTENSION POLICY, METHODS, AND RELATED CONSIDERATIONS

It is no use outsiders knowing what needs to be done, unless, the farmers who will have to take up the ideas are convinced of their value for them and their communities. The methods advocated must have credibility for the local situation, which means that unless influential local people have seen the methods working well in a very similar situation, they must first be demonstrated locally. Sometimes it is difficult and risky in terms of social relationships for people to try something different, however when people have been convinced, they have "the courage of their own convictions", and so are more likely to try the new method. It is always best if somehow a culture of consent is developed early on.

For encouraging soil and water conservation and soil improvement as with any rural development work we have to be able to understand the local situation and people and fit the approach to the situation,or we will have little if any lasting success.

For a thorough coverage of policies, methods, approaches and other considerations see "Quality Rural Development" by this author which will I hope be published in 2003.

B6.1 Policies
    B6.1a Impatience
    B6.1b. Development Fashions
    B6.1c. Participation
    B6.1d. Towards a better understanding of a multi-disciplinary approach
    B6.1e. How feasible is it to introduce new crops, livestock or other enterprises?
    B6.1f. Trials should normally be successful over at least two seasons before wide use and acceptance is encouraged
    B6.1g. Identifying where the power for change lies
    B6.1h. Social or religious considerations
    B6.1i. It is very important to be able to hear the views of each socio-economic group separately
    B6.1j. Change is often threatening, so proceed slowly, humbly, and carefully
    B6.1k. Quota's
    B6.1l Analysis of what will be the long term result of actions being planned
    B6.1m. It should be realised that it may be necessary to help with marketing, perhaps by encouraging a cooperative
    B6.1n. Should people be given inducements?
    B6.1o. Careful planning
B6.2 Methods and approaches
    B6.2a. Attitudes
    B6.2b. Credibility
    B6.2c Giving the farmers a voice, and participation from the start
    B6.2d If possible avoid alienating any group
    B6.2e Meeting felt/perceived needs
    B6.2f Local trials
    B6.2g. The importance of showing we care for the real interests of the people and not just for filling government quotas or gaining promotion.
    B6.2h. Helping with cooperatives can lead to a change of attitude and general improvement programs
B6.3 How to run extension trials/training/demonstrations
    B6.3a A careful approach is needed

B6.1 Policies

B6.1a Impatience

Policies are often made too quickly, without a long careful look at the whole local situation. They may be made by a person with interest in one discipline, whose particular interest could be the need to plant trees. The local vocals may indeed speak out for such work in public or private meetings, however it may be that those local vocals are the people who own the timber yard, or the trucking company, or they are the people with the good land, who know that planting trees will not harm them as it would not be done on their land.

It may be that the poor people are not present at the public meeting, or dare not express their opinions; so that their side is not appreciated; until, the trees which were planted, and the fences round the new forest disappear. Then, it may be realised that the land planted to trees is the common area where the poor people collect fodder for their animals, and that the needs of livestock come into the equation. It may be realised that the manure from that livestock is needed for the crops, that if there is no manure the crops are thinner, the soil structure is weakened, and serious erosion and perhaps landslides may occur. In short that rural development should be multi disciplinary, and people from each socio-economic grouping must be carefully consulted.

It may be that rather than spending a lot of time and money on extension personnel who the farmers are likely to ignore, or who undermine the confidence of the farmers, it may be better to use different methods. One experienced advisor has said that it would be more effective to be a travel agency than involved in extension work. He explained that if groups of farmers were taken to see other farmers who are using different and successful methods under similar conditions to their own; the farmers being free to see and feel the results, ask questions, and discuss with the other farmers, they were much more likely to take up the ideas. However, with such visits care should be taken that the farms visited really do have similar constraints and possibilities to those of the visitors; and that what they are doing has been tried for a long enough period of time for it to be appropriate under different weather, pest, or other conditions.

An example of is of a visiting group of Nepali farmers being impressed by large imported goats and wanting to buy some. The importer who of course was happy to sell to the farmers at a high price insisted that they were just what was needed. In fact they were very unsuitable for the conditions in the part of the country the visitors came from. Under those conditions the goats had very poor reproduction rates, and in addition needed regular veterinary attention and better food than was normally available. Though the trip organiser knew this he did not say anything. The goats were bought and subsequently failed, and with it the interest in alternative technology.

Impatience can lead to:

- trying to move too quickly, starting before the target people are really ready, and moving at too fast a pace;
- by passing community participation, and involvement; upsetting people;
- putting too much money into a project in an attempt to speed things up;
- working in too wide an area or trying to do too many things at once;
- using too many people from outside the area rather than training local people or using local knowledge and experience.
- It can lead to using a technology that is not really appropriate and cannot be continued without unfeasible outside support; using methods that are not sustainable.[175 chapter 4 p4.]

B6.1b Development Fashions

When going into a situation we should not do as is too often done, follow the latest fashion. Fashions are ideas which have merit for one place, time or situation; the idea is then spread by people who cannot think for themselves but like to appear clever and up to date. We should *look at the particular situation, and fit the plan to the situation, not the situation to the plan or fashion.* I have heard of many instances where organisations wanting a program, or project funding for development work, have paid people to look through the proposal and pepper it with the fashionable buzz words such as participatory, and participation. They find it greatly increases the chances of funding, whereas if the fashionable words are not included there is little chance of funding from the donor organisations.

Of course participation is essential for real and lasting development. Unless we show we respect the people, listen to them and involve them we cannot expect them to respect us. That should always have been obvious and courteous, unfortunately it was not, ideas were imposed from above, many mistakes and much resentment was the result. However, the pendulum has swung too far the other way as more and more people are recognizing. In the early 1980's "Base line surveys were the fashion', every self respecting project had to do baseline surveys. Projects since find that unless they scatter 'participation' in their reports and hold PRA (Participatory Rural Development) meetings they find it hard to attract funding. The meetings are often a sham or the only people participating are the 'local vocals' rather than the busy or less vocal people. Technical ability is ignored in favour of sociology some of which is dubious. See also 1c.

> For agriculture in the tropics the pendulum of aid agencies' emphasis appears now to be swinging away from the strong earlier stress on the scientific aspects of what is needed to get agriculture moving and towards a strong relative stress on the social science panacea. This swing is perhaps understandable, because results of trying to apply purely technical solutions to problems of inadequate plant production - which have a complex of technical, social and economic causes - have not been as successful as hoped. In recent years it has become abundantly clear that rural families motivations and constraints affect their decisions about how best to manage land to improve their families conditions, and that purely-technical recommendations for improved plant production may be unacceptable, inappropriate or even damaging in this wider human context. Growing sensitivity to, and understanding of. Farm families conditions and aspirations makes easier the growth of confidence - of farmers in their advisers, and of advisers in the farmers they serve. Information and understanding of every sort can then move more easily and credibly in both directions. While this can greatly improve the *rapport* in rural areas we should not see this as and end in itself, an alternative to the technological approach, but rather as a valuable complement to it, for the melding of both sets of 'disciplines' and the best use of their different but interlinking sets of specific knowledge and skills. [195.]

Another example is *Women in development:* The present fashion is that women are the key to development. In many cases they are and always have been, in many cases women are the real though perhaps hidden power in the home. In most places women are the key to influencing the future generation. We often find that the men are away from home most of the time, or, at the key periods of the year for the particular aspects of development concerned.

Sometimes the men appear to be lazy, leaving the women to do most of the work, but it is hard to do things differently. Perhaps traditionally the men would go hunting to feed the family or fight in battles, and adjustment to a changed situation comes slowly. People need friends, if people do things differently they tend to lose friends until/unless the different way proves to be better, then it becomes fashionable and the person who made the change becomes a leader. If it fails he makes a loss, he loses friends, respect and cooperation from his neighbours, in many societies the risk is too great.

Women are very important and ignoring women has been both unjust and unwise, but by concentrating on the women we may alienate the men.

If we ignore the women we may get nowhere, but equally if we ignore the men, or the men think we are ignoring them we may get nowhere. In some situations ignoring the men may stimulate them to start doing something, but it may be that what they do is not what we feel is best for the community but rather opposing what we are attempting.

To concentrate on the presently fashionably correct stress on working with women may be unsuccessful as:

   1. The women who need to be involved may have not have the time or energy to be seriously involved.

Those who do attend training programs may be quite unsuitable.

2. The women may not want to alienate the men.

3. The men may get suspicious and/or jealous, and forbid the women to attend the meetings or action involved.

4. It may just add to the long list of work which is classed as women's work without other activities being taken over by the men.

5. It will usually be better if the work/training can be with or for both the men and the women. Perhaps with the women first but not far ahead, to encourage the men to do it better, with the result that all are involved or even that the men take it on as a work they can do best.

B6.1c Participation

PRA (Participatory Rural Development) including PRA (Participatory Rural Appraisal) are popular requirements for rural development. In fact anything with participatory in it is regarded by many organisations as an essential requirement for a project proposal. Yet it needs very careful consideration, it can be bad as well as good, it can result in the opposite results to those hoped for.

What is PRA?

PRA developed from RRA, (Rapid Rural Appraisal) as people began to realise that they could learn more about rural situations, and learn in a more effective and accurate way, if they used a more participatory approach. Listening and learning and involving the people in working out appropriate responses to situations. Rather than asking questions, recording the answers and coming up with plans to help.

Peoples participation is not new, it has of course always been used by good rural development workers. We should always show people respect, and involve the people in working out how to improve their livelihoods. As we show respect, mutual respect and relationships develop, together improvements can be made, and the communities assisted to help themselves to a better future. Sadly in the past and in fact still in some places too many development projects did not respect the people and properly involve them, and so unsurprisingly were not successful. The PRA and PRD discovery that 'they can do it' has been a startling revelation to many outsiders. "Many of the activities which we thought only outside professionals can do, can be done by local people, and done better".

The newer aspects of PRA are the improved techniques of analysing the local situation. Drawing information from the community through participatory methods, and, hopefully by their involvement in the analysis, encourage them to share their needs; and be motivated to discuss what could be done to improve their situation.

Some ideas may have been mentioned previously, some people have been to see other places before; but the suggestions for change should come from the people, or, be freely accepted as desirable, or worth trying, by most of the community.

It may be that preliminary pilot work may be needed to show that alternatives could be feasible and worth developing before there is acceptance by the people, then the work can develop with the people.

PRA methods can be used to recognize problems. The problems may have been accepted previously, but now they are brought out and analysed. This can be useful in helping people find the real cause of a problem, then they can be helped to solve it.

As a very simple example: Why has the boy an infected toe? Because he got a thorn in it. Why? Because he has no shoes. Why? because his father has no money to buy shoes. Why? Because his crops are too poor. Why? Then the people, with or without the projects help, work out a strategy to deal with problems.

PRA-Participatory Rural Appraisal as any participatory work has a lot to commend it, it is very important to have the participation of the people in appraising the situation. This should start from the point where it has been decided to start working with the community or members of it. However, I have been told by people who work with PRA and PRD (participatory Rural Development) methods, that they found that while in general the resulting development seemed effective, the poorest were not helped or if they were helped it was only indirectly and was of less help than the benefit received by the 'go for it' people. So the result can be a widening in the gulf between the poor and the more prosperous members of the community.

Downtrodden people find it difficult or impossible to share in public that the real problems are the local big wigs, who take more of the water; who cheat and browbeat them, but on whom they may need to depend for work and loans when in need. It is hard to help downtrodden people to stand up to other people whom they detest, but know that they might need help from which they perhaps could not get from anywhere else.

As with previous methods of surveying needs, in some societies courtesy demands that people give the answers they think the outsiders want to hear. In some situations as I have personal experience of the people are primed by their leaders beforehand as to what they should say.

It is vital to realise that PRA needs skill, or as with the baseline surveys, previously used to plan development work, the answers can be misleading. The hopes of the people may be raised and then dashed, the poor end up poorer and disillusioned.

PRA can be used to manipulate people, and once people sense rightly or wrongly that they are being, or may be, manipulated, they turn against the work.

PRA can also be used to manipulate donor agencies. An agency in Nepal asked for two VSO volunteers to help run a project for helping poor hill farmers improve their livelihood. They told of how they and the staff had lived in the villages and had held PRA meetings to find out what the people felt were their problems. They had then helped the community to work out how to change their situation. The project was designed to work with the communities to develop a sustainable future.

After being in the project for a few months the volunteers discovered that the whole thing was in fact a way for the project organisers to receive money for doing very little. The project office was on the plains not where the villages were. The staff had not been to the villages, they had merely asked what one or two merchants thought were needed and developed the project plan themselves. Furthermore the project was being fully funded by two different donors who had been impressed by the project proposals which fitted their fashionable criteria. The volunteers both resigned, but it is quite possible that the project organiser is still working the same way of attracting money from other donor organisations.

PRA can be useful but it tends to be used as part of a quick fix, real development takes time. PRA used as a part of the process, at the right time in that process, is good and has an important part to play in many situations. It is not the magic formula we are often expected to believe.

The point of development is to motivate and enable the people to see the needs for themselves and plan to meet the needs themselves.

> Participation is not always constructive. In some programs, a single leader emerges and takes control; everyone else learns to be submissive rather than to participate. In other cases, a lack of experience at making decisions as a group causes disagreements. Factions develop and organisations disintegrate. Even well-made decisions can lead to failure, causing disappointment and mutual casting of blame. Many cultures have no acceptable method of correcting the inappropriate or dishonest actions of leaders. When leaders misbehave, people merely sit back and gradually become convinced that organizations are ineffective, or even dangerous, and very often, too little is known about handling money. Financial losses, because of either insufficient planning, poor decisions, graft, or nepotism, will also cause division and mutual recriminations. Even if these more noticeable problems do not occur, programs may merely fail to produce much recognizable success. As people become convinced that little is going to improve, whatever enthusiasm they had wears off. The best motivated and most talented leaders may go elsewhere. Those who remain do so for the only reasons left, - their salaries, or graft. Tremendous pressures for deceit and manipulation can be produced by situations in which the continuation of salaries depends upon superiors believing that successes exist where, in fact, they do not.
>
> These kinds of participation teach people that other villagers are not trustworthy, that getting involved in organizations only causes them problems, and that village people are not capable of solving their own problems. These kinds of participation teach manipulation, deceit, exploitation, individualism, hopelessness, and dishonesty. They are destructive rather than constructive. They do not *produce* development; they *preclude* it." [187 p28.]

Constructive participation does not develop merely by having meetings and forming a committee it has to be learned and requires new skills, integrity and openness.

Another point to be noted about participation is the degree of and speed of participation appropriate in the running of programs to help the communities to help themselves for instance:

> Don't flaunt the money bags. Programmes that inform the villagers early on that, say, $100,000 are available for the program will confront a good number of problems. Such sums, astronomical by village

standards, tend to attract those villagers interested in graft. They can also produce considerable pressures for everyone to receive inflated salaries. Costa are thereby inflated and voluntarism reduced or eliminated---. High salaries and low voluntarism reinforce also the feeling that outside money, not the people's own efforts, has made the program successful. Thus, the growth of pride and enthusiasm is stunted. And very likely, when the money is spent, the work will come to a screeching halt.[187.]

In a meeting held by a project chairman to discuss a possible drinking water project, the representative people from one part of the area said that they were not interested, they could manage well enough with the sources they had. As was usual in that area where there was no real village as such, but only small clusters of houses scattered over a wide area, few women attended, they were too busy. There were no women from the group which stated that they did not need the water project. I later spoke to the women and children of that group; *they* were very sorry that their group would not be part of the water project; *they desperately needed the water.* The men did not want the extra work involved in digging trenches, carrying materials etc, fetching water was not their job!

It may be easy for outsiders to go in, see the problem/s and have a meeting about it. However, it is normally best to take time, to get people thinking about the situation, discussing, and working out how the problem can be faced together. 1. It could be that the outsiders do not realise the whole picture. 2. The people in the community may in fact be able to see a better solution or way to achieve it than an outsider.

Participation is vital yet has to be applied with patience and preparation. It is important to prepare very carefully, there is need for good training before PRA is used and the right attitude among the staff.

It is important to first go into the area quietly as an Indian national working for OXFAM told me he did. Get to know the people, particularly the needs of the poor and oppressed, build relationships with them. He would not think of holding PRA meetings until he had studied the situation over a six months period or longer if necessary. Only then think of having PRA meetings, otherwise the poorest will consider that they do not have time to attend the meetings and in any case the more confident and vocal will come into prominence. We should also remember that we can get the weaker members into trouble if we encourage them to be frank in public meetings. PRA can be dangerous for people low in the pecking order, in the excitement of a PRA meeting, explicit details of political interference, of bribery and corruption etc, can emerge. While it is important for the project staff to know about the real situation, such openness can be a threat to those involved. The more influential, will not like it, they may refuse to give loans or may give preferential treatment to the ones who kept quiet. The poor may find that the more influential will even use more violent means to punish those who spoke out. Until we have had time to really sense the situation and gain confidence of the weak members and plan accordingly we can harm the ones we want to help.

The PRA approach can easily tend to work with those who seem keen and cooperative, who are very often the crafty ones who know what they want and how to get it. Unless we have first established a relationship with the least confident and less vocal those people and their opinions will be left further behind.

Let's not forget technology!

I have been concerned for some time by several of the changes that are taking place in soil and water conservation. Quite correctly, it has been recognized that soil conservation programs are unlikely to succeed unless the landusers themselves are closely involved in the whole process of identifying the problems, working out the solutions, and implementing the programs. Certainly, programs like the Landcare Program in Australia have made substantial progress and farmer groups have achieved impressive results.

Unfortunately, this has led some to the conclusion that all governments need to do is provide the facilities for farmers to work together and they will then be able to develop their own solutions and implement the required measures largely by themselves. I think it is partly due to this approach that so little research is currently being done, in spite of the fact that land degradation is now recognized as a growing problem in so many parts of the world.---.

While farmers can do much innovative work on their own and are constantly coming up with new solutions, there are limits to what they can achieve without outside help. I believe there is now a pressing need to support them with more and better scientific research into the processes of land degradation and its control. Also, if extension workers are to win the respect of the farming community in which they live and work, they must be given the necessary technical training and be equipped with the skills and equipment that field work needs to be able to operate effectively. Unfortunately, some now see the role of soil conservation extension workers only as "facilitators" whose job is to create the conditions for the landusers to be able to do everything themselves.[194.]

The sort of statement which is fashionable is that the xx xx xx program 'aims to facilitate sustainable development by building on communities 'own experience, capacity and skills'. The primary goal of xxx is to promote healthy communities, and so enable them to undertake their own sustainable development activities. The programme ---. Focuses on organizational and local human resources development among the poor and marginalised. The main areas of work include community organisation and motivation, non formal education xxx, yyy, and agriculture.

I know of several such programs none of which include soil and water conservation which is basic to healthy communities. Privately it is admitted that the long term effect helps the better off more than the marginalised. In fact sometimes it is admitted that afterwards there was more of a gap between the rich and the poor and often the poorest those who for various quite reasonable reasons do not attend the meetings are even more demoralised and poor afterwards.

Of course it is good to work on health and literacy programs, building on communities ' own experience, capacity and skills and enabling people to organise to improve their situation but people cannot pull themselves up very far by only using their own resources. There is a need to help them to use other resources and experience *so people are needed who can do that.*

Why should people have to go about re-inventing the wheel, should not the methods and skills which have been found to work in similar situations be shared and tried? Should not the mistakes that have been made also be made known so that others can avoid them? The currently fashionable swing that a social approach to development is the most important ignores the contribution of the sharing of knowledge and skills in a non negatively paternalistic way. As with most things a balanced approach is the best.

The techniques of PRA may be essential to get people cooperating to enable better methods to be used. For sustainable land use there are many situations where a joint approach is needed. For instance I taught farmers to grow fruit trees and fodder grasses and legumes instead of maize on eroding slopes. However when it came to marketing the fruit the local merchants paid less than the cost of picking the fruit and so people started to cut down the trees in order to grow maize to feed themselves. Very reluctantly I had to get involved in setting up a marketing cooperative. The cooperative proved successful, the standard of living improved dramatically and people started to plant more fruit trees. Without the peoples ability to cooperate all the earlier work would have been wasted, and would have been another negative example of what happens when new things are tried.

It may be that by cooperating a pressure group can be formed to send representatives to talk to government about improving the road and transport situation so that produce can be sold, and with a better standard of living fewer bright young people will be interested in moving elsewhere. Teachers and health workers will be happy to live and work in the area. See also B6.2h.

PRA type methods like most things can have both good and bad aspects, so care and consideration is needed.

B6.1d Towards a better understanding of a multi-disciplinary approach

From the previous chapters it will be clear that a multi-disciplinary approach is needed to meet the vital challenge of sustainable soil and water conservation, and soil improvement.

However many people will point out that a multi-disciplinary approach does not work in practice. People can quite rightly speak of multi-disciplinary garbage, as does a letter in the Tropical Agriculture Association Newsletter September 1993. It quotes from a book as follows:- 'Individual creativity - must be subjected to the needs and ambitions of the team. Individuals participating in interdisciplinary research who do not embrace the common goals may find themselves in opposition to the groups objectives- However the administrators overseeing such projects must serve as facilitators and must suppress the divisive tendencies of undisciplinary thinking'. The letter writer goes on to say this infers that team members will have to ingratiate themselves with the bosses by getting specially genned up on *'special techniques of group interaction'* and the *'psychosocial sciences of group dynamics'*.

I can appreciate what the writer is getting at. The pursuit of fashion and buzz words, to appear to be 'with it'; is a typical way in which good concepts are ruined.

Another criticism of the multidisciplinary approach is that it is understood that there needs to be discussion about every aspect of work carried out. To facilitate reasonably meaningful discussion, long explanation of points from each discipline needs to be given. The result is that team members spend so much time in meetings, that they have insufficient uninterrupted time to get on with their own work; the bureaucrats are happy but the doers are frustrated, the whole program becomes stifled and bogged down.

I still maintain that the key people in rural development are the farmers, if programs fail to help farmers the programs are failures. Too many programs are failures. A major reason is that *the farmers have to be multi disciplinary, the components of their agriculturally related work have to fit together.*

Too much research, development, and extension does not address this fact, so that the result is like having a car which may look very nice, it may even have electric windows and stereo sound but if it does not have a drive shaft or any fuel it is not going to get anywhere unless it is towed. Sometimes programs are towed, they appear to be working but once the support which is towing them is withdrawn movement ceases. The structure like a disabled car is vandalised, it rusts and falls apart.

Unless we first see what is needed and supply and fit all the component parts properly together we are wasting our time.

I believe two points should be seriously considered.

1. We need input from the farmers and others involved in the process at the beginning and throughout. These may include merchants, transporters, processors etc. Not, in long time consuming meetings, but asking for and putting proposals to the people concerned, encouraging them to be an important part of the process. Letting them mull over the suggestions and come up with points for and against. Do not expect rural people to be ready with instant answers they often know better than project people how unwise that is.

2. At the risk of appearing to push myself forward I believe that my kind of background is more appropriate for appropriate rural development; and so training and recruitment for this work should take this into account. My background is in mixed farm management followed by training appropriate to the work I was involved in, rather than a narrow training for an academic degree, so it is nearer the situation of farmers. It has therefore been easier for me to understand problems and constraints likely to occur; how things must fit together, and if they are likely to fit together. I know that farming is a business and is more complex than many in R and D understand. I speak the same language as farmers, the farmers appreciate that I realise more of their difficulties, I can work with them in the field. I am a multi-disciplinary person, though not an expert in one field. I have had a much wider training. I can call on advice from experts and act as an intermediary.

Of course we need the experts but between them and the actual practitioners there should be multi-disciplinary people NOT multi-disciplinary committees.

B6.1e How feasible is it to introduce new crops, other enterprises, or new methods of managing land? If there are constraints can they be rectified to make it feasible?

While it may be clear that alternative crops are needed, it is important to look at how those crops will be transported and marketed. As an example, it was realised that it would be good to grow temperate fruit in Jumla a higher altitude area of Nepal. The sale of the fruit would enable farmers to buy in other food so that they would not grow cereals on land deteriorating under cereal growing, and at the same time have a better standard of living. Insufficient consideration was given to how the fruit would be transported to markets. The result was that most of the fruit was wasted, transport costs were too high, and so the farmers standard of living was further depressed and the credibility of alternatives to the unsustainable system was lost.

Another example is of farmers growing maize even though they know growing maize is spoiling the land. The reason- fear of being at the mercy of merchants and money lenders, if they have to buy maize their stable food, instead of growing it. The solution could be a purchasing cooperative, or that part of the payment for the fruit was made in maize. In another situation a bridge may be needed so that with a jeep trail being made by the community, produce could be transported to a sales point or market.

B6.1f Trials should normally be successful over at least two preferably three seasons before wide use and acceptance is encouraged.

Rural people are so often made to feel inferior by people with more academic education, or by outsiders, because of this they are quick to notice mistakes made by people who appear to think themselves superior. The news of mistakes travels very rapidly, while that of successes only slowly. That is why it is so important to test new methods thoroughly before recommending them. See also B6.3c.

B6.1g Identifying where the power for change lies.

We should always realise the importance of power. The real power in a locality may not be comprehended easily, particularly by outsiders. If we alienate the power or some aspects of the power we may get nowhere, what we are attempting to do is doomed to fail. Aid agencies working with a national government, may by doing so alienate themselves from the real power for change in a district; unless they find and establish good relations with those with real influence in the area who will often be members of a political party in opposition to the government. Effective work can be quite tricky to balance.

B6.1h Social or religious considerations.

These may not be obvious, but must be considered. If the local priest tells the people that it is because they are displeasing the gods that the land is falling down the hill, he may be angry if teaching on soil conservation contradicts this. We may need to point out that if the people have better crops they will have more income and crops to support the temple and the priest.

Another example is that in an area without a road it was decided that nuts would be an appropriate crop, they would not spoil, and could be carried to market whenever people were available to carry them. They planted nut trees and things looked well until the nut trees attracted monkeys to the area. The monkeys not only ate nuts but also other crops such as peanuts and corn cobs. As the monkeys were regarded as sacred they could not be killed or removed and so the people were much worse off than before. From this experience, before encouraging nut trees it is important to have a clear decision that if monkeys become a problem they will be removed or killed.

From these examples it is clear that wise and representative local people should be involved from the start, and careful consideration given to the real long term sustainability of what is being planned and how it fits in with other local considerations.

B6.1i It is very important to be able to hear the views of each socio-economic group separately

to discover the real situation. This must be done discreetly, though it may be appropriate to admit that separate meetings discussions were held to enable each group to be frank about their views.

If only main village or area meetings are held, the views expressed may be mainly those of the local vocals, that is the people who are more familiar with the language used, or better educated, which usually means more wealthy. Or, the more powerful members of the community such as landlords, money lenders, religious leaders or the local mafia. The less important members of the community may feel intimidated in a large open meeting.

"For any project to succeed the local people must always retain power over their livelihoods and local resources."[186.]

B6.1j Change is often threatening, so proceed slowly, humbly, and carefully

There are so many factors involved, it is important to avoid things which are needlessly threatening; or, make them less so. When people feel uneasy they tend to reject that which causes the feeling. That is why a slower approach, showing respect, being ready to be humbled and risk failure, and persevere until the best way is found, is usually more effective. Failure by humble sincere people can also bring sympathy, and when perseverance pays off- respect.

As already mentioned, it could be that the best way to help development is to organise visits by farmers to see and talk with other farmers who have discovered a better way. We still need to remember and perhaps remind farmers, that what works in one place may need to be modified or may not be suitable at all in another place.

B6.1k Quota's

Policy makers love to set targets and quota's; these can be good but can also be dangerous. One of the risks is that quality of work is sacrificed for the need to fulfil the quota.

In one year the suitable weather, or the labour for tree planting may be very limited, due to the pressure to do other agricultural work at that time. Consequently it may be better to plant fewer trees but to plant them correctly and successfully than to rush the job to fulfil the quota and so have few of the trees survive.

If quota's are issued from a distant place with little or no chance that the project work will be visited and checked properly, the quota's may be filled by unscrupulous means. For instance it is much easier and quicker and so cheaper, to plant an area which was prepared and planted the year before. If tree seedlings are taken up a hillside and thrown like grenades to the places where they are to be planted, they are less likely to survive. So the area can easily be planted again next year, the quota met, the policy makers happy, and the full payment per tree made, so it seems everybody is happy, but the money has been wasted. I have seen a hillside being planted four times in four years. 10 years later there are no more trees on the area than there were before the program started.

It may be good to have a flexible approach so that if programs go slower than planned the budget allocated can be extended over a longer period, rather than if it is not used in the planned period it is lost to the project. The saying slow and steady wins the race is very often true of rural development. Because the money is not used, or the quota's not met in the time, it should not be assumed that it is due to laziness or inefficiency on the part of the staff. The writer has known of budget money being wasted on unhelpful work and other expenses, due to a policy which would cut the future budget if the money allocated was not used.

The situation should be examined to see if there were good reasons for the spending being less than forecast, and if the overall need would still be as great as had been planned. It is important to be sure that the project is not going slow merely to extend the period of employment of the staff.

B6.1l Analysis of what will be the long term result of actions being planned

It may be that it is realised that fruit trees rather than crops should be planted on steep hillsides. The extension agency does good work in researching and supplying the most suitable fruit trees; and also in teaching how to manage and care for them. After several years work and investment the trees start to bear good fruit, now there are masses of peaches or apples in the area, in fact there is a glut. The fruit rots on the ground, while the people have no food to eat. There is no economic way to transport the produce to other markets, no marketing and no processing facility. Perhaps there are merchants but they keep the prices so low that the farmers start to cut down the trees. Eventually the situation may balance out, but much time effort money and credibility has been lost. This is poor planning but happens too often.

It is important to look to the possible future results of what is planned now rather than waiting until there are undesirable results.

Another example is that of changing cropping systems or management to one which initially brings in extra yield or a greater profit which may lead to imbalance and exhaust the soil, such as growing cassava; or encourage the build up of pests and/or diseases, in other word unsustainable systems.

B6.1m It should be realised that it may be necessary to help with marketing, perhaps by encouraging a cooperative.

In Taiwan, after training and encouraging farmers to plant fruit trees rather than growing maize on steep hillsides I found that farmers were starting to cut down healthy trees. On investigation I found that the payment received for the fruit was less than the cost of harvesting the fruit. After cooperative marketing was established the return to the farmers over and above the needed expenses, was 20 times more than before, then far from cutting down the trees, fruit growing was very much in favour.

It is important to follow up the work done to conserve soil, if I had not taken on the job of organising the cooperative my previous work would have been wasted. Donor organisations and development organisations must consider the possible less obvious future needs of the area. *Economics are an important consideration for effective and lasting soil and water conservation.*

B6.1n Should people be given inducements?

Sometimes it may be necessary to give inducements to people to attend extension activities. This would be particularly at the beginning of a program to introduce the program and the staff; and to meet and learn the reaction of the farmers to what was planned. However the aim should be to earn credibility so that the people want to see what we are doing, and are interested in trying out methods we suggest because they have confidence in us. Obviously this means that there should in most cases be longer commitment to the agriculturally related needs of an area, than say in water programs or even health work. The treatment of illness, vaccination programs, improving infant nutrition etc can achieve credible results in a comparatively short time. The treatment of worms or TB in people is the same in UK or Taiwan, Nepal or any other country.

Some programs pay people to attend meetings and give out T shirts etc, this writer does not believe this should be the normal situation. If after a time the credibility of the work is still so low that people have to be bribed to attend meetings, it is probable that the extension work is not going to be effective anyway. It could be that the open meetings were being held before the methods had been properly tried out; or under conditions which were better than those of the ordinary farmers of the area, and so the farmers did not consider the trials really relevant to them. If inducements are always needed there should be a careful consideration of why? Farmers are wary of institutions which seem to have too much money. They ask 'why are they using so much money, they must be getting something out of it'. There will be likelihood of mistrust of the motives and so mistrust of the message.

Six years ago I was invited to help improve the livelihood of people in an area where the people were selling their daughters for prostitution. There was potential to produce fruit, vegetables, poultry products and milk to tourist areas nearby. As it takes time for fruit trees to mature I suggested planting a range of trees which would probably be suitable so that the people could discover which were best so that they could then plant more of those kinds and varieties. I asked the families concerned which kinds they would like. The people were eager to have the tree seedlings but when I said that there would be a subsidised very nominal price they lost interest. Other development organisations had paid them to do anything. Similarly when I gave a vegetable training they were angry that they were not paid to attend, and they lost interest in vegetable growing. So called 'development' in the past had spoilt them from real development. It had taught them to be beggars rather than helping them to help themselves.

B6.1o Careful planning

There should be maximum appropriate participation of the target people at each stage. Without the active involvement of the communities involved there is little or no real commitment. However, the degree of participation appropriate at different stages will vary with each situation---.

In order to be able to achieve the specified objectives of the project, we need to establish what are the real long, medium and short term needs of the target people and their environment. The key major objective must be to help the target people to help themselves to a sustainably better future.

For this future we must help them with systems which are appropriate to local circumstances, considering such things as their constraints of soil, climate, economic, social and religious factors.

These systems must be practical, integrated , and sustainable. The requirement is for development to be an enabling process that will be continued by the community after the project is completed. They must also take account of both the obvious, and, the subtle relationships, between living things including people, and their environment.

Minor objectives are those which if they prove inappropriate or no longer needed can finish. Sometimes experience may show that a minor objective even though perhaps a felt need of the people, is not feasible, or, is not sustainable and so dependent on unreliable outside funding, minor objectives may be such as prepare the way for the major objectives but are not ongoing.

It may be appropriate to set intermediate objectives and so it is important that the objectives of the program are specific as to whether major, intermediate or minor. If the objectives are only stated in a general way, more like an aim, for example to reduce soil loss, it is not clear exactly what should be accomplished and so the programs cannot be evaluated properly and modified (changed) if needed to ensure that the specified objectives are met---.

For effective evaluation and modification we need to have a means to compare the situation before and after a particular set of events. The evaluation should involve the people in each case. It should be the target community who evaluate, it is the work of the project - target people, the project staff, and supporters, - to facilitate. (To make it possible).[175.]

B6.2 Methods and approaches

B6.2a Attitudes

Too much of what is supposed to be encouraging extension work results in further demoralising the farmers.

One example is the extension worker who goes to visit farmers in smart clothes, perhaps with an umbrella and dark glasses to protect him or her from the sun. A person who obviously does not intend to get dirty if he or she can possibly avoid it. The dirty farmer in his work clothes feels inferior, and feels he is in a dirty, inferior job. This will be further emphasised if as often happens the extension worker spends quite a lot of time in the local tea shop, and generally seems to take life easily; while the farmer has to sweat in the sun and rain.

Another example is the use of an affected smart accent, and a condescending attitude to the farmers.

It may also be the agency itself has a wrong attitude to the farmers.

In Britain during the second world war the government appointed people as ministry officials who had power to tell farmers what they should do and what they should plant. The farmers resented this particularly as many of the officials were people who had failed as farmers, or were incapable of being farmers, yet were telling practising farmers what to do. Many mistakes were made such as ordering farmers to plough good pasture and meadowland and plant cereals in areas of high rainfall, where it was too wet to harvest cereal crops. As soon as the war was over the resentment was shown, farmers did not want the advice of the ministry men even if it was good. In some cases farmers even pointed guns at ministry men and told them to get off their land.

Similar resentment will be found elsewhere when the extension people are people who would not be able to farm themselves, yet feel that they know better than farmers who can. Or if they can blackmail the farmers by blocking or delaying needed inputs or the use of equipment such as sprayers.

When extension agents show respect to the farmers, the farmers are much more likely to respect the extension people, particularly if the ideas the extension agents suggest have credibility. The extension people must realise that the key people are the farmers, not themselves.

If the farmers self respect and confidence is built up, they are more likely to try different methods. If as often happens the reverse occurs, the more intelligent and progressive young people will leave farming, and improvements will be even harder to bring about.

Respect for the farmers includes not teaching alternative methods until they are well enough proven. As an example of what can happen; a new extension person who was a good communicator told the farmers in his area that they should not be growing millet; sorghum which was not grown would give much higher yields and tasted better. The villagers grew a small amount the first year and liked it, and the next year almost all their land was planted with sorghum. Unfortunately a small insect, the sorghum midge, arrived in the area and chewed the sorghum flowers so that few grains developed. The quantity harvested was about the same as the amount sown. The people had little other food except for some old millet which kept well and never had serious problems, but they needed that for seed; there was starvation!

B6.2b Credibility

It is both foolish and arrogant to believe that all agriculturally related practices are suitable for all conditions, yet government and other extension agencies often act as if they are. An example is that the wheat and rice research institutes in Nepal are on the plains at about 450 ft above sea level. They would give recommendations for varieties to be grown in the Nepali hill areas which comprise about 68% of the total land area and have average height of 4,500 feet above sea level; they also have different soil and rainfall to the plains.

Wise farmers are not going to believe the credibility of advice unless they feel reasonably confident that it will be suitable under their own conditions; yet, when farmers ignore such advice the advisors often describe such people as being stupid and stubborn.

There should be local trials carried out under normal conditions with no better treatment than normal farmers could use.

The author knows of one extension program which encouraged a new variety of maize. They had a demonstration comparing the local maize with the improved variety. To make sure that the results were what they wanted the new seed was pre germinated. Also, fertiliser was secretly applied to it at night but not to the local variety. The farmers found out and all credibility was lost.

The trial should be carried out in a public place where all the management and the mistakes and problems can be seen, as well as how the problems are overcome, or other methods tried. When the new ways are convincing, farmers are encouraged to try on a small scale, then on a larger scale. Once farmers have success it is better if interested farmers are directed to the local farmers who use the technique. Farmers will generally learn better from other farmers than from non farmers. Of course if the extension worker is himself a farmer his work will be much more effective.

Extension agencies usually work with leader farmers, but these are usually wealthier farmers who can afford to take risks. They have better soil and better access to timely inputs, and advice, than the ordinary farmers. They may have other advantages such as a better education, or more influence where it matters. While it is often more difficult to work with ordinary farmers and we may need to give some assurance of compensation in case of genuine failure, it is much more effective. While the new method may work with the best farmer in the village, it does not always follow that it will work under the poorer conditions of the average farmer. On the other hand if it works for the poorest farmer in the village it will work for everybody and is more likely to be taken up.

B6.2c Giving the farmers a voice, and participation from the start

In Taiwan there are two agricultural extension systems, the combination seems to work well. One is government controlled, the other is mainly farmer controlled, although the government has some say in it. The first is a typical government system with research institutions and county livestock improvement stations which are linked with the research institutions but also have demonstrations, some trials and extension staff.

The Farmers association is funded by a levy on agricultural produce.

Each local association has a cooperative for the purchase of agricultural inputs at wholesale prices. Some also sell articles of clothing soap and other necessities. They have their own extension workers and may have their own trials and demonstration areas. The extension workers are appointed, and the direction of the extension work and research work decided by the local farmers, or their appointed representatives. Their extension workers have a different attitude towards the farmers, this in turn brings about a better attitude to the extension workers by the farmers. The farmers may decide that a farmer will make a better extension worker and so may appoint and pay the farmer. If he proves to be good he may be sent away for extra training or to see other areas to gain ideas for possible use in his own area. This seems to bring about a much more dynamic situation and the government of Taiwan has stated that while the governments research and development has played a large part, the rapid development of agriculture has been mainly because of the work of the farmers cooperatives. I think a similar two sided approach would be useful in other countries.

B6.2d If possible avoid alienating any group

We should try to be on good terms with all the people. For instance, in a Hindu society working with the high caste people or a progressive group is likely to cut us off from the low caste or the less able group. If we work with the less able group the situation is reversed. The same effect could occur if working with one tribe or group; another group could be resentful and cut water pipes or otherwise cause problems. We may have to work with all groups to some degree in order to help the most needy.

B6.2e Meeting felt/perceived needs

At least some of what we do should be what the people want, even if we see that other needs are really vital for long term survival. Trying to meet the felt needs of a farming community is very important, but care needs to be taken.

a). Care is needed to know if the apparent felt needs of the community, are the real needs; and not just the needs of the 'local vocals'. The local ruling group or the "educated"!

b). Are all the felt needs, in the real interest of the community ?

The community, or members of it, may be convinced that bringing in a new crop will be a good thing. It could in fact be a tragedy. We can become quickly popular by meeting needs and leave in a blaze of glory; later someone else has to pick up the pieces.

c). When a village community has no knowledge of alternatives they will not mention those alternatives as felt needs. There may well be a need to make trials of potential felt needs.

Potential felt needs may also be addressed if we proceed in a quiet, friendly, humble way. For instance in the area of Nepal in which I worked, while farmers recognised that both soil and soil quality was diminishing and the future looked very uncertain or worse, soil conservation was not expressed as a need. When asked why? they said that there was nothing to be done about it. Fertiliser could be applied, if it could be bought; it would help for a while, but in the end it made the situation worse.

It was not until we started transforming poor land into fertile land that there was belief that anything worthwhile could be done.

When after trial and error we find a better way, we can encourage our friends to try and see if they can improve on our results. Then they become the experts and others learn from them. This is non threatening and more easily acceptable. (See also giving the farmers a voice B6.2c.)

B6.2f Local trials

It is often said that extension is not working, the farmers are not interested in change. The extension workers do not tell of alternative methods because they feel it is a waste of time.

I would suggest **firstly**, perhaps it is because there is no facility for the extension worker to try out the new practice, under similar constraints and conditions to those of small farmers. **Secondly**, it could quite possibly be because the extension worker couldn't use the new practice himself. Too many extension workers can only tell what they have been told, they are not very good at *doing*. That is more likely why the new technology is not demonstrated. It may be a good idea to employ a local farmer to demonstrate the new or alternative method.

Thirdly, too little or no attention is given to the vital credibilities needed before wise small farmers try new practices, unless they are bribed into trying them.

1st Needed Credibility. Will it work consistently in conditions like mine ?

2nd. Could I make it work in my conditions ?

3rd. If I switch to the new technique, will I lose, financially, socially, or in other ways ?

Unless those criteria are met it would not be wise for farmers to change methods. I believe that there is need for much more respect to be shown to small farmers. We must learn to be small farmers ourselves, working under similar constraints to theirs. Though not dependant for our livelihood on how we succeed; our real credibility to advise them, will depend on whether we can succeed. We should earn credibility. We should aim to build up the farmers self respect. To work with them, and show respect for them. Few of us could survive given the constraints they have.

B6.2g The importance of showing we care for the real interests of the people and not just for filling government quotas or gaining promotion.

Several years ago at an international seminar the Ghanaian Minister of Agriculture told of the way agricultural missionaries were more effective at extension than the government extension staff. He said that they first asked the farmers what was the most important need for them. The farmers replied that "it was to grow enough of the staple crop to feed their families, otherwise we may starve". The agricultural missionaries researched ways which improved the yields of the staple crop and the people were very thankful. Because they knew that the missionaries cared for them and had helped them the people asked "now we have spare land what should we plant? The agricultural missionaries advised them that the cocoa crop was now a good crop to grow and was profitable so they should plant some, so the people planted cacao trees.

The government workers had been saying that all along, but the people did not believe that the extension workers were really interested in their welfare, or that the government would feed them if the cocoa crop failed. How would they fed their families? The result was that where the mission worked, more cocoa was grown than where the government extension workers were urging the people to grow the crop. When the farmers realise that we are on their side, that we do not have a hidden reason for encouraging them to do things differently the situation normally improves.

B6.2h Helping with cooperatives can lead to a change of attitude and general improvement programs.

An example of this occurred in Nepal. As there was little work for the men of a village they had to spend most of the time away seeking work. Jobs which had been the responsibility of the men were not being done. The path to the funeral area was in bad repair and eventually a funeral party slipped and they and the dead body slid down the hillside. This was a great disgrace and so the women decided to work together to repair the path. This was successful and they decided on other work which should be done, a government health worker helped them to set up a cooperative. Next they decided that instead of selling their goats to the merchant who came to the village they would send the goats with some of their members to Kathmandu and sell them direct. They were so pleased with the result that they put some of the extra money they earned into having a literacy class. Eventually they hired an extension worker to help improve their agriculture. When people learn or relearn how to work together in one aspect of soil conservation, it can trigger more cooperative work on improving the soil and water management, or the other way round, first soil or water improvement then soil conservation.

B6.3 How to run extension trials/training/demonstrations

B6.3a A careful approach is needed

If the approach is wrong we will achieve at best little, and at worst a negative result; careful preparation is important.

We should involve the people from the point when it is decided to consider work in the area, not when we think we are going to introduce a program for them. Ideally the people realise that it could be of benefit to have the opportunity to consider other ways of doing things and ask for some sort of program. However care needs to be taken not to raise false expectations, development of the program should depend on the interest of the people. It may be that there is a need to start with a pilot project to try out some possibilities, but the people should be informed and involved. If there is interest in the results of the pilot project a program should be discussed with the people.

As mentioned earlier it could be that the local people realise that the soil is losing both fertility and soil but they do not ask for help in reversing the situation because they do not believe that anything can be done. In that case I believe we should say that we would like to carry out some trials to learn about the local conditions. We do not want to do anything secret so would like to use some land near a road or path so that anyone can see what is being tried. The trial work should be carried out using materials and methods which the local people have, or could have, and use.

The ideal is when local people see what we are doing and say they would like to try it. We then work with them on their land to adapt the method to their situation. Then we can refer other farmers to those farmers, build up the peoples self respect and involve people in helping other people to help themselves.

B6.3b Care needs to be taken when starting trials See also B3

One example of this is that of a farmer who was asked if he had a piece of land at the side of the path or road which was fairly uniform and so could be used to compare two different methods. "Yes", he said and showed a field which looked suitable. At harvest the new method performed better than the old method so it seemed that the trial and demonstration had been useful. Then the farmer said in public "oh well that part of the field is always the best"; so the whole project was a complete waste.

It is important to have previous records or that the farmer states publicly that the field is even. It is also important to check for yourself beforehand that one side is not near a place from which insects, disease, weather, or neighbours animals etc may affect that part more than the rest of the field.

A problem I experienced was when planning a trial to try out a method of rice growing which gave higher yields of better quality rice, while using less water. Originally one field was to be used, but then the farmer changed to another which seemed suitable. However he had not said that the particular small field had been used as the threshing and winnowing floor, consequently it had masses of weed seeds which grew and choked the crop, when the weeds were removed the rice seedling were damaged ,and very quickly other weeds germinated to take the place of those removed. The result was that the trial was a failure the weed problem was just too great.

Another pitfall may be to ask people beforehand or part way through a trial what they think the result will be? The person is then committed and will not want to be proved wrong and so will look silly. Rather than that they may interfere with the trial in order not to lose face. An example of that was when a farmer secretly put extra fertiliser on the part of the trial he had said would be best, the traditional method. As that made that part more affected by disease the introduced method still did better, but the farmer would not admit it. Again the whole project was a waste, in fact it had a negative result as neither the farmer nor the agency wanted to have a trial again, the relationship had been spoilt.

The author had a policy of encouraging soil conservation trials. In each area in the first year the farmer would supply half the labour, counting a ploughing team as equivalent to three men; the project would supply the other half. The second year if the method had been shown to be successful and more farmers wanted to start it, the farmers would need to supply more than half the labour. This started well, a ploughing team and two men could make an average of 10 metres per hour of properly constructed 2 metres wide terrace, a perfectly viable system.

The next year we had training in soil conservation on cooperating farmers land in other areas, trainees came from other areas. *Sadly other people than the author were directing the work.* Although those in charge said that they had arranged that the owner farmers supplied a ploughing team to loosen the soil and helped with the work; in fact the farmers either did nothing but supply tea, or practically nothing but watch the trainees doing the work. The correct way was that the soil was ploughed and the men moved the loose soil, they did not have to dig it up as well.

The farmers made excuses that the bullocks were not available or that they were not needed as there were so many people to do the digging. The result was that the training was of negative value. People agreed that terraces had been built but that it took too much labour, even though I explained that it did not need so much labour, what they had experienced was that it did. I heard later that the people rejected the method as needing too much labour. Rather than the organisers putting on a properly organised demonstration, and so correcting the mistaken view, a great opportunity had been lost. A careless approach to setting up the demonstration had spoilt what could have made a great impact for good. I am told that because of those failures the good method has been abandoned, and instead making contour strips is now being taught in those areas; even though the organisers know that contour strips are not really good for the local conditions. The method was blamed rather than the managers.

B6.3c Selecting workers

One of the common problems with extension programs is that the extension agents are often from the dominant groups themselves or they are dominated by the dominant groups.

B6.3d Wrong priorities

When as is common, village meetings are dominated by the wealthy and/or the more powerful people, the extension activities normally reflect the interests of the wealthy. The result is that the wealthy or powerful become more successful and the gap between them and the ordinary farmers is increased. The debt problem of the poor increases and so more of the land moves into the hands of the rich. The poor may still farm the land but it is owned by someone else who receives most or all of the profit. There is no incentive for the farmer to farm better and so the soil deteriorates. The farmer puts in less effort knowing that the harder he works the more goes to the landlord, if he increases the yield the rent will be increased. The result of such extension is exploitation rather than care of the land, it is better to help all groups so that the farmers can keep and care for their land. Helping the poorer less vocal farmers may not be welcomed by the powerful but in the long term it is better for all the population.

B6.4 "Six ingredients of a sustainable soil and water conservation project

- ❏ Suitable systems- to combat erosion, there are many systems to choose from but not all will suit each site.
- ❏ Training and motivation- villagers are more likely to become involved in better management of soil and water where the methods are shown to work, they are then interested in training which helps them understand about loss of quality and quantity of soil and water, and how to control it.
- ❏ Use existing institutions- where traditional groups exist, working through them can greatly enhance the progress of a project.
- ❏ Flexibility- projects may push activities that local people do not want. Staff must listen to local people, and respond by changing how the project works.
- ❏ Traditional systems- these techniques are an important starting point for any project's work and can often be greatly improved.
- ❏ Village land use management- an integrated approach to village resource management is essential if individual conservation activities are to have any long term benefits" [107].

B6.5 Some final points

B6.5a Looking to The Future

Farmers' continued interest in being farmers rather than moving to other possible occupations depends on many factors. Even if the farmer has planted fruit trees on land too steep for sustainable crop production he will not be able to continue unless he can get an economic price for the fruit.

Decline of the lands productivity or other disincentives such as crop theft, or his altered way of farming spoiling relationships with neighbours, can tip the balance towards abandonment of the better methods and so sooner or later migration.

B6.5b Improving land

The good farmers philosophy in any country is to hand the farm over to his son in at least as good a condition as he found it, unless life is so unpleasant and precarious in farming that the children only want to get away. Marketing cooperatives to give more income to the growers; and organisations such as 4H and young farmers clubs, may be needed to provide social and educational facilities, and to encourage go ahead young people to stay in the rural communities

Good farmers who farm their own land will try to improve it, farmers farming land which they cannot hand on to their family do not have this incentive, their incentive will be to get as much as possible from the land for the least input. Unless closely controlled by the owner of the land, the land loses soil fertility and texture.

For this reason **there needs to be a fair system of tenant rights, fair compensation for improvements made when a farmers lease ends or more security of tenure so that if a farmer invests in soil or water conservation or soil improvement he will be able to stay on that land for a long enough period to be compensated for the investment of time or other factors he has put into the improvements. Without this soil will continue to deteriorate, so this situation deserves close attention by those concerned that there be sustainable agriculture.**

The land to the tiller program made a huge difference to agriculture in Taiwan. The rents were first reduced to a reasonable percentage of the crop, then the government bought land from the landowners with shares in government owned enterprises. The land was then leased to the tenants, but belonged to the tenants after 10 years. The landowners were encouraged to be involved in the running of the government enterprises which they owned shares in and that improved the efficiency, of those enterprises, and so everyone was happy. It is a system that if modified to fit individual situations might help result in a more sustainable agriculture.

B6.5c The capacity of the target people to adapt to changing circumstances by adopting methods new to them will vary greatly

People who have always been cheated by outsiders will be reluctant to believe and act on suggestions from outside. They will be reluctant to take up introduced ideas unless there has been very convincing evidence of the benefits. People who walk a narrow path where survival is not assured will be reluctant to take risks. People who do not know whether they will have sufficient resources to live from this year to next will not be as interested in the longer term future, as, how to survive in the present?

Too much development work ignores such history which effects peoples reactions to change. *There must be more appreciation and understanding of the need to build confidence in the communities themselves, and credibility that things can change for the better. That the programs and program staff have not come to cheat or abuse them, but have their interest at heart.*

For much more on the subject of rural development see the soon to be published book 'Quality Rural Development' by this author.

TO CLOSE

B7.1 The importance of good soil management
B7.2 Is genetic engineering the answer to food shortages?
B7.3 Water
B7.4 The future

B7.1 The importance of good soil management

The most important requirement for sustainable food production is that soil, a non renewable natural resource, the basis of food, is protected from degradation.

Derek Balinsky of the Zimbabwe Agricultural Society has I think rightly said:

Without doubt the real wealth and the greatest heritage of any nation is the soil.
It takes nature countless ages to produce productive agricultural soil. Conversely we have evidence that men can completely destroy the same productive soil in a matter of a few years, leaving it not only unsightly, but rendering it useless to both man and beast, neither of which can survive without soil. In essence, therefore the wanton destruction of the soil is a more serious crime against humanity than the most potent weapon used during wartime. Soils in the tropics are very young in terms of soil formation processes and low in organic matter. The maintenance of soil structure, therefore, is one of the most important factors to be considered in it's utilization.

A book entitled "More people less erosion: Environmental recovery in Kenya" by M Tiffen, M Mortimore and F Gichuki, John Wiley /ODI; describes a change in what was considered an environmental disaster.

Soil erosion had declined, new agricultural technologies had been introduced, and a predicted wood-fuel crisis came to nothing. A terraced and tree planted landscape has replaced eroded pastures and gullied arable plots- and there was an *eleven - fold* **increase** in food output per hectare---.
There were many contributing factors. Machakos district retained a strong social structure at village level which helped people to adapt to modern circumstances. Coping with unpredictable rainfall, small farmers adapted and invested in new methods- and their investment was especially crucial. Government support, research and extension played a positive role. Attention was paid to water conservation. The rate of population growth, although high, was not inimical to sustainable environmental management and did not exceed the capacity of the system to evolve.

Soil degradation and erosion are inter-related; the cumulative effect has a compounding result. The rate of increase of loss increases with time.

A) As erosion takes place, the most fertile elements and the most water retaining constituents, go first. This results in the protecting effect of plants being less, as the plants are poorer and thinner, and have a shorter growing period.

B) It also means that there is more moisture loss, which results in even poorer plant growth, and so less organic matter available to reduce soil fertility and texture degradation.

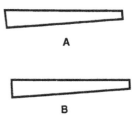

C) With the soil bare for longer periods the soil structure is spoilt, making it much more susceptible to erosion.

D) Fertiliser may be used to cover/lessen the effect of fertility loss but this only works in the short term. It increases acidity, which is a problem with most tropical soils. This in turn results in a fixation of plant nutrients in a form unavailable for the plants. At the same time, it discourages microorganisms which would release the plant nutrients. Furthermore it spoils soil structure and makes the soil harder to cultivate, less friendly to plant roots and more easily eroded.

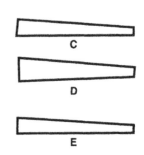

E) Reduced protective effect given by growing plants from Sun, Rain, and Wind:

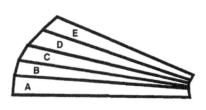

The compound result is deterioration of all aspects needed for sustainable agriculture.

The people who would be needed to work to improve the situation being absent or malnourished, with reduced energy, time, skill, intelligence and confidence.

In all a vicious cycle. The situation in so many places is rushing towards catastrophe. Indeed in some areas it has already reached catastrophic proportions. As someone has said "it has passed the threshold beyond which limiting factors become so severely operative that recovery, in periods meaningful in the human time scale, becomes impossible."[2]

B7.2 Is genetic engineering the answer to food shortages?

There is now much optimism about new varieties and new crops through genetic engineering. We are told that it is going to save the future by producing super plants which have all sorts of qualities built in, high yield, disease resistance etc. It sounds hopeful but we need to think about it. Suppose a new high yielding variety of cassava produces three times the yield of traditional varieties. Surely that means that it is taking more from the soil than the traditional varieties; so how long will it be possible to maintain the soil fertility. Hill farmers in Taiwan were convinced that cassava exhausted the soil and so if possible only grew it when they particularly needed money and only grow once in 7 - 10 years.

We need to remember the saying "You can't both have your cake and eat it." Unless we can reverse loss of soil and loss of soil fertility the genetic engineering will only bring about a short lived increase by further mining the soil; followed by worn out soil, famine, disease and war, as desperate people use desperate measures.

High yielding varieties still need soil, water and good soil even more than indigenous varieties. No crops can grow in subsoil or without water.

B7.3 Water

The other great need is unpolluted water. More and more water is polluted by chemicals or in other ways made unfit for drinking. In "The State of the World 2000" published by the Worldwatch Institute Sandra Postel, Director of the Global Water Policy Project draws attention to the serious problem of falling water tables.

Of all the vulnerabilities characterising irrigated agriculture today, none looms larger than the depletion of underground aquifers. Across large areas, farmers are pumping groundwater faster than nature is replenishing it, causing a steady drop in water tables---. If so much of irrigated agriculture is operating under water deficits now where are farmers going to find the additional water that will be needed to feed the more than two billion people projected to join humanity's ranks by 2030?[165]

B7.4 The future

It is our children's future we are building now.

A Kikuyu proverb should speak to all of us.

God will not ask your race, nor your sex, nor your birth. But he will say to you: What have you done with the land I gave to you ?

The experience from which the book is based was gained while I was an agricultural missionary, from 1963-1990. And then 1994-1996 with Voluntary Service Overseas, I am not ashamed of having been an agricultural missionary, only that there are not more of us.

Contrary to a common misconception the Judeo Christian tradition does not only say that mankind should "have many children, so that your descendants will live all over the earth and bring it under their control" Genesis 1.28; some have wrongly thought that meant the earth is ours to do what we like with it.

God also said we are "to cultivate it and guard it" Genesis 2.15

> God created the world a wonderful place. He created people. He wants us to live in harmony with himself, with his creation and with each other, we are not doing that.
> We are first to love God, and that includes what is his. The second most important commandment is to love our neighbours, including other nationalities and people we disagree with; as ourselves. That is the only way to a better and a sustainable world.

It is of no use trying to act as gods, we are not. Whether we believe God created the world or nature happened to develop a mostly fertile planet; we have a responsibility to care for it for future generations, or be cursed by them. To be a blessing or to be a curse? To act responsibly or turn our backs on the future?

What should we choose? What will we choose? NOT TO DECIDE - IS TO DECIDE!

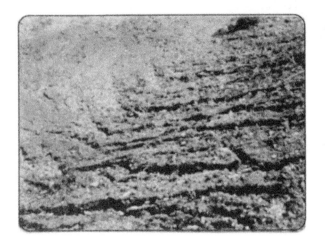

Unless we can heal the wounds and restore it's health the bleeding will increase. When the bleeding stops the earth will be dead, but before it dies the people who caused the wounds which killed it will themselves have died.

The earth is bleeding! It's blood is brown.

The future is flowing away

We are all contributing to the coming disaster by our greed, selfishness and short-sightedness. The time to act is now!

ASSESSING SOIL MOISTURE CONTENT

General relationship between soil moisture and the feel and appearance of the soil.

FEEL   OR   APPEARANCE

Moisture between wilting point and field capacity	Coarse soil	Light soil	Medium soil	Heavy and very heavy
0	Dry, loose, single-grained, flows through fingers.	Dry, loose, flows through fingers.	Dry, sometimes slightly crusted but easily breaks down to powdery condition.	Hard, baked, cracked sometimes has loose crumbs on surface.
50% or less	Appears dry, will not form ball * with pressure.	Appears dry, will not form ball.	Somewhat crumbly, holds together from pressure.	Somewhat pliable, balls under pressure.
50-75%	Same as coarse under 50% or less.	Tends to ball under pressure but seldom holds together.	Forms ball, somewhat plastic, will sometimes slick ** slightly with pressure.	Forms ball, ribbons out between thumb thumb and forefinger.
75%-field capacity	Tends to stick together.	Forms weak ball, breaks easily, will not slick.	Forms ball, very pliable, slicks readily if high in clay.	Easily ribbons out between fingers, has slick feeling.
At field capacity	Wet outline of ball is left on hand upon squeezing.	Same as coarse.	Same as coarse.	Same as coarse.
Above field capacity	Free water appears when soil is bounced in hand.	Free water released with kneading.	Can squeeze out free water.	Puddles and free water forms on surface.

*Ball is formed by squeezing a handful of soil very firmly. ** Slick means having a smooth surface, or forming a smooth surface.

Prepared by Max Jensen and Claude H Pair. In [52 p719].

GLOSSARY/DICTIONARY OF TERMS, WORDS, AND PARTS OF WORDS WHICH HELP US TO UNDERSTAND THE MEANING OF THE WORDS

This glossary will not only be useful for this book, it is prepared to be useful to people in third world countries who often do not have the resources, or time, to search for definitions of terms found in books and articles related to soil. It should enable easier translation of books and articles by non specialist translators, so speeding the spread of information on this important subject. I would have found it useful, I believe and hope that many of my readers will.

NOTE:The names underlined are the names of soils, there are many other names for soils other than those listed here. I do not think this is the place for more than those listed. I do think it is useful to have a list of the sources of the names of soils, and some other scientific names and terms used, as this knowledge is helpful in understanding the terminology of the subject. It is not easy for the field worker to discover these meanings and so much of what he sees written will be more difficult to understand than it need be. When we see we have no idea what it relates to. However when for instance we know that ferr is from latin ferrum, iron, so ferr or fer indicates the word has something to do with iron; and when we know that al comes from Latin for aluminium., it is easy to understand that Ferralsols are soils containing iron and aluminium. When we know that morph comes from the greek morphe and means form, body, it is easy to see and to remember that morphology is related to form or structure. For that reason these definitions are included in this glossary.

A Horizon, the top layer of undisturbed soil, see also horizon.

Acidity, soil acidity represents the excess of hydrogen ions (H+) over hydroxyl ions (OH) in the soil solution. Soil acidity is expressed (described) in terms of pH units by which pH 7 represents a neutral soil. The lower is the pH number of a soil than pH 7, the more acid that soil is. The higher it is above pH 7 the more alkali the soil. Plants vary in the pH they like best but on average the best pH is about pH 6.5.See also base.

Acr, from Greek akros, at the end, very much weathered.

Acre, measurement of land, 4840 square yards, about 0.4 hectare (0.404 of a hectare.)

Acri from Latin acris, very acid.

Acrisols, (from Latin acris, very acid). Soils having an argillic B horizon and of low base saturation.

Actinomycetales, see nitrogen fixing.

Actinomycetes, a group of soil micro-organisms which produce an extensive threadlike network. They resemble moulds in some respects but are more like bacteria in size.

Adsorb, see adsorption.

Adsorption, the attachment of compounds or ionic parts of salts to a surface or another phase. Nutrients in solution (ions) carrying a positive charge become attached to (adsorbed by) negatively charged soil particles.

Aeration, in soil, the exchange of air in the soil with air from the atmosphere. The composition of the air in a well aerated soil is similar to that in the atmosphere; in a poorly aerated soil, the air is considerably higher in carbon di-oxide and lower in oxygen than the atmosphere above the soil.

Aeric, browner and better aerated soil.

Aerobic, conditions with oxygen as part of the environment, living organisms which need to have air in order to live.

Aggregate, a collection of individual particles into clumps or structural shapes.

Aggregates of soil, Fine soil particles held in a single mass or piece of soil such as a small clod, or a clump, or a crumb of soil.

Al, from Latin aluminium.

Albi, bleached, from Latin adjective albus, white.

Albic, bleached, (from Latin albus adj, white).

Alfisols, soils of medium age, "formed under a hardwood forest become alfisols. "[16] With grey to brown surface horizon, medium to high base supply and subsurface horizons of clay accumulation, usually moist but may be dry during the warm season.

Alkali soil, soil with a pH of higher than 7, particularly if more than 8.5. see pH and base.

Allophane, "comprises amorphous hydrated aluminium silicates of rather variable composition as the first weathered product of some kinds of fresh volcanic material." [18]

Alluvial, deposited by streams, surface water.

Amendment, any material, such as lime, straw, manure, which is added to soil to make it more productive. Fertilizer is also an amendment but the term is not normally used for it.

Ammonification, the formation by organisms of ammonium compounds from nitrogen containing organic materials.

Amorphous, from Greek not + morphe = form, without definite shape or structure, shapeless, without crystalline structure.

Anabaena, see note on Azolla.

Anaerobic, living forms which live where there is no air or free oxygen.

Ando from Japanese an - dark, and do - soil.

Andosols from Japanese an - dark, and do - soil; connotative of soils formed from materials rich in volcanic ash and commonly having a dark surface horizon. Weakly developed soils, rich in allophane ("comprises amorphous hydrated aluminium silicates of rather variable composition as the first weathering product of some kinds of fresh volcanic material."][18] and having a low bulk density.

Anion, a negatively charged ion; when the element is written a negative (-) sign is added to it.

Annual, a plant that lives for only one year. Each year new plants must grow from seed.

Antagonistic crops or plants, do not grow well together.

Antibiotic, from Greek anti, opposite, against, instead, rivalling, reverse of in exchange; and Greek bios, life. So the meaning is against life, substances used against bacterial infection.

Aqu, from Latin aqua, water.

Aquifer, a water bearing material in the ground through which water moves more easily than in the surrounding materials. Water springs run from Aquifers.

Aquults, from Latin aqua, water, seasonally saturated with water.

Ar, from Latin arare, to plough.

Areno, from Latin arena, sand.

Arenosols, from Latin arena, sand. Strongly weathered sandy soils of tropical and subtropical areas.

Argil, from Latin argilla, clay.

Argilic, from Latin argilla clay, soils with a horizon of clay accumulation.

Arid, see arido.

Arido, from Latin aridus, dry, parched.

Aridosols, from Latin aridus, dry parched. Soils of medium age with separate horizons, low in organic matter, and dry for more than six months of the year in all horizons.

Auxins, organic substances which promote growth in plants

Azolla, from Greek azote, former name of nitrogen. "A tiny water borne fern, used as a nitrogen rich green manure in rice production. It has a nitrogen fixing blue green alga Anabaena axollae in its fronds and can fix 100-150 kg N/hectare/year." [22 P13]

"**Azospirilium brasilense**, from Brazil is a bacterium which has been tried in India and has shown benefit in increased yield from inoculation on wheat, rice, sorghum, maize, barley and oats. Saving on the application of inorganic nitrogenous fertilizers from 20-30 kg N/ha". [22 P12]

Azotobacter Chroococcum, a soil micro-organism. "Inoculation of rice, cabbage and eggplant (aubergine = Solanum melongena) with this organism gave significant increases in yield."[22 P11]

B horizon, the horizon or layer of soil just below the top layer of undisturbed soil.

Bacillus megatherium var. phosphaticum converts bound phosphates into forms which are easily assimilated by plants Gerretsen 1948. Sperber 1957. Sethi and Subba Rao 1968. [22 p11]

Bacteria, from Greek bakterion, stick or rod, kinds of one celled organisms which group together in various shaped rods. They are classified according to the shape that they group together in. 1. Cocci; from Greek kokkos a grain, are individually round or oval like grain, eg streptococci. Strepto from Greek strepho, turn. These group into what look like rows of round beads. Staphylococci which grow in clusters. 2. Bacilli; rod-like or rather cylindrical, straight or slightly curved. 3. Vibrios; rigid, curved rods usually with a flagellum (whip) at one or both ends like a tail. 4. Spirilla; rod-like curved in a spiral with tufts of flagella at the ends. 5. Spirochaetae. spiral or wavy without flagellum.(Based on Hutchinsons 20th Century encyclopaedia 1961 definition.)

Base, a substance which reacts with an acid to form a salt and water; a positive ion usually as calcium in soils that acts as a plant nutrient and neutralizer of soil acidity. Any excess of base over and above that needed to neutralise the soil (See acidity) is present in the form of carbonates, mainly calcium carbonate. By determination of base- status we understand any determination which has the object of answering the questions

1. Does the soil need lime? 2. If it does how much does it need? Base- status and pH are directly related. pH can be described as the power of the Hydrogen ion.

The relation of hydrogen-ion concentration to pH is seen in the following table:-

HYDROGEN-ION CONCENTRATION	pH
0.1	1.0
0.01	2.0
0.001	3.0
0.0001	4.0

Neutrality is represented by pH 7, pH values less than 7 represent acidity and higher than 7 represent alkalinity.

A high base status is alkaline and a low base status is acid.

Bedrock, the underlying solid rock, which may be exposed.

Bench terraces, see terraces.

Benefit/cost ratio, the relationship between the benefit of say terracing and its cost spread over a defined period of time.

Biennial, a plant that lives for only two years.

Bloat, this is a condition of cattle sheep and goats and other animals with four stomachs (ruminants). It can occur for two reasons. Either 1) because the animal has swallowed something like a plastic bag or string which blocks the exit of one of the stomachs usually the first or second stomach. Or, 2) because the animal has eaten food which has a bad chemical action in the stomachs. Some plants can stop the animal from belching up gasses which form in the stomach. Other plants have the affect of making gassy bubbles which stop the gasses from getting out. The result is that the stomachs particularly the fist stomach blows up and pressure on the lungs and heart can cause the animal to die.

"**Blue green algae**, there are many species of these water borne algae, they have the ability to photosynthesize and also biological nitrogen fixation. They also excrete vitamin B12, auxins, and ascorbic acid, which may also contribute to the growth of rice plants." [22 p113.]

Bog, wet spongy soil.

Bor, from Greek boreas, northern, cool.

Broad base terraces, terraces which have a gentler slope than the area had before terracing but are not level.

Broadcasting, spreading seeds by throwing by hand or by machine.

Buffer, Buffering, substances in the soil which have a neutralising effect against changes in acidity.

Buffer strips, of erosion resisting plants which stop or slow down the force of water or wind which would otherwise carry away soil. They also increase the absorption of water into the soil.

Bund, a bund is a marked strip of land, contour means horizontal/level lines, so a grass bund is a level strip of grass crossing a piece of land, normally about two feet wide.

C horizon, the third layer or type of undisturbed soil counting from the top or surface layer.

Calcareous soil, a soil containing enough calcium carbonate to fizz when treated with dilute hydrochloric acid.

Calci from Latin calx, lime.

Calcic from Latin calx- lime. Soil containing lime/chalk calcium Ca CO_3

Cambi from Latin cambiare, change.

Cambisols from Latin cambiare, change; connotative of soils in which changes in colour, structure and consistence have taken place as a result of weathering eg loss of Ca CO$_3$. Soils formed by a weak alteration of the parent material.

Carbon - Nitrogen ratio. The ratio of the weight of organic carbon to the weight of total nitrogen in a soil or organic material. This ratio is important when making compost.

Catalyst, a substance which changes other bodies chemically without being changed itself.

Cation, a positively charged atom or ion in soil. When written a positive (+) sign is attached to it, eg K+.

CEC. See below.

Cation exchange capacity, a measure of the soils capacity to store nutrients. See also Ion and Base. Cation exchange capacity is a term now used more than it was before.

To explain simply, it is used where pH was often used in the past. It is more complex than just thinking of soil being more or less acid, and this being measured in terms of the pH of the soil, which can be increased if needed by adding lime; it gives a better understanding of the situation. For more information on this subject see appendix 10.

Chern from Russian chern, black.

Chernozems from Russian chern, black and zemlja, earth land. Soils of the grassland steppes showing strong accumulation of organic matter in the surface horizons and an accumulation of calcium carbonate at shallow depth.

Chinampa, this is a way of irrigating similar to that used with hill paddy fields but can be used for structures like swales, with a zig zag construction. Water is diverted into the upper end of a terrace or ditch, it flows along slowly, wetting the land as it goes. When it reaches the other end it flows down to the strip below in the opposite direction. When all the land is wetted any excess is diverted into a safe ditch or watercourse. This may carry it to another chinampa system or to a water soakage or storage area. [44 p480.]

Chrom from Greek chroma, colour, meaning soils which have relative purity or strength of colour.

Clay loam, soil that contains 27-40% clay.

Clay soil, contains 40% or more of clay less than 45% sand, and less than 40% of silt.

Clay pan, a compact slowly permeable layer of soil rich in clay and more or less different from the soil on top of it. They are hard when dry and soft when wet. Most plant roots do not like to penetrate them and so they are a barrier to growth.

Clay mineral, soil particles which are less than 0.002 mm. in diameter.

Clear fell, cut all the trees on an area at one time. see also strip fell.

Colloids from Greek kolla, glue. Soil colloids are very fine particles which though they may be suspended in water and carried by it, do not go into true solution and so will not pass fine filters. Compound particles of soil are held together by a kind of binding material consisting of humus and these finest clay particles, or as it is sometimes called the soil colloid.

Because the soil colloids act as anions in the soil it follows that they can attract and hold the mineral cations in their vicinity. The process is known as ION ADSORPTION. Were it not for this process of ion adsorption by the clay and organic colloids, mineral cations of the soil would be lost rapidly by leaching, as would also the soluble cations added in fertilizers---.It is a fact that the organic colloids (humus) have a high ion fixing capacity- about 5 or 6 times more than an equal weight of clay colloid, that has partly to do with density, or surface area in relation to weight. That portion of the ion fixing capacity of soils due to the organic colloids varies from 10% to 50% or more of the total. Thus it is obviously in the farmers interest to maintain the soil organic content in any farming system. " [21 p12.]

Compacted, pressed tightly or consolidated by pressure normally from above.

Companion crop, friendly crops, crops or plants which grow well together. They help each other in some way which may or may not easily be explained, and one at least of the crops assists the others in one or more of several ways in which different crops can help others. For example shading from too much sun, deterring harmful insects, or encouraging helpful insects. Providing substances which help the growth of the other plants, improving the soil, providing a frame for weaker plants to climb on. Some encourage soil antibiotics which destroy disease organisms in the soil which would otherwise infect the companion plants.

Compost, the product of composting.

Composting, this is the term applied to the rotting down of plant and animal remains in heaps or pits before the material is added to the soil. It has advantages over adding the material directly to the soil. The composted material can be a better balanced mixture of materials, and if kept at the correct moisture and air level will heat up, disease organisms and weed seeds killed, and the materials rot sooner than they would normally do in the soil. In addition micro-organisms can be added, and perhaps animal manure, and/or urine, or wood ashes added, all of which speed up and improve the breakdown into plant foods. The finished product can therefore be of better value than if waste organic material was applied directly to the soil. In addition when we apply organic matter to the soil directly this can have bad side effects, particularly if our soil has a low population of micro-organisms. The available organisms in the soil set to work on the slow job of breaking down (rotting) the raw organic matter. There is less effort spent on the later stage of releasing plant foods in the soil to a state where they are available to the plant roots, this results in sickly plants. When we apply well composted material, sufficient and well balanced plant food is soon available; the plants are more healthy and more resistant to attack by pests and diseases.

Confined, kept within limits.

Contour, an imaginary or actual level line connecting points of equal height on the surface of the ground.

Contour cultivation, cultivation carried out across the slope, level cultivation; not up and down the slope.

Creep, when related to soil. The downward mass movement of sloping soil; it occurs when the lower soil is very wet and soft. The movement is slow and irregular at first but may turn into a landslide if the water continues to wet the soil.

Crumb structure, very porous granular structure in soils.

Crust, a thin, brittle layer of hard soil that forms on the surface of some soils when they dry out following wetting.

Cry, from Greek Kyros, cold.

Culm, from Latin culmus, stem of plant usually grasses.

D Layer, any layer of material under the soil profile that is unlike the material from which the soil has been formed.

Debilitate, from Latin debilitatum = weak. To become feeble = not strong.

Decomposition, from French decomposer, separate into it's elements.

Deep percolation, a general term for the downward movement of water beyond the reach of plant roots.

Deflocculate, see flocculation, of which it is the opposite, to disperse, separate.

Degrade/degradation, to become of lower value, for example the change of one kind of soil to a more highly leached (less fertile) kind of soil.

Denitrification, the process by which nitrates or nitrites are changed to ammonia or free nitrogen by bacterial action and then escape into the air and are wasted.

Diffusing, spreading out.

Duff, the matted partly rotted organic surface layer of forested soils.

Dur, from Latin duros, hard.

Duripan, cemented.

Dust mulch, a dusty surface created by cultivation carried out with the idea of reducing the loss of soil moisture. Not a good practice.

Dystr, from Greek dys, ill; from dystrophic, infertile.

Ecology, from Greek oikos, house, a branch of biology dealing with living organisms' habits, modes (ways) of life, and particularly their relations with their surroundings.

Ecosystem, from ecology, describing the system formed by the relationship and interactions of organisms with their environment including with each other.

Eelworms, see nematodes.

Elecrolyte, a substance that conducts electric current through ionisation.

Elements, what things are made of.

Eluvial soils, are those which have lost material to other areas, not dissolved in the water but in suspension. The movement of the material can be downward or sideways according to the direction of the water which carries it.

"Endophytic fungi, known as vesicular-arbuscular (V-A) fungi, are common inhabitants of roots of several plants and help in the mobilization of phosphorus into plants. Such a type of beneficial association between a plant root and it's fungal associate is known as V-A Mycorrhiza". [22 p157]

Enti, from Greek entelkh, the becoming of potential.

Entisols, from Greek entelekh the becoming or being actual of what was potential; young soils which have not yet differentiated clear horizons, as they mature they develop into inceptisols. Their agricultural value and use depends on the parent material, if alluvial or volcanic origin they will be potentially fertile soils.

Enzymes, protein like organic substances produced by living cells which produce or speed up chemical changes. A plant may contain many hundreds of different enzymes that are active in different processes. Soil enzymes are derived mainly from soil micro-organisms.

Equilibrium, from Latin aequalis, even. A state of steadiness or balance.

Erode, from French eroder from Latin rodere, to gnaw away, destroy, gradually wear out.

Erodibility, from erode but a recent more precise term, it refers to the extent to which a soil is vulnerable to erosion, the ease of detachment of soil particles.

Erosion, from Erode see above, gnawing away, destroying, gradually wearing out. See also Erosivity and Erodibility.

Erosivity, from erode but refers to the intensity or effectiveness of the eroding agent, the ability to erode the soil. Heavier rain or large droplets would be more erosive than gentle or fine rain. Rain with wind would be more erosive. Rain falling on bare soil would be more erosive than rain falling on a crop.

Establishment, when related to crops, means the early growth period until the plants are well developed and can stand difficult conditions such as grazing or lack of rain reasonably well.

et al, from Latin et aliae/alia, and other people or things.

et sequens, from Latin and what follows.

Eutr, or Eu from Greek eutrophic, fertile.

Evapotranspiration, the loss of water from a soil by evaporation and plant transpiration (breathing out).

Excreted, put out.

Exist, from Latin existentia, continue to be.

Exotic, from Greek exoticos, exo-outside, ticos- come from or related to. = introduced from somewhere else.

Fallow, land rested from agricultural use for a period.

Ferr from Latin ferrum, iron.

Ferralsols, from Latin ferrum and aluminium; connotative of a high content of sesquioxides. Strongly weathered soils consisting mainly of kaolinite, quartz and hydrated oxides.

Fertility, the ability of the soil to supply plant foods in proper amounts for plant growth when weather and moisture levels are suitable.

Fertilizer, any natural or manufactured material added to the soil to add plant foods, it is normally only applied to mineral fertilizers, but not including lime or gypsum.

Fibr, from Latin fibra, fibre, least decomposed.

Field capacity, the amount of water a field can hold without excess water draining away; or the amount of water left after the excess has freely drained away.

Fixation in soil, the conversion of a soluble material from a soluble form in which it is available to the plant roots, to a form which is not available to plants. This process is sometimes called ionization.

Flocculation, Latin floccus tuft, to aggregate or clump together individual tiny soil particles, especially fine clay, into small groups or granules.

Fluvi, from Latin fluvius, river.

Fluvisols, from Latin fluvius, river; connotative of floodplains and alluvial deposits. Weakly developed soils from alluvial deposits in active floodplains.

Forage/fodder, food for grazing animals.

Frag, from Latin fragilis, brittle, presence of fragipan.

Fragipan, a dense impermeable subsoil layer.

Friable, from Latin friabilis - crumble, easily crumbled (broken into crumbs or very small pieces).

Fungi, form of plant life which does not contain chlorophyll and are unable to make their own food.

Gley, from Russian gley, mucky, dirty.

Gleysols, from Russian local name gley' meaning mucky soil mass; connotative of reduced or mottled layers resulting from an excess of water. Soils in which the hydromorphic processes are dominant.

Glossic, tongued eluvial and illuvial horizons.

Gravitational water in soils. The water in the large pores of the soil that drains away with good drainage. Well drained soils have this water only during and just after heavy rain or irrigation. In badly drained fields this water stays longer in the soil displacing air, as plant roots also need to breathe, plant growth is affected. Reeds, rushes and other plants of little or no agricultural value start to appear.

Greyzems, from grey and Russian zemlja, earth, land. Soils of the forest steppe transition, rich in organic matter and having a grey colour caused by white silica powder on structure faces.

Ground water, water below the water table.

Gypsic, containing Gypsum = Hydrated Calcium Sulphate, $CaSO_4 . 2 H_2O$.

Gypsum, see gypsic.

Ha, short for hectare = 10.000 square metres, about 2.5 acres. (2.471 acres).

Hal, from Greek hals, salt.

Halo, from Greek halos disc of sun or moon.

Hapl, from Greek haplous, simple, as minimum horizon.

Hardpan, a hardened or cemented layer of soil, it may be sandy or clayey and cemented by various chemicals in the soil. It may also be caused by ploughing or cultivating at the same depth and the ploughs compressing the layer they skid over. When this happens it is sometimes called a ploughpan.

Hem from Greek hemi, half, intermediate state.

Hista from Greek histos, tissue; therefore has organic matter.

Histasols, from Greek Histos, tissue. Organic soils. "If the parent material is highly organic, such as a filled-in bog or swamp, the soil that develops, regardless of climatic type and age, is called a histosol. Because of their high organic matter content, histasols can be highly productive, although they tend to be difficult to manage".[16]

Horizon, when used with soil we mean a layer of soil of similar description seen when we look at a vertical cut through the soil from the side.

Humu, from Latin humus, earth, but meaning rich in organic matter.

Humults, From Latin humus earth, soils with high or very high organic matter content.

Humus, is the end point of the microbial decomposition of plant and animal residues. Its value lies in its coagulating effect on the soil, it's water holding capacity and that it is a complex material which has biological, chemical and physical properties. It contains many of the nutrients which were used to build the plants which in turn built the animals before the respective products or by products were returned to the soil. It is a transition stage between one form of life and the next."It used to be thought of as dark coloured but it varies in colour. Some types are indeed very dark, but the humus of red tropical soils does not appear to be so dark in colour as the humus of temperate humid soils" [19 p142,]

I personally have a feeling that humus loses colour in the presence of UV light. At least one authority believes it is colourless.

Hydr, from Greek hydro,water.

Hydrophyte, from Greek hydro - water, and phuton - plant, so a plant which can tolerate wet conditions.

Hydrous, containing water,

Hyp, from Greek hypnon, moss, indicates the presence of hypnon moss.

Illuvial, soil which has received material which has been moved to it in suspension from other soil.

Immature soil, soil which does not have clear soil horizons because of the relatively short time for the soil to separate into layers.

Impair, to damage, weaken, spoil, reduce.

Impermeability, when a soil or other material does not allow penetration of water, air or plant roots.

Incepti, from Latin inceptio, beginning.

Inceptisols, from Latin inceptio, beginning, young soils. Soils that are usually moist with soil horizons showing alteration of the parent materials, not by accumulation. Their value for agriculture depends on the parent material.

Indigenous, from Latin indigena, be born, native, belonging naturally.

Ingested, taken in.

Inhibited, to hold in or back, to keep back, to restrain or check.

Inhibit, from Latin in, against, and Latin hibere/hibit, hold, to restrain, hinder, repress (keep down or under).

Inoculate, to add micro-organisms to soil or plants, this may be done by various methods. The simplest is to take soil which has the desired micro-organisms and add it/them to the soil which does not have it/them. Another way is to make a soup containing the micro-organism and wash seeds in the soup just before sowing. Another way is to add the micro-organisms to compost heaps well before the compost is applied to the soil; or we may add it to water and water on to the soil.

Inorganic, of mineral origin; where organic relates to substances produced by living organisms.

In situ, in place, formed in that place. As it is before being moved.

Interactions, act on each other.

Interface, between faces, a surface forming a common boundary. A meeting point or area of contact between objects, systems subjects etc. The connection or junction between two systems or two parts of the same system.

Intermittent, occurring at intervals, interrupted.

Internode, the part of a stem which is plain (without nodes) it is between the nodes.

Intertilled crop, one which requires cultivation during growth.

Inundation, being flooded or covered over.

Ion/Ions, molecules (tiny particles) of dissolved substances are separated into electrically charged *ions*.

If an atom loses one of its electrons there is no longer the right number to balance the positive charge of the nucleus. In this condition the atom is called a positive ion. If on the other hand, an atom gains an electron there are more than enough to balance the positive charge, and the atom is called a negative ion. The positively charged ions are called cations, these include K^+, Mg^{++}, Ca^{++}, Fe^{+++}, Mn^{++}, Zn^{++}, and Cu^{++}. The negatively charged ions are called anions, these include NO_3, $H_2PO_4^-$, SO_4^{--}, Cl^-, HB_4O_7, $HMoO_4^-$. For example, all or part of the potassium chloride (muriate of potash) in most soils exists as potassium ions and chloride ions. The positively charged potassium ion is called a cation, and the negatively charged chloride ion is called an anion.

When an electric current is passed through such a solution, the ions are attracted to the oppositely charged electrodes. Chemical and biological processes occurring in the soil solution (the dissolved substances) and at the interface of solid soil particles create the ions necessary for plant nutrition." [20 p81.]

Ionization/Ion adsorption. Sometimes the term is used for fixation. Were it not for this process of ion adsorption by the clay and organic colloids, mineral cations of the soil would be lost rapidly by leaching, as would also the soluble cations added in fertilizers---.It is a fact that the organic colloids (humus) have a high ion fixing capacity- about 5 or 6 times more than an equal weight of clay colloid, that has partly to do with density, or surface area in relation to weight. That portion of the ion fixing capacity of soils is due to the organic colloids varies from 10% to 50% or more of the total. Thus it is obviously in the farmers interest to maintain the soil organic content in any farming system. [21 p12.]

Kasstano, from Latin castaneo, chestnut.

Kasstanozems, from Latin castaneo, chestnut and from Russian zemlja, earth, land. Soils of the semi arid steppes showing an accumulation of inorganic matter in the surface horizons, often calcareous throughout.

Keyline, when fields to be terraced or contour bunded have irregular slope the terraces or contour strips will be irregular in width. This makes cultivation and planting difficult with short rows needed to fill in the wider parts. To avoid this the keyline system is used. The terraces or strips are made parallel in width and wedge shaped pieces of the slope which do not fit the parallels are left. These are either planted separately or left as grass, or planted with trees.

Lacustrine, soil which was or is still being deposited at the bottom of a lake.

Landscape, a view of the environmental scenery.

Lateritic, see lato.

Lato, from Latin later a brick, soil that can be made into bricks; with weathering water and sunlight it goes brick-like.

Latosols, from Latin later, a brick, because blocks of it dug out set so hard as to be able to be used as building material. Found in the Humid tropics (also in areas with monsoon rain but otherwise dry, PJS) Such soils have a high content of iron and aluminium, in these soils leaching tends to be considerable.They contain a high percentage of clay but are not as heavy working as would be soils with the same clay percentage in temperate latitudes. They are acid and have a clay fraction dominated by kaolinite or iron and aluminium oxides. Their red colour is due to the iron hydroxides formed under tropical conditions of weathering; their yellow colour is from aluminium hydroxides.

Anthony Young pointed out in a paper "The nature and management of the poorer tropical latosols" that:-

> Within the latosols there are some 10-12 distinctive soil types. The classification employed in the FAO-Unesco Soil Map of the World(1974) contains eleven classes which fall within the group; the six classes of ferralsols and three of nitrosols, plus the ferric luvisols and ferric acrisols---. In the US Soil taxonomy

(USDA 1975)the whole order of oxisols together with Ustic and Udic suborders of both alfisols and untisols are included. In a recent review by the present author (Young 1976) six main types of latosols with eleven subdivisions are recognized.

By no means all of the latosols are agriculturally poor; those derived from basic rocks, the ferrisols (FAO nitosols) can be highly productive, with a capacity to sustain continuous cultivation. The ferruginous soils (FAO class as ferric luvisols) have a potential for fairly continuous arable use, given a reasonable level of inputs. Many of the humic latosols of higher altitudes, with higher organic matter contents associated with the cooler temperatures, are not problem soils except where steep slopes give a high erosion hazard

Leaching, the removal of materials *by dissolving* and moving away with the water usually down through the soil.

Legumes, are leguminous plants, the seeds of which are carried in a pod. They are divided into three main subfamilies of the family Leguminoseae, they are Caesalpinoieae, Mimosoideae and Papilionoideae.

Leguminosae. The peas and beans are members of this family, the flowers are similar to those of the pea and all members have their seeds in pods.

Leguminous, from Latin legumen pod; of the botanical family - see legumes.

Lept, from Greek leptos, thin.

Leptic, thin soil horizons.

Level terraces, see terraces. A level terrace suitable for paddy fields or when the soils are permeable enough to allow all the water accumulating on them to soak away.

Lime, generally lime for agricultural use is ground limestone -calcium carbonate; hydrated lime- calcium hydroxide, or burnt lime-calcium oxide. It is applied to reduce the acidity or the pH of soils.

Limitations, soils with slight limitations are relatively free of problems, or the problems are easily overcome. Soils with moderate limitations have problems which can be overcome with good management and suitable use. Soils with severe limitations have problems which are so serious that it is unlikely that it can be used for the intended purpose.

Limnic, organic soil with basal layer of marl, diatoms (Coral) or sedimentary peat.

Lithic, soil with hard rock within 50 cm's of the surface.

Litho, from Greek lithos, stone.

Lithosols, from Greek Lithos, stone, shallow soils over hard rock.

Loam, a soil with a fairly even mixture of sand silt and clay.

Lodging, when crops fall over. This is sometimes caused by strong wind, sometimes because of too much, or unbalanced, plant food, causing tall, soft or weak stems. Sometimes by disease or insect or animal attack weakening the stems or roots. Sometimes by rats or other creatures digging under the roots.

Loess, a soil which has been deposited by wind.

Luvi, from Latin luo, to wash.

Luvisols, from Latin luvi, from Greek luo, to wash, connotative of illuvial accumulation of clay. Soils having an argillic (claylike) B horizon and of low base saturation.

Lysis, from Greek lusis, loosen dissolving the outer membrane.

Macro, from Greek - long, large, big or great. Sometimes interchangeable with mega. Macro to do with computing means a bigger more complex rather than a simple instruction.

Macro organism, bigger organism. Soil macro organisms are those which can be seen with normal eyesight such as worms, insects, and sometimes small animals such as moles and mice are included.

Mature soil, a soil which has well developed horizons or layers of soil, produced over time by the natural division which occurs due to the passage of water, and chemical and biological reactions.

Mellow soil, a porous, softly granulated soil which is easy to cultivate and does not become compacted easily.

Metabolism, from Greek metabole, change; the building up or changing of material into living matter.

Micro-nutrients see trace elements.

Microbes/Micro-organisms/micro life, see also soil organisms.

An essential component of soil is the enormous population of beneficial micro-organisms, which are microscopic (they can only be seen with a microscope) plants and animals that thrive in the top soil layer. There are millions of them in just a few crumbs of soil. [23 p37.]

The decomposition of organic matter in the soil is almost entirely the work of micro-organisms principally bacterial and fungi. These decompositions are important for the maintenance of soil structure; they are also of the highest significance in rendering available for the plants the reserves of plant food in the soil organic matter.[19 p27.]

These organisms are responsible for the break-down and decay of organic waste materials, converting them into forms which plants can utilise for growth, eg:

They convert cellulose and carbohydrates into simpler substances useful to plants.

They change nitrogen from the air and the soil into soluble nitrates which roots can absorb.

They lock up and store nitrogen surpluses and then release them only as required by the plants. (Compare this with the action of chemical nitrogenous fertilizers which stimulate excessive leafy growth and render the plant liable to disease.)

They absorb insoluble phosphates, and after they die, plants can then utilise these phosphates from their decayed bodies.

Many species prey on harmful disease organisms and pests such as nematodes (microscopic worms that eat roots).

Micro-organisms secrete a type of glue which binds soil particles loosely together and coats them with nutrients, improving both soil structure and available plant food.

They continue to feed on composted organic matter even after it has been re-incorporated into the soil, and thus slowly make nutrients available to plants as they need them.

They prevent valuable nutrients from being washed down through the soil ('leached') beyond the reach of the plant's roots. [24 p28.]

"Microbial decomposition of organic matter not only releases its carbon and nitrogen, but also the many other contained minerals as well---. Microbes also affect the availability of various minerals in their organic combinations. Iron, manganese, and sulphur are transformed from unavailable to available forms by microbial oxidations and reductions."[20 p161.]

Microbes also affect the availability of various minerals in their inorganic combinations. Iron, manganese, and sulphur are transformed from unavailable to available forms by microbial oxidations and reductions.---. Sulphur applied to soils as a corrective for excessive sodium salinity is ineffective until it has been oxidised by the soil flora (microscopic plants).

The soil microflora provides nutrients in available forms and also conserves those present from fixation (locking up) or flocculation (binding together) in unavailable forms. In the case of phosphorous, biological interference with mineral fixation is accomplished in part by the elaboration (production from its elements) of organic acids. The acids combine with iron and aluminium and thus keep them from forming insoluble iron phosphate and aluminium phosphate. In soils that have little available iron, the tendency of microbially produced organic acids to combine with iron and to hold it in solution benefits plants. [20 p161.]

The main divisions of soil micro-organisms are Bacteria, Actinomycetes (something between a bacterium and a fungus) including Streptomyces, Mycorrhiza (meaning fungal root), Fungi including yeasts and moulds. Algae, the Protozoa, Nematoda (Eelworms), and microscopic arthropods (tiny insect like soil creatures.) They also liberate non organic plant foods and help to break down rock particles into soil particles. They can cause plant disease but they can also prevent it and help healing. Some produce secretions which stimulate growth; some transform nitrogen from the air into nitrogen the plants can use for food. For more information see appendix.

Micro-organisms, see microbes.

Micro life, see microbes.

Mineral soil, a mineral soil is one which contains mostly minerals, little organic matter.

Mineralisation, the release of mineral matter from organic matter, this occurs especially through the action of soil microbes decomposing the organic matter.

Modules of rupture, modules, standard unit for measuring rupture, break relationship with, so together it is the measurement of the force needed to separate. In the case of soils, it is the effort needed to cultivate the soil.

Molli, from Latin mollis, soft supple.

Mollisols, from Latin mollis, soft supple, soils of medium age with nearly black, organic rich surface horizons and high base supply. They develop under conditions of prairie, steppe or savanna.

Monosaccharide's, from Greek monos, alone, only, single, simple sugars.

Mor, occurs in forests, a layer of organic matter which forms a layer of its own and does not easily mix with the soil under it.

Moribund, from Latin moribundus/mori die, at the point of death.

Morph, from Greek morphe, form, body.

Morphology, morphological, from Greek morphe, form or structure.

Mottle, patches of different coloured soil in an area or cross section of a particular soil for instance as we look at the sides of a trench. It indicates intermittent wetness or flooding.

Muck soil, peat which has become very well rotted soft organic material and has mixed with soil.

Mulch, a layer of material on the surface of the soil. It reduces drying of the soil, protects the surface from the weather and helps keep the soil temperature more regular. It also reduces weeds.

Mycorrhiza, from Greek mukes, mushroom and Greek rhiza; root microscopic fungi which live in the soil usually in association with plant roots. The relationship is usually helpful to both the fungus and the plant. See also Endophytic fungi and microbes.

Natric, from Arabic Natrun, containing Sodium.

Natric, from Arabic Natrun - Native sesquicarbonate of soda; soil containing sodium.

Nematodes, microscopic worms in the soil also called eelworms. Many of them attack plant roots, but there are others which feed on the harmful ones. When nematocides are used (chemicals like special insecticides), all the nematodes are killed along with other micro-organisms. Harmful nematodes quickly return when there are plants to feed on which is most of the time. Helpful nematodes only return slowly when there is a big enough population of bad nematodes for them to feed on. The balance is in favour of the harmful nematodes and so more chemicals are needed. It is best to control nematodes by other means.

Neutral soil, is neither significantly acid nor significantly alkaline. In theory this is soil with a pH of 7. In practice it is between 6.6 and 7.3.

Nito, from Latin nitidus, shiny, bright, lustrous.

Nitosols, from Latin nitidus, shiny, bright, lustrous; connotative of shiny ped faces. Soils having deeply developed argillic B horizon and showing features of strong weathering.

Nitrates and nitrites, both are salts of nitric acid. Nitrogen in the soil is released as ammonium salts. Nitrosomonas convert it into Nitrites. Next as plants need nitrogen the nitrite (in which form the nitrogen is held tight) is converted by nitrobacter into nitrates (the form in which it can be 'ate' by the plants). The form in which plants can use it. So to remember which is available to the plants remember what rhymes:- Nitrites hold tight so the N is not available, but the plant ate the Nitrates as they were available.

> **Nitrification**, consists of two steps. First the ammonia is oxidised to nitrite by a specialised group of bacteria represented mainly by the genus Nitrosomonas. The nitrite is further oxidized to nitrate by other specialists, such as Nitrobacter. Some commonly occurring heterotrophic bacteria and fungi can oxidise ammonia to nitrites and nitrates to a limited extent.[20 p161.]

Nitrogen fixation, in a general sense it means the conversion of free nitrogen to nitrogen combined with other elements. In soils we mean the conversion of nitrogen from the air by micro-organisms to nitrogen plant foods. Apart from Rhizobium it is also carried out by members of actinomycetales which can also fix large amounts of nitrogen.

If this is done by nitrogen fixing organisms which live in close association with plants the organisms are called symbiotic nitrogen fixing organisms. If the organisms do not live in association with plants they are non symbiotic.

Node, the part of a stem from which leaves flowers or side shoots grow. See also internode.

Nodulation, from Latin nodus-knot knob or band, the forming of a small roughly round lump, these form on the roots of legume plants like peas, beans, and clover. They are caused by special bacteria which take nitrogen from the air and make it into plant food which is stored in the lumps or nodules. This process is helpful to the plants as it supplies them with nitrogen plant food. It also improves the soil for following crops or neighbouring plants. The process cannot happen unless there are suitable bacteria in the soil. Some of the correct bacteria are better at producing plant food than others so it is sometimes best to supply some of the better bacteria.

Non symbiotic nitrogen fixation, occurs without the need for association with plants or another host.

Northern hemisphere, the half of the world north of the equator.

Nutrients, plant foods.

Ochr, from Greek ochros, pale.

Optimum, from Latin optimus = best. Most favourable conditions.

Organic, produced by living things, having or having had life.

Organic soils, are those which contain mostly organic matter such as peat, muck, or bog soils.

Organic matter, the remains of plants or animals. "Organic matter affects the availability of certain plant nutrients An ample supply of organic matter may increase the supply of available phosphorus and zinc--. peaty soils high in organic matter are frequently deficient in the trace element copper" [3 p41.]

Orth, from Greek orthos, true, common or typical.

Oxi, from Greek oxus sharp or acid from the understanding that oxygen was essential for acid formation, so oxides or oxidation.

Oxisols, from Greek Oxus sharp or acid, oxidisation, to cause to combine with oxygen. This comes from the understanding at first that oxygen was the essential principle in the formation of acids.

Oxisols as the order name suggests, are highly weathered, nutrient poor soils with an oxic (showing oxidation) layer. They occur on some of the oldest landscapes on earth such as the Brazilian and Guyannan Shields in the Amazon Basin and the African shield of the Congo Basin. High Annual rainfall leaches the bases from the soil, yielding relatively high levels of iron and aluminium oxides (sesquioxides). As a result, oxisols are acid in reaction. Extreme weathering removes even the silica component producing Kaolinitic clays with a cation exchange capacity of only 2-4 meq/100 g soil. Base saturation levels of oxisols in the Amazon Basin typically fall below 15%.

Oxisols range in colour from yellows to reds to dark browns. Unlike with temperate zone soils, however, colour has little to do with organic-matter content and fertility. Rapid decomposition and uptake of organic matter nutrients leaves little chance for clay- humus complex formation, and excessive weathering results in weatherable minerals occurring at levels below 1%. Deep weathering of some oxisols places the hard rock parent material 10-40 metres below the surface[16.]

Pachic soil, from pakhus, a thick dark surface horizon.

Pach, from Greek pakhus meaning thick.

Palatable, pleasant to the taste. hence palatability being pleasant to the taste.

Pale, from Greek paleos, old, eg paleosol-old soil.

Paleosol, old soil.

Pan, a layer of soil which is firmly compacted or is very rich in clay and which is hard when dry; eg Plough pans, claypans, traffic pans. In the construction of paddy fields the land is cultivated and puddled to form a waterproof pan.

Parasite, from Greek para, beside, beyond, wrong, irregular. and Greek sitos, food. A plant or animal living in or upon another object and drawing nutrient directly from it. It includes plants or animals growing on in other plants, animals or walls etc.

Parasitic, see above, living on, feeding on, or otherwise using or climbing on the thing it is parasitic on.

Parent material, the rock or peat from which the soil has developed.

Pathogenic, from Greek pathos, suffering, disease, or passion, and Greek gen, become or originate, as genesis in the bible. The meaning is source of suffering or disease or the problem.

Peat, partly decomposed (organic matter which accumulates in boggy (wet) conditions.

Ped, an individual soil aggregate (piece of soil).

Pell, from Greek pelos, dusky.

Perched, above where it would normally be. For example a perched water table is caused when poor drainage holds soil water from rain or irrigation above relatively dry soil, instead of it passing through the soil or rock or other barrier to the real water table below.

Pellet, a small round or roundish piece of material such as a grain, a pill, a small ball of paper, or a small piece of soil.

Pelleted, made into pellets, this is sometimes done to seed. A paste or thick soup is made of lime, dung, soil or another material, sometimes containing microorganisms helpful to the seedlings or the soil in which the seeds are to be sown. The seeds are coated with the paste and then the paste is dried, so that the seed can be sown. Lime is used for plants which do not like acid soil but are to be sown in such soil. Quite often they grow well enough in the acid soil if they have a help from lime to start with. When seed is broadcast sown it is sometimes pelleted with mud and animal dung or urine, this is to deter birds from eating the seed.

Perennial, a plant which lives for several years.

Permeability, the ability of water, air and plant roots to penetrate the soil.

Permeate, penetrate, saturate, spread through.

Petro-calcic, containing indurated Calcium Sulphide.

pH, see acidity, and base.

Phaeo, from Greek phaios, dusky.

Phaeozems, from Greek phaios, dusky, and Russian zemlja, earth. Soils of the forest steppes showing a strong accumulation of organic matter in the surface but a deep leaching of calcium carbonate.

Plac, from Greek plax, flat stone, presence of a thin pan.

Plag, from German plaggen, sod, indicating the presence of

Plaggen (sod), horizon.

Plano, from Latin planus, flat, level.

Planosols, from Latin planus, flat, level. Connotative of soils generally developed in level or depressed topography with poor drainage.

Pod, from Russian pod, under.

Podzols, from Russian local name derived from pod, under and zola, ash. Soils with a strongly bleached horizon having B horizons with iron or humus accumulation, or both.

A zonal group of soils having an organic mat and a very thin organic mineral layer above a grey leached layer which rests upon an illuvial dark brown horizon, developed under the coniferous or mixed forest , or under heath vegetation in a temperate to cold, moist climate. Iron oxide and alumina, and sometimes organic matter, have been removed from the A and deposited in the B horizon.

Podzoluvisols, (combined Podzol and Luvisol). Soils having an argillic B horizon but also showing features of Podsols.

Pollard, a tree cut off at an appropriate height, so that new branches will sprout from the top of the remaining trunk. Not all kinds of trees will survive pollarding particularly if the cut is lower down.

Polysaccharide, from Greek polus, many, and Greek saccharon, sugar. They are natural carbohydrates derived from the condensation of several or many molecules of simple sugars (monosaccharide's). The polysaccharide's include cellulose, hemicelluloses, starch and pectic substances. In the soil most come from microbial action, they are gummy and help the soil crumb structure, they also help the soil to resist erosion.

Porosity, means the amount of pore, spaces between the soil particles (very small pieces). The more the proportion of the soil mass which is pores or spaces, the more is the porosity, and the easier for water, air, and plant roots to penetrate the soil.

Precipitation, in agriculture, moisture condensed and fallen to the earth from damp air, mist, or clouds, as dew, rain, snow, or hail.

Preponderance, to exceed in number quantity or importance.

Prevalence, being widespread, superior strength or influence, preponderance.

Prohibit, to forbid, refuse to allow, or exclude (keep out).

Proliferation, to grow or to reproduce by multiplication of parts, to increase in numbers greatly or rapidly.

Psamm, from Greek psammos, sand.

Puddled soil, Paddi or Paddy fields, are purposely puddled to make them waterproof. If we cultivate or walk or drive on land when it is wet, particularly if it is a clay soil it becomes pressed and solid, when it dries it becomes hard. It is then difficult to cultivate and difficult for plant roots to enter.

Quartz soil, from German quarz, quartz, indicating a high quartz content.

Raceme, from Latin racemus= grape bunch, a flower cluster with the separate flowers attached by short equal stalks at equal distances along the central stem.

Radical, from Latin radicolis = root. Of the root, naturally inherent, essential, fundamental, forming the basis of. In chemistry - a group of 2 or more atoms behaving like a single atom and passing unchanged from one compound to another.

Rangeland, is land which grows native pasture plants.

Rankers, from Austrian rank, steep slope; connotative of shallow soils. Soils which develop a surface horizon enriched in organic matter over siliceous materials.

Recuperate, recover.

Reddish brown lateritic soils, a zonal group of soils with dark reddish brown granular surface soils, red friable clay B horizons, and red or reticulately (honeycomb-like) mottled lateritic (capable of being made into bricks) parent material; developed under humid tropical climate with wet-dry seasons and tropical forest vegetation.

Regasols, from Greek rhegos, blanket; connotative of mantle of loose material.

Regeneration, bring or come back into renewed existence.

Rego, from Greek rhegos, blanket, connotative of a mantle of loose material.

Rendzi, from Polish rzedzic, noise, connotative of plough noise in shallow soils.

Rendzinas, from Polish rzedzic, noise; connotative of plough noise in shallow soils.

Replicate, from Latin plicare fold, and re again. To repeat/make a copy of, to reproduce an identical result.

Replication, see replicate, make a duplicate copy/copying. To reproduce an identical result. To repeat a number of times in the same way. Make an echo.

Replications, repeats, the number of times a thing is repeated, this is used for experimental trials

Residual fertilizer, the amount of fertilizer remaining in the soil after the crop for which it was mainly applied has been cropped or harvested.

Residues, French residu remainder, what is left over.

Resource, a quality or ability that can be used for the benefit of people or other organisms in the environment.

Reticulate, Latin rete, net, divide in fact or appearance into a network.

Retiform, from Latin rete, net, netlike.

Rhizobia, the bacteria that can live in symbiotic relationship with leguminous plants, inside nodules on the roots.

Rhizobium, from Greek rhixa, root. It has been known for many hundreds of years that legume plants improve the soil for following crops. Experiments about 100 years ago proved that nodules (lumps) on the roots of legumes were the reason for this as they fixed nitrogen from the air so that it could later be used by the plants. The nodules are caused by the micro-organisms Rhizobium. Some of this family are more efficient than others and so are sometimes inoculated into the soil or onto the seeds of legume crops to increase the nitrogen fixing ability, and so the benefit to following crops. Rhizobium may also be inoculated if there is no naturally occurring rhizobium in the particular piece of land the legume crop is to be grown in.

Rhizome, from Greek Rhizoma- take root. Rootlike stem which spreads out from the parent plant just below the surface of the soil. It sends out roots and normal stems and can be used as planting material.

Rhizosphere, the soil space surrounding and including the surface of the roots, in which the number and kinds of soil micro-organisms are influenced by the roots.

Rhod, from Greek rhodon, rose, dark red colour.

Rockiness, the amount of a soil surface where the underlying rocks show (USA definition). In British terms the amount of rocks 5-24 inches (13-60 cm's) in diameter.

Runoff, the water that runs off the surface of an area.

Ruptic, from Latin rupt = break, soils with intermittent horizons.

Saline soil, a soil containing so much salt as to effect the growth of normal plants.

Sapr, from Latin sapros, rotten.

Saprophytic, Greek sapros, rotten, an organism living on dead or dying organic matter.

Scarify, (to cut, scratch, or graze.) This is done to hard coated seeds to allow moisture to enter the hard waterproof coating which some seeds have to allow them to survive for long periods before germinating. Some times it is done by making a little cut in the seed coat, sometimes by cutting of a piece of the seed-coat. In other cases the seeds are put through a mill or ground between stones, or treated with acid to remove part of the waxy coating.)

Scrub/Scrub country, in regions where the annual rainfall is between 30-60" (76-152 cm's) at the upper limit of which it merges from monsoon forest into rain forest. The trees are smaller and thinner and do not grow as dense and yet the grasses under them are also sparse and poor.

Sedimentary, rock or soil, is composed of particles deposited by water. The chief groups of sedimentary rocks are conglomerates from gravels; sandstones from sand; shales from clay; and limestones from soft masses of calcium carbonate.

Seepage, water moving slowly through the ground and emerging from a spread area rather than one place as springs do.

Self mulching soils, soils with a high content of clay which swells and becomes very sticky and impervious (water proof) when moist, then crust over. When very dry deep, large, cracks form. Surface material fills the cracks in the dry season. Mulches help to conserve moisture but in this case if the soil is on a sloping surface the moisture is likely to run off the sealed surface and so have less to conserve. This kind of soil is normally only good for paddy fields, grass or bamboo.

Sesquioxides, from Latin semis-que, one and a half, in chemistry it refers to compounds where there are three equivalents of the named element to two of others. So it is 1.5 to 1. eg sequioxides or sesquisulphides. Sesquioxides are oxides of trivalent cations such as iron or aluminium.

Shale, a kind of soft flaky stone.

Sider, from Greek sideros, iron, eg siderite indicating the presence of free iron oxides.

Silt, soil mineral particles 0.002-0.05 mm in diameter deposited by water. Sedimentary soil.

Slip, the down the slope movement of soil when some or all of it is soaked with water. It could develop into a landslide or to the start of rills or gullies.

Slope, "the incline of the surface of a soil. It is usually expressed in percentage of slope, which equals the number of feet of fall per 100 feet of horizontal (level) distance." [20 p767.]

Sod, from Latin and Greek sode, turf, upper layer of the soil of grassland including the plants and the soil surrounding the roots.

Soil profile, a sideways view of the soil such as at the side of a trench where we can see the different layers from the top down.

Soil conservation, the efficient use of soil to ensure the protection, preservation, improvement, and stability of soil by sustainable means.

Soil organisms, these are micro-organisms or micro life; soil bacteria, and tiny fungus like plants and moulds and mosses which can only be seen with a microscope; and macro organisms such as ants, worms, and other soil living creatures. See also microbes/micro-organisms/ micro life.

Sol, a solution in which the solate is present in the colloidal state, (a substance in a state in which, though apparently dissolved, it cannot pass through a membrane; a substance that readily assumes this state. Colloidal system- a dispersed substance plus the material in which it is dispersed.

The solvent for example water, is termed the dispersion medium, and the dispersed substance as the disperse phase.)

Solic, from Latin Sal, salt. Salty, containing Sodium Chloride, Na Cl .

Solon, from Russian sol, salt.

Solonchaks, from Russian sol, salt. Soils showing strong salinity.

Solonetz, from Russian sol, salt. Soils developed under the influence of high sodium saturation.

Solum, the upper part of a soil profile, above the parent material, in which the processes of soil are active. The solum in mature soils includes the A and B horizons. Usually the characteristics of the material in these horizons are quite unlike those of the underlying parent material. The living roots and other plant and animal life characteristic of the soil are largely confined to the solum.

Sombr, from French sombre, dark.

Southern hemisphere, the half of the world south of the equator.

Sparse, thinly scattered.

Sphagno, from Greek sphagnos, bog, indicating the presence of sphagnum moss, which grows best in cool humid regions.

Spindly, weak and thin.

Spodo from Greek spodos, wood ashes.

Spodosols, from Greek, spodos, wood ashes," middle aged soils which develop under coniferous forest"[16.]

With accumulations of amorphous (without definite shape or structure) materials in sub surface horizons. They are not normally very good for agriculture.

Stand, a growing crop.

Standover, a crop left uncut/ungrazed to die back and provide winter or dry season feed.

Stolon, a prostrate surface stem which roots at the nodes.

Stoniness, the amount of stones 10-24 inches (25-60 cm's in diameter in or on the soil (American definition, in the UK would be called Rockiness.) In the UK stoniness would be the amount of stones 1/2-5 inches in diameter in or on the soil.

Stratified, from Latin stratum spread thing or blanket, arranged in strata or layers.

Strip felling, when forests on hillsides are felled (cut) it is usually better if they are cut in strips across the slope. When the strips cut have been recovered perhaps replanted and the soil is again covered with plant growth instead of being bare the alternate strips are cut. This reduces the erosion compared with clear cutting which leaves all the surface bare and exposed to wind sun and rain.

Strip cropping, crops planted in level strips across the slope instead of up and down the slope. The strips are planted with different crops, crops which give better protection from erosion alternating with those which give less protection.

Subsoil, the B horizons of the soil, the soil below the depth where most plant roots live. The less fertile underneath soil below the ploughed or dug soil.

Surface soil, the more fertile ploughed or dug soil, the living soil, about 5-8 inches.(13-20 cm's).

Susceptible, capable of receiving, easily affected by.

Suspension, carried, hanging in, something else, such as soil suspended in water being carried down a stream.

Swale, a barrier of stone, soil or plant matter made across the slope with a slow slope to a ditch to carry excess water away. The idea is to divert and spread the water flowing down a slope so that it can soak into the soil, it is very useful where rainfall is not regular, but when it does come is heavy, and normally quickly runs away with little penetration. It also reduces the erosion of the water pathways. [44 p480.]

Sward, the plant material covering the ground, usually for animal feed.

Symbiosis/symbiotic, from Latin sym, harmony and Greek bion -ountos, life -part of. Permanent union between organisms each of which depends for it's existence on the other.

Symptoms, noticeable change in a body or object or its functions; indicating disease or a sign of some other problem or abnormality, (not normal situation).

Synthesize, to combine simple substances to form another substance.

Terraces, strip fields made on sloping land, the strips are made level and the land in between the strips slopes steeply so that the area is made into big steps.

Texture, this is the feel of the soil and depends on what the soil is made up of, it may be a sticky soil with a high proportion of clay, a sandy soil, a gravelly soil etc but the exact definition of where one texture becomes another varies between countries.

Thapto, a buried soil.

Tillage, the cultivation of land by ploughs and other implements.

Tilth, this is basically soil structure but other considerations should be included. For instance the best tilth for winter wheat in England would be a soil with a mixture of sizes of the pieces of soil, including larger clods about 1.5-2.5 inches (4-6 cm's). If it is too fine it will run together when wet and this will become a crust when the soil dries. This forms like a tight collar round the plants, when frost or drying weather causes the crust to lift the plants are damaged. The larger clods help to stop the crusting and also help against frost injuring the base of the plants. The cold air seems to keep at the level of the clods rather than at the soil surface.

There is a subtle difference within tilths caused by the way the tilth was obtained.

MECHANICAL TILTH. If a sticky (clayey) soil is ploughed when wet and dries out it will become hard and lumpy. It is possible to cultivate the soil till it is broken and sliced into the correct mixture of sizes of the pieces of soil this is called a mechanical tilth. It is not as good an environment for plant roots as :-

NATURAL TILTH. If as the soil is cultivated while in a state of change, (becoming wet after being dry or more usually starting to dry or thaw after being wet or frozen), it breaks easily and naturally. As well as requiring less work the natural tilth produced is much better for plants. In areas where frost occurs it is best to dig or plough the soil before frost comes. The frost will have a good effect on the soil, so that when cultivated for the next crop a good tilth may be obtained easily. The frost will also reduce soil pests and diseases.

Tolerate, withstand, exist with, sustain the use of without harm, eg If we were in a cool country we could tolerate having nothing to drink for a day much better than if we were in a hot dry country and had nothing to drink for a day.

Topsoil, the surface layer of the soil, the living part of the soil, containing organic matter, and soil organisms.

Torr, from Latin torridus, hot dry, usually dry.

Toxin, poison. hence Toxic means poisonous.

Trace elements, the old term for micro-nutrients. The plant foods which though very important, are only needed in very small amounts; also other elements found in plants which may not be important to the plants but or may not be important to animals.

Transpiration, from Latin trans - across, beyond, on or to the other side, through, into a different state or place. And Latin spirare - breathe. So to emit through excretory organs of skin or lungs or plant tissue.

Trifoliate, three leaves coming together from a single stalk.

Trop, from Greek tropicos, of the solstice, continually warm.

Udults, from Latin udus, humid, damp; soils of low organic matter content, temperate or warm and moist.

Ulti, from Latin ultimare, the last, come to an end.

Ultisols, from Latin ultimare, the last, come to an end. Older more leached soils that are usually moist with horizon of clay accumulation and a low base supply. As they are not able to store plant foods they are normally very infertile unless managed well.

Two common features of ultisols are plinthite and fragipans. Plinthite is a subsoil phenomenon dependent on alternating wet and dry periods, like a fluctuating water table. The drying out of a wetter iron rich layer can harden irreversibly into plinthite.(A kind of brick red clay). Fragipans occur in poorly drained ultisols, where they restrict water movement and are sometimes responsible for perched water tables."[16.]

Umbr, from Latin umbra, shade, dark colours.

Unaggregated soils, soils which do not clump into pieces of soil but stay separate such as sandy soil or shaley soil.

Ust, from Latin ustus, burnt, of dry climates with summer rains.

Verm, from Latin vermes, worm.

Vermiform, from Latin vermis, worm, worm shaped.

Vermisol, indicating a soil which is wormy or mixed by ants moles etc.

Verti, from Latin verto, turn, connotative of a turn over of surface soil.

Vertisols, from Latin verto, turn; connotative of a turn over of surface soil, self mulching soils, with a high content of clay which swells and becomes very sticky and impervious (water proof) when moist, and forms deep, large, cracks when dry. Surface material fills the cracks in the dry season, this causes an inverting process and is the reason for the name. This kind of soil is not good for crops other than paddy crops or pasture, or perhaps some species of Bamboo.

Virgin soil, is one which has not been disturbed from it's natural state.

Water table, the upper limit of the part of the soil or underlying rock that is wholly saturated with water (wet). In some places an upper or perched water table may be separated from a lower one by a dry zone.

Water holding capacity, the capacity (ability) of a soil to hold water. In other words how much it can hold without the water draining out. Sandy soils have a lower water holding capacity (ability) than clayey soils. This may be described in inches of water per foot depth of soil. or centimetres per metre.

Waterlogged, a situation where all the pore spaces in the material are filled with water. It is full of water.

Watershed, sometimes called a catchment area; the area from which a river or stream receives its water.

Weathering, the physical and chemical breaking down of rocks.

Xero, from Greek xeros, dry.

Xerophyte, from Greek xeros - dry, and phuton - plant, so a plant which can tolerate dry conditions.

Xerosols, from Greek Xeros, dry. Soils of semi arid environments. Dry with winter rains.

Yermo, from Spanish yermo, desert.

Yermosols, from Spanish yermo, desert, derived from Latin eremus, solitary, desolate. Soils of desert environments.

Yield, the amount of growth a soil can produce.

Zem, from Russian zemlja, earth, land.

Zol, from Russian zola, ash.

THE EFFECTS OF SOIL MANAGEMENT ON SOIL EROSION

1. The Effect of Different Crop Rotations on Soil Erosion

For sustainable agriculture we need to reduce soil loss so that the loss is not more than the new soil which has developed over the period, and to farm in such a way that the fertility of the soil does not decline. If possible we should be adding to the soil and improving it's fertility. It is useful to have a comparison of the effect of different patterns of cropping.

The table below is from research in Iowa in the United States. It will give an idea of differences, though as there are so many variables on steep land, long uninterrupted slopes, soil with poor structure, or in places with very heavy rainfall the differences would be much greater than this table shows.

Crop Rotation.	Measure of amount of Erosion
Continuous Corn (Maize).	4.0
C-C-O	3.0
C-C-Osc	2.0
C-O	2.0
C-Osc	1.5
C-C-O-M	1.4
C-O-M-C-Osc	1.2
C-O-M	1.0
C-C-O-M-M	0.9
C-C-O-M-M-M	0.8
C-O-M-M	0.6
C-O-M-M-M	0.4
C-O-M-M-M-M	0.3
C-O-M-M-M-M-M	0.2

C refers to inter- tilled crop.(Such as Corn, Sorghum, Tobacco. PJS.)
O refers to small grain crops.
M refers to Legume or Grass.
sc refers to Sweet Clover. (Presumably sown under the small grain crop it is listed with. PJS.)[8] p88.

2. Long Term Effects of Organic and Conventional Farming on Soil Erosion

Conventional, intensive tillage farming systems have greatly increased crop production and labour efficiency. But, serious questions are being raised about the energy- intensive nature of these systems and their adverse effects on soil productivity and environmental quality (Bidwell, O. W. Journal, Soil Water Conservation 41, 317-320 [1986]) (Papendick R I; Elliot L F; & Dahlgren R B; Amer. Journal Altern. Agric 1, 3-10 [1986])

This concern has led to an increasing interest in organic farming systems because they may reduce some of the negative effects of conventional agriculture on the environment. (US Department of Agriculture report and recommendations on Organic Farming 1980 US Govt Printing office. Washington).

We compare the long term effects (since 1948) of organic and conventional farming on selected properties of the same soil. The organically-farmed soil had significantly higher organic matter content, thicker topsoil depth, higher polysaccharide content,(gummy materials which help soil structure and are evidence of soil micro organism activity) lower modules of rupture (easier to cultivate) and less soil erosion than the conventionally farmed soil. This study indicates that, in the long term, the organic farming system was more effective than the conventional farming system in reducing soil erosion and, therefore, in maintaining soil productivity. [34.]

An article in the important journal 'New Scientist' listed further work on the same subject as condensed here.

Among the rolling hills of the Palouse region of eastern Washington State (in the United States) lie two neighbouring farms. The two have much in common; they lie on the same type of land and share the same soil; both have been worked since the early years of the century. But there is a crucial difference. One is an organic farm; it has relied on green manure crops, crop rotation and the natural fertility of the soil since first ploughed in 1909. The adjoining farm, slightly larger and cultivated a year earlier, became a "conventional farm", nourished by fertilisers from 1948 and protected by pesticides since the early 1950's. Together, these two farms offer the opportunity to settle some of the many arguments about the pros and cons of each system of farming---.

What we describe as conventional farming is of recent origin. With the appearance of cheap fertilisers at the end of the second world war, and pesticides in the early 1950's advanced countries quickly abandoned traditional or organic methods of farming and became heavily dependent on both agrochemicals and labour saving machinery. Farmers relinquished organic methods not because they did not work but because they could not compete with the new type of agriculture. But the modern concept of organic farming is not a return to the past, rather a marriage of scientific advances with many traditional practices.

The US Department of Agriculture considers organic farming to be a system that avoids or excludes the use of synthetic fertilisers and pesticides and relies upon practices such as crop rotation, the use of animal and green manures, and some forms of biological pest control. Most organic farmers use modern machinery, recommended varieties of crops, certified seed, sound methods of livestock management, prescribed practices for conserving soil and water, and innovative methods of managing crop residues.---.

Our study is one of the few systematic comparisons that gives a detailed picture of the effects organic and conventional have on the quality of soil and on erosion on working, commercial farms.---.

The study area (100 metres by 50 metres) lies on a 4 degree (about 9%) slope at the juncture of the two farms. The validity of this research is based on the assumption that all environmental conditions and soil properties at the site were similar until 1948. Any differences today can be attributed to the fact that after 1948 one farm was managed organically and the other conventionally.

Soil microorganisms enrich the soil as they live and die and are indispensable in their influence on crop production. They are a kind of hidden workforce that provides many key benefits. The most important of these are the breakdown of organic matter in the soil to humus and the release of nutrients held in organic combinations, which plants can then use. Microbes also help to stabilise soil aggregates, fix nitrogen and break down some pesticides. In 1983, Bolton found that the soil on the organic farm contained a much higher mass of microbes and showed much more enzyme activity than the soil on the conventional farm. Soil enzymes are derived primarily from micro organisms in the soil. These results indicated that the organic farm had larger and much mor active populations of microbes in its soil.

Elliot and I later found almost 60 per cent more organic matter on the surface of the soil on the organic farm. these results support the findings of other researchers that organic farmers can, and generally do, achieve higher concentrations of organic matter in their soils than conventional farmers. They also help to explain the greater microbial activity in the organically farmed soil; the more decayed organic matter there is, the more microbes it will feed.

Organic matter has a profound effect on the quality of the soil; it encourages mineral particles to clump together to form granules, improving the texture of the soil; it increases the amount of water the soil will hold and the supply of nutrients; and the organisms in the soil are more active. All in all, organic material makes the soil more fertile and productive. In our experimental plots, we found that the soil on the organic farm was well granulated, the best structure for most ordinary crop plants; the other soil was not. The organically farmed soil also contained much more moisture; its cation exchange capacity (a measure of the soils capacity to store nutrients) was greater; the total amount of nitrogen and available potassium were also much larger. Most of the extra organic matter came from green manure.

The presence of microbes brings other benefits. As the microbes break down organic material they produce polysaccharides, gummy substances that can stabilise the soil by binding particles into aggregates. Aggregates are less vulnerable to breakdown and erosion. Soil organisms also break down polysaccharides, however. So farmers must continue to maintain a supply of these stabilising substances.

The organic soil had a significantly lower "modulus of rupture" an index related to the hardness of crust that forms on the surface of the soil. In general, the lower the modulus of rupture, the easier it is for seedlings to emerge. The combination of all these factors gives the organically farmed soil better tilth than the soil on the conventional farm. The better the tilth, or physical condition of the soil, the easier it is to till and the easier it is for plants to germinate and push out shoots and roots.---.

Andrea Weilgart Patten studied water erosion in our test area. She measured rill erosion on the two fields when both fields were growing winter wheat. The results were impressive: water erosion removed 32.4 tonnes per hectare on the conventional field, but only 8.3 tonnes on the organic field. Interestingly, the rate of erosion on the conventional farm is close to the average annual rate for the area, which is farmed almost entirely by conventional methods.---.

Erosion is one of the greatest threats facing American farmers in many regions of the country. The US Department of Agriculture's Natural resources Inventory for 1982 showed that about 44 per cent of arable land is eroding faster than the soil conservation service considers tolerable. Interestingly, the National Soil Map, published by the Soil Survey of England and Wales in 1986, also indicated that 44 per cent of the arable soils in England and Wales are at risk from erosion by water and wind.

Alternative Futures

Though the two fields in this study have the same type of soil and were probably identical 40 years ago, the topsoil is now eroding more rapidly on the conventional field. At this rate, all the topsoil on typical Naff (fine - silty, mixed, mesic Ultic Argixerol; a dark coloured, well drained soil which formed under grass. Typically these soils have a silt loam surface A1 horizon 20-46 cm's[8" - 18"] thick, overlaying an A2 horizon of heavy silt loam over a strong silty clay loam) and similar soils will be lost in another 50-100 years.---.

Technological advances in the form of new fertiliser, pesticides, and plant varieties mask the decline in productivity as a result of erosion. Intensive farming has produced record breaking yields year after year, but production could fall dramatically in the coming decades. *Research has shown that technological advances have the greatest potential for increasing yields on deep, relatively uneroded topsoils. If erosion continues at its present rate, the topsoil will finally become so thin that fertilisers will fail to increase yields, and then they will begin to decline.---.*

As to the economics another landmark study by William Lockeretz and his colleagues from Washington State University, in St Louis, Missouri; made economic comparisons of numerous matched pairs of organic and conventional farms in the American Midwest.

Between 1974 and 1978, the organic farms in the Lockeretz study produced lower yields than the conventional. But operating costs on the organic farms were also lower by about the same amount.

As such, the income from each hectare of cropland was about equal for the two types of farm. In addition, the amount of energy spent to produce the crop was dramatically different on the two types of farm; the organic farms required about 40 per cent of the fossil fuel to produce crops of the same value.

Because of decisions two farmers made four decades ago, the two farms in Palouse today illustrate how different methods of farming affect one of the country's most precious, though often overlooked, natural resources -its topsoil. In the late 1940s and 1950s, the conventional farmer, like almost all farmers in the US, abandoned traditional farming methods in favour of intensive, high- technology agriculture, the organic farmer stuck to tradition. and from the measure of topsoil and other soil characteristics, he appears to have made the right decision. [35 pp49-52.]

3. During the BBC morning farming program on 19th May 1992 it was announced that a new survey had stated that for every bushel of maize grown in the U.S. one and one half bushels of soil was lost.

THE GOOD NEWS BULLDOZER

See also Chapter 6.

When I first lived in Taiwan, labour was very cheap and we made terraces using only human labour. After a few years the situation changed and it was difficult and expensive to get labour for such work.

With the shortage of labour in rural areas there were now walking tractors for cultivating the rice fields and other work in most villages. I realised that in many areas in developing countries, there are walking tractors with rotary cultivators used for rice field cultivation. Some are normally used for pulling trailers, but the cultivators are available for fitting to them. I wondered if we could make a bulldozer conversion which people could hire and fit on their local tractor and use it as a bulldozer, first using the cultivator to break up the soil, then reversing, lowering the dozer blade, and pushing the loose soil to where needed. Such a bulldozer could be used for levelling land, making terraces and local roads or tracks.

I drew up a simple plan and sent it to an organisation for agricultural engineering research, for their advice and opinion. They replied that they thought it should work, they would be interested to know if it did.

The Good News Bulldozer Conversion. After the rotovator has dug up the soil it is raised and the bulldozer lowered to push the loose soil to where needed.

However, when I had it made I found that it did not work. I had thought that when the depth wheel was lowered to raise the cultivator the bulldozer blade would come down into the loose soil, then as the tractor moved forward the soil could be pushed to where needed. The reason why it did not work was that there were now three weight bearing points, the depth wheel, the drive wheels, and the bulldozer. As the tractor tried to push the soil the drive wheels slipped and turned without being able to move the tractor.

I found that the solution was to make cranked push bars, they were fastened to the drive shaft casing for the cultivator drive. When the tractor moved forward and the bulldozer blade pushed into the soil the push arms pushed down on the axle, and so the more the soil was pushed the more the pressure on the wheels. Instead of the wheels slipping they gripped better and so if the bulldozer depth wheels were too high it was possible to stall the engine.

The bulldozer parts were made by a local welding shop, the wheels were paddy field wheels, with spuds

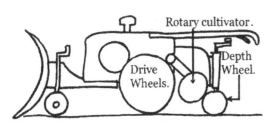

Rotary cultivator.

Depth Wheel.

Drive Wheels.

When the blade is attached like this the driver wheels slip.

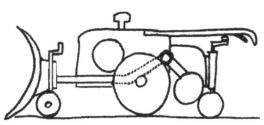

The more the dozer pushes soil, the more weight on The tractor wheels.

The Good News Bulldozer conversion can fit on most two wheel rice tractors. First the cultivator is lowered to loosen the soil, then the loose soil is dozed.

welded on instead of the normal paddles.

The conversion kit was designed to be able to fit a range of walking tractors, so that it could be hired out and fitted to a farmers own tractor.

We found that it was better not to use the depth wheel on the tractor to control the bulldozer action. Instead we used the depth wheels on the bulldozer to control the depth, and adjusting the angle of the blade till it easily entered the soil. We used scrap broken truck springs as weights attached below the tractor handle bars to act as counterweights to the bulldozer when we wanted to lift the bulldozer out of work. The bulldozer worked very well as can be seen by the pictures in chapter 6. B1, B2, B3 and B4.

We were very pleased with the performance particularly when starting to make terraces when we were driving across the slope. We found that unlike a normal bulldozer the good news bulldozer could go crabwise (diagonally), across the slope using the steering clutches to control the direction , so that while the tractor wanted to slip down the slope we could aim it slightly up the slope to compensate and still angle doze the soil from the top of the slope down to where it was needed. Only once did we have it roll over, it was when the lower wheel slipped into a hole where a rock had been. However the two of us easily rolled it back onto it's wheels.

The bulldozer was designed to push the soil either forward, or to either side. It was pivoted in the centre and the straight parts of the push bars (which were made of tubular steel used to make tripods for lifting logs of wood,) telescoped with pins through them to lock each position.

The bulldozer depth wheels and their screws were scrap from something else but could be simply wheels attached to a sliding upright bar with ways of fixing it at different positions, perhaps by welding a bolt to the top which could be raised or lowered and locked by two lock nuts.

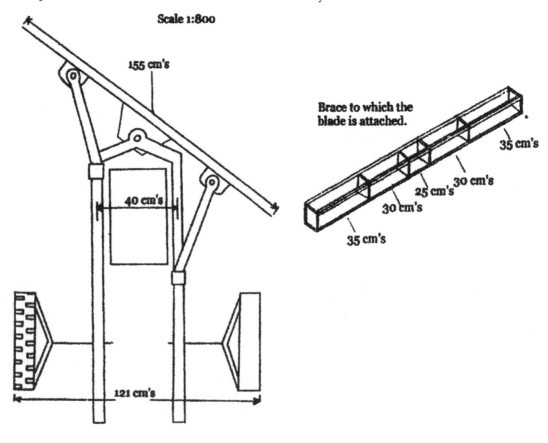

Scale 1:800

155 cm's

40 cm's

121 cm's

Brace to which the blade is attached.

35 cm's

30 cm's

25 cm's

30 cm's

30 cm's

35 cm's

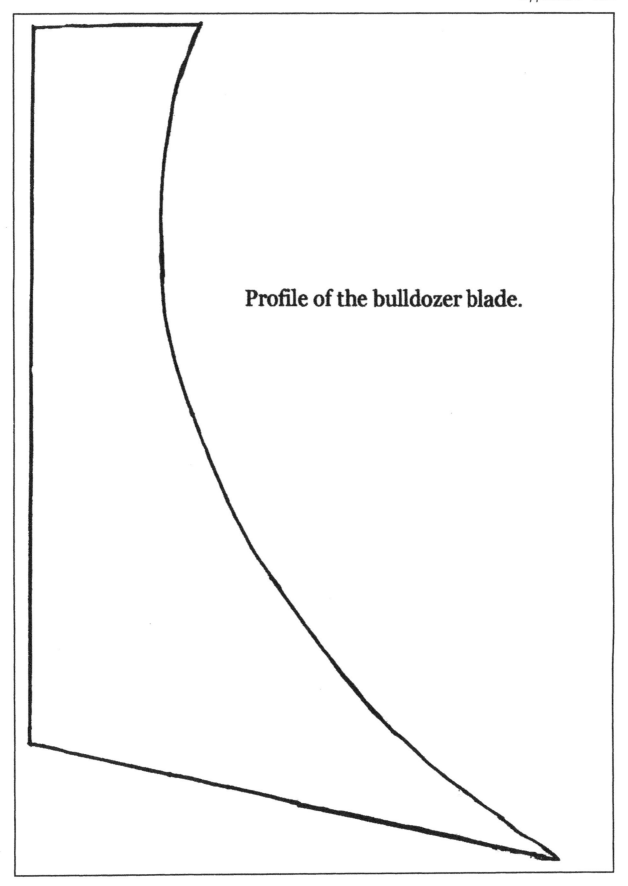

Profile of the bulldozer blade.

SOME GRASSES AND LEGUMES FOR USE IN SOIL CONSERVATION

Grasses and legumes are very important for soil and water conservation, and also for improving the soil quality. This includes covering the soil surface from the harmful effects of sunlight and weather; and their use in providing fodder and bedding for livestock which in turn provide manure with which to enrich and improve the soil. They may also provide material to be ploughed in as green manure or used as a surface mulch to conserve moisture.

Books on soil conservation often say that excess surface water should be diverted to grassed waterways, some say that suitable grasses and legumes should be grown on banks and terrace walls; however I have not found any that give much description of suitable species for this purpose. It may well be considered that it is too big a subject. The author believes it is a very important subject and so it is covered in detail here.

Though the following does not describe all the possible species it does give a good selection. The information is gleaned from many sources, together with the experience of the author.

The grasses and legumes described are suitable for a wide range of conditions; the hope and expectation is that readers working in widely differing situations will be able to find some species suitable for their needs. They should then conduct trials including selected local species, to discover those best for soil conservation in the conditions they meet with.

Grasses

AP5.1   Qualities which should be considered in grasses before their use in soil conservation.

AP5.2   Index of grasses which are suitable for soil conservation and their qualities as enumerated in the list above.

AP5.3   Selecting grasses which may be used for soil conservation and which meet specific conditions

AP5.3a   Section one. Grasses which are particularly tolerant of/or can survive certain conditions.

AP5.3b   section two. Grasses tolerant of a wide range of conditions.

AP5.3c   section three. Grasses suited to difficult conditions.

AP5.3d   section four. Climatic conditions.

AP5.3e   section five. Particularly for erosion control.

AP5.3f   section six. Values related to grazing/fodder.

AP5.3g   section seven. Growing conditions.

AP5.4   Description of the grasses.

Legumes for soil conservation

AP5.5   What is the role of legumes in soil conservation ?

AP5.6   Qualities of pasture/fodder legumes which may be used for soil conservation.

AP5.7   Index of the qualities of 35 legumes which are suitable for soil conservation.

AP5.8   Selecting pasture/fodder legumes which may be used for soil conservation and which meet specific conditions.

AP5.8a   section one. Legumes which are particularly tolerant of/or can survive certain conditions.

AP5.8b   section two. Legumes which are tolerant of a wide range of conditions.

AP5.8c   section three. Legumes suited to difficult conditions.

AP5.8d   section four. Climatic conditions.

AP5.8e   section five. Particularly for erosion control.
AP5.8f   section six. Values related to grazing.
AP5.8g   section seven. Growing condition.
AP5.8h   section eight. Special features.
AP5.9 A description of suitable legumes.
AP5.10 A list for readers who would like further information, or to obtain seeds or other propagating material for local trials.

Note 1. It is important to realise that there can be quite a difference in performance between different varieties or selections of the same species; also that the selection which is best in one location or season may give much different results in another location, or another season with different weather conditions.

Note 2. It is often best for sustainability to use a mixture of legume and grass species; they complement each other and there is less chance of the soil becoming exhausted, and so more prone to erosion. When this happens weeds will start to take over the area; these may be a nuisance to surrounding crops and may not be as good for soil conservation, even though they may in the longer term improve the soil fertility.

Note 3. It is important to give introduced plants time to adapt to particular soils and conditions. Sometimes plants which at first do not show well, will, after one or two seasons adapt themselves and do very well. Sometimes we need to learn how to manage the introduced species; an example of this is the authors experience with Greenleaf Desmodium. Where I first worked in Taiwan this species had been sown and had hardly grown. It was difficult to find any of it among the razor grass type of Imperata cylindrica which grew everywhere if it had a chance. We cut the grass down twice to weaken it prior to digging it up. Then the Greenleaf Desmodium which was also cut sprang into life and soon we had a three feet (1 m) high crop of high quality fodder which choked out the razor grass. We then found that if we planted as cuttings, lengths of stem about 9" (23 cm) long, at an angle of 45° = 100% slope; with 2/3 in the soil and the base end butting against soil, we had much quicker and better development.

When I went to Nepal, I asked about Greenleaf Desmodium, and at two research stations was told it was of no use. However, I tried Greenleaf and it seemed to perform as in Taiwan. I asked one grass nursery if I could have more planting material, and they told me I could have it all it was of no value. Later after we had shown it was in fact very suitable they asked for planting material back. I think they had been too gentle with it. If left it is weak and straggly, if cut it becomes vigorous and produces much high quality feed, palatable to all classes of farm livestock except dogs. It is very useful for soil conservation as it both protects and binds the soil even on vertical surfaces.

AP5.1 Qualities which should be considered in grasses and legumes before their use in Soil Conservation.

0. Acid soil tolerant.
1. Alkaline soil tolerant.
2. Cold tolerant.
3. Drought tolerant.
4. Flooding tolerant.
5. Heat tolerant.
6. Poor soil tolerant.
7. Salt tolerant.
8. Shade tolerant.
9. Wide range of pH tolerant.
10. Wide range of temperatures tolerant.
11. Hardy.
12. More temperate.
13. More tropical.
14. Good for erosion control.
15. Good for protecting soil.
16. Good for binding soil.
17. Good for improving soil.
18. Good for waterways.
19. Good grazing/fodder.

20. Good for summer grazing.
21. Good for winter grazing.
22. Can be grazed.
23. Heavy grazing tolerant.
24. Combines well with legumes.
25. Vigorous grower.
26. Quick growth after rain.
27. Responds well to winter rain.
28. Responds well to summer rain.
29. Long growing period.
30. May become a weed.
31. Grows on a wide range of soils.
32. Grows on light soil.
33. Grows on heavy soil.
34. Grows on badly drained soil.
35. Grows on waterlogged soil.
36. Long lived.
37. Sod forming.
40. Wide adaptation.
41. Survives fire.
42. Sensitive to waterlogging.
43. Survives frost.
44. Produces a lot of fodder.
45. Good seedling vigour.
46. Grows on acid soil.
47. Does not tolerate salinity.
48. Not good with legumes.
49. Grows with a wide range of rainfall.
50. Does best in moist conditions.
51. Can exhaust the soil.
52. Prefers well drained land.
53. Can tolerate heavy treading.
54. Encourages rats.
55. Good pioneer crop.
56. Suffers from frost.
57. Produces good standover feed (uncut hay) for winter.
58. Can stand light frost.
59. Exceptionally drought resistant.
60. Very tolerant of low fertility.
61. Can grow at high altitudes.
62. Has early spring growth.
63. Gives worthwhile winter feed.
64. Good for contour strips.

AP5.2 Index of Grasses which are suitable for Soil Conservation and their Qualities a enumerated in the List Above

African Star grass *(Cynodon plectostachyus)*.18 31 4 22 30 13 16 59.
Bahia grass *(Paspalum notatum)*.3 4 6 10 11 13 14 15 16 18 22 23 32 37 49 52 53 59 60 61 64.
Birdwood grass *(Cenchrus setigerus)*.6 13 3 25 27 19 32 24 26.
Blue grama *(Bouteloua gracilis)*.31 10 3 9 19 14 38 40 43.
Buffalograss *(Buchloe dactyloides)*.37 15 38 19 40 10 23 14 31 53 43.

Buffel grass *(Cenchrus ciliaris)*. 13 3 23 16 41 26 21 10 42 31 25 44 58 6 45 29 14 59.

Canary grass see Phalaris.

Carpet grass *(Axonopus affinis)*.46 32 4 47 12 5 37 16 48 0 50 34.

Centipede grass *(Eremochloa ophiuroides)*.0 46 47 6 31 16 8 4 2 25 30 50 43.

Crested wheatgrass *(Agropyron cristatum)*, see wheatgrasses.1 2 3 5 61 59 23 62 29 19 31 40 20 11 45 10 22 44 49 43.

Elephant grass. see Napier grass.

Eragrostis obtusiflora. See lovegrasses.3 16 1 44 19 5 2.

Green panic *(Panicum maximum* var trichoglume).5 25 12 24 8 3 11 49 27 41 23 6 53 55 43 59 62.

Guinea grass *(Panicum maximum)*.13 5 11 3 16 25 50 8 41 20 24 51 31 9 46 19 44 26 43 21 59 58.

Harding grass see Phalaris.

Kikiyu grass *(Pennisetum clandestinum)*.20 38 16 14 44 10 49 12 52 39 19 23 37 3 2 29 50 18 30 53.

Kus, see Vetiver grass.

Lovegrasses.*(Eragrostis species)*.6 44 15 14 55 60.

Makarikari grass, *(Panicum coloratum)* var. makarikariense.4 16 3 43 11 20 21 33 63 34 59 23 31.

Mat grass *(Axonopus affinis)*.30 6 13 55 60.

Mesquite grasses *(Hilaria)*.1 7 32 3.

Molasses grass *(Melinis minutiflora)*.14 15 16 37 13 39 5 6 31 40 46 42 3 10 20 64 62 29 58 55 18.

Napier Grass, *(Pennisetum purpureum)*.51 44 54 19 13 18 16 14 31 50 42 25 20.

Paspalum grass *(Paspalum dilatatum)*.37 23 53 20 26 33 3 58.

Phalaris/canary grass/Harding grass. *(Phalaris tuberosa)*.12 36 29 19 3 2 4 18 62 63 59 34 7 31 21.

Plicatulum *(Paspalum plicatulum)*.11 19 6 55 3 4 20 24 35 56 29 57 5 13 58 44 62.

Rhodes grass *(Chloris gayana)*.11 25 13 16 55 39 38 41 23 58 3 14 42 7 24 31.

Setaria *(Setaria sphacelata)*.9 10 13 6 34 31 40 19 25 44 29 24 64 4 58 18.

Signal grass *(Brachiaria decumbens)*.60 13 44 19 23 58.

South African pigeon grass see Setaria.

Vetiver grass. *(Vetiveria zizanioides)*.14 2 4 35 7 9 31 6 40 49 10 62 64.

Weeping Lovegrass *(Eragrostis curvula)*, see Lovegrasses.32 10 44 45 9 31 5 2 3 38 55 43 62 60.

Western wheatgrass *(Agropyron smithii)*, see Wheatgrasses.3 11 37 9 40 31 1 25 33 55 18 38 14 15 16 64 62 19 44 45 5 10 22 49 43 61.

Wheatgrasses (Agropyron spp).11 3 40 19 44 61 5 2 38 31 9 45 22 49 62 55 16 14 10 43.

AP5.3 Selecting Grasses which may be used for Soil Conservation and which meet specific conditions

This table aids in choosing grasses suitable for particular conditions. Select the most important condition and look to see which grasses are listed as suitable for that condition, then compare their other qualities to select possible species for the particular situation being considered.

Note:- The recommendations below are derived from careful analysis of the description of the grasses given in this chapter. The descriptions are drawn from a compilation of many sources. The author can therefore only use his judgement based on those sources. Information may in certain cases be incomplete or differ from a readers own experience; that is unfortunate but cannot be helped. For instance if no source mentions that a particular grass is sensitive to frost, sensitivity to frost will not be noted. The grass may in fact be sensitive to frost but never normally be grown in areas where frost occurs. The reader must consider that if it is a tropical grass it is likely to suffer from frost unless resistance to frost is noted.

INDEX OF SECTION TITLES:

SECTION 1. GRASSES WHICH ARE PARTICULARLY TOLERANT OF/OR CAN SURVIVE CERTAIN CONDITIONS. NOTE:-There are other grasses which are adaptable to a wide range of conditions. (SEE SECTION 2.)

1.-13.

SECTION 2. GRASSES TOLERANT OF A WIDE RANGE OF CONDITIONS.

14.-18.

SECTION 3. GRASSES SUITED TO DIFFICULT CONDITIONS.
19.-30.
SECTION 4. CLIMATIC CONDITIONS.
31.-36.
SECTION 5. PARTICULARLY FOR EROSION CONTROL.
37.-53.
SECTION 6. VALUES RELATED TO GRAZING/FODDER.
54.-74.
SECTION 7. GROWING CONDITIONS.
75.-105.

AP5 3a Section One. Grasses which are particularly tolerant of/or can survive certain conditions.

NOTE:-There are other grasses which are adaptable to a wide range of conditions. SEE SECTION 2.

0.   Grasses which are particularly tolerant of acid soil.
        Carpet grass (Axonopus affinis).
        Centipede grass (Eremochloa ophiuroides).

1.   Grasses which are particularly tolerant of alkaline soil.
        Eragrostis obtusiflora. See under Lovegrasses
        Lovegrasses.(Eragrostis species).
        Weeping lovegrass *(Eragrostis curvula)*. See under Lovegrasses.
        Mesquite grasses (Hilaria).
        Wheatgrasses *(Agropyron spp)*.
        Crested wheatgrass *(A.cristatum)*.See under Wheatgrasses.
        Western wheatgrass *(Agropyron smithii)*.See under Wheatgrasses.

2.   Cold tolerant grasses.
        Centipede grass *(Eremochloa ophiuroide)*.
        Crested wheatgrass *(Agropyron cristat)*.
        Eragrostis obtusiflora. See lovegrasses.
        Green panic *(Panicum maximum var trichoglume)*.
        Kikiyu grass *(Pennisetum clandestinum)*.
        Vetiver grass *(Vetiveria zizanioides)*.
        Wheatgrasses *(Agropyron spp)*.
        Western wheatgrass *(Agropyron smithii)*.

3.   Drought tolerant grasses.
        Bahia grass.*(Paspalum notatum)*.
        Birdwood grass *(Cenchrus setigerus)*.
        Blue grama *(Bouteloua gracilis)*.
        Buffel grass *(Cenchrus ciliaris)*.
        Crested wheatgrass *(Agropyron cristatum)*.
        Eragrostis obtusiflora. See lovegrasses.
        Green panic *(Panicum maximum var trichoglume)*.
        Guinea grass *(Panicum maximum)*.
        Kikiyu grass *(Pennisetum clandestinum)*.
        Makarikari grass, *(Panicum coloratum)* var. makarikariense.
        Mesquite grasses *(Hilaria)*.
        Molasses grass *(Melinis minutiflora)*.
        Paspalum grass *(Paspalum dilatatum)*.
        Phalariscanary grass Harding grass.
        Plicatulum *(Paspalum plicatulum)*.
        Rhodes grass *(Chloris gayana)*.

Weeping Lovegrass *(Eragrostis curvula)*.
Western wheatgrass *(Agropyron smithii)*.
Wheatgrasses (Agropyron spp).

3a.   Exceptionally drought resistant.
African Star grass.*(Cynodon plectostachyus)*.
Buffel grass *(Cenchrus ciliaris)*.
Crested wheatgrass *(Agropyron cristatum)*.
Green panic *(Panicum maximum* var trichoglume)
Guinea grass *(Panicum maximum)*.
Makarikari grass *(Panicum coloratum)*.
Phalariscanary grass Harding grass.

4.   Flooding tolerant.
African Star grass.*(Cynodon plectostachyus)*.
Bahia grass.*(Paspalum notatum)*.
Carpet grass *(Axonopus affinis)*.
Centipede grass *(Eremochloa ophiuroide)*.
Makarikari grass, *(Panicum coloratum)* var. makarikariense.
Phalariscanary grassHarding grass. *(Phalaris tuberosa)*.
Plicatulum *(Paspalum plicatulum)*.
Vetiver grass. *(Vetiveria zizanioides)*.
Napier Grass, *(Pennisetum purpureum)*.

5.   Heat tolerant.
Buffel grass *(Cenchrus ciliaris)*.
Carpet grass *(Axonopus affinis)*.
Crested wheatgrass *(Agropyron cristatum)*.
Eragrostis obtusiflora. See lovegrasses.
Green panic *(Panicum maximum var trichoglume)*.
Guinea grass *(Panicum maximum)*.
Molasses grass *(Melinis minutiflora)*.
Plicatulum *(Paspalum plicatulum)*.
Weeping lovegrass *(Eragrostis curvula)*.
Western wheatgrass *(Agropyron smithii)*.
Wheatgrasses *(Agropyron spp)*.

6.   Poor soil tolerant.
Bahia grass.*(Paspalum notatum)*.
Birdwood grass *(Cenchrus setigerus)*.
Buffel grass *(Cenchrus ciliaris)*.
Centipede grass *(Eremochloa ophiuroides)*.
Lovegrasses.*(Eragrostis species)*.
Mat grass *(Axonopus affinis)*.
Molasses grass *(Melinis minutiflora)*.
Plicatulum *(Paspalum plicatulum)*.

6a.   Very tolerant of low fertility.
Lovegrasses *(Eragrostis species)*.
Mat grass *(Axonopus affinis)*.
Signal grass *(Brachiaria decumbens)*.
Weeping Lovegrass *(Eragrostis curvula)*.
Eragrostis obtusiflora. See lovegrasses.

7.   Salt tolerant.
Mesquite grasses *(Hilaria)*.

Phalariscanary grassHarding grass.
Rhodes grass (*Chloris gayana*).
Vetiver grass. (*Vetiveria zizanioides*).

8. Shade tolerant.
Centipede grass (*Eremochloa ophiuroides*).
Green panic (*Panicum maximum var trichoglume*).
Guinea grass (*Panicum maximum*).

9. Survives fire.
Buffel grass (*Cenchrus ciliaris*).
Green panic (*Panicum maximum var trichoglume*)
Guinea grass (*Panicum maximum*).
Rhodes grass (*Chloris gayana*).

10. Survives frost.
Blue grama (*Bouteloua gracilis*).
Buffalograss.(*Buchloe dactyloides*).
Centipede grass (*Eremochloa ophiuroides*).
Green panic (*Panicum maximum var trichoglume*)
Guinea grass (*Panicum maximum*).
Makarikari grass, (*Panicum coloratum*).
Weeping Lovegrass (*Eragrostis curvula*).
Kikiyu grass (*Pennisetum clandestinum*).

10a. Can stand light frost.
Buffel grass (*Cenchrus ciliaris*).
Guinea grass (*Panicum maximum*).
Molasses grass (*Melinis minutiflora*).
Paspalum grass (*Paspalum dilatatum*).
Plicatulum (*Paspalum plicatulum*).
Rhodes grass (*Chloris gayana*).
Signal grass (*Brachiaria decumbens*).
Napier Grass (*Pennisetum purpureum*).

11. Can grow at high altitude.
Crested wheatgrass (*Agropyron cristatum*).
Western wheatgrass (*Agropyron smithii*).
Wheatgrasses (*Agropyron spp*).
Kikiyu grass (*Pennisetum clandestinum*).

12. Tolerant of heavy grazing.
Bahia grass.(*Paspalum notatum*).
Buffalograss.(*Buchloe dactyloides*).
Buffel grass (*Cenchrus ciliaris*).
Crested wheatgrass (*Agropyron cristatum*).
Green panic (*Panicum maximum var trichoglume*)
Kikiyu grass (*Pennisetum clandestinum*).
Makarikari grass, (*Panicum coloratum*).
Paspalum grass (*Paspalum dilatatum*).
Rhodes grass (*Chloris gayana*).
Signal grass (*Brachiaria decumbens*).

13. Tolerant of heavy treading.
Bahia grass.(*Paspalum notatum*).
Buffalograss.(*Buchloe dactyloides*).
Green panic (*Panicum maximum* var trichoglume)

Kikiyu grass *(Pennisetum clandestinum)*.
Paspalum grass *(Paspalum dilatatum)*.

Ap5.3b Section Two. Grasses tolerant of a wide range of conditions.

14.   Wide range of pH tolerant.
Blue grama *(Bouteloua gracilis)*.
Guinea grass *(Panicum maximum)*.
Setaria *(Setaria sphacelata)*.
Weeping Lovegrass *(Eragrostis curvula)*.
Western wheatgrass *(Agropyron smithii)*.
Wheatgrasses *(Agropyron spp)*.

15.   Wide range of temperature tolerant.
Bahia grass.*(Paspalum notatum)*.
Blue grama *(Bouteloua gracilis)*.
Buffalo grass.*(Buchloe dactyloides)*.
Buffel grass *(Cenchrus ciliaris)*.
Crested wheatgrass *(Agropyron cristatum)*.
Kikiyu grass *(Pennisetum clandestinum)*.
Molasses grass *(Melinis minutiflora)*.
Setaria *(Setaria sphacelata)*.
Weeping Lovegrass *(Eragrostis curvula)*.
Western wheatgrass *(Agropyron smithii)*.
Wheatgrasses *(Agropyron spp.)*.

16.   Wide range of soils tolerant.
African Star grass.*(Cynodon plectostachyus)*.
Blue grama *(Bouteloua gracilis)*.
Buffalograss.*(Buchloe dactyloides)*.
Centipede grass *(Eremochloa ophiuroides)*.
Crested wheatgrass *(Agropyron cristatum)*.
Guinea grass *(Panicum maximum)*.
Makarikari grass, *(Panicum coloratum)*.
Molasses grass *(Melinis minutiflora)*.
Phalariscanary grass Harding grass.
Rhodes grass *(Chloris gayana)*.
Setaria *(Setaria sphacelata)*.
Vetiver grass. *(Vetiveria zizanioides)*.
Weeping Lovegrass *(Eragrostis curvula)*.
Western wheatgrass *(Agropyron smithii)*.
Wheatgrasses *(Agropyron spp)*.

17.   Wide adaptation.
Blue grama *(Bouteloua gracilis)*.
Crested wheatgrass *(Agropyron cristatum)*.
Molasses grass *(Melinis minutiflora)*.
Setaria *(Setaria sphacelata)*.
Vetiver grass. *(Vetiveria zizanioides)*.
Western wheatgrass *(Agropyron smithii)*.
Wheatgrasses *(Agropyron spp)*.

18.   Wide range of rainfall tolerant.
Bahia grass.*(Paspalum notatum)*.
Green panic *(Panicum maximum* var trichoglume)
Kikiyu grass *(Pennisetum clandestinum)*.

Vetiver grass. *(Vetiveria zizanioides)*.
Wheatgrasses *(Agropyron spp)*.

AP5.3c Section three. Grasses suited to difficult conditions.

19. Hardy.
    Bahia grass.*(Paspalum notatum)*.
    Crested wheatgrass *(Agropyron cristatum)*.
    Green panic *(Panicum maximum* var trichoglume)*.
    Guinea grass *(Panicum maximum)*.
    Kikiyu grass *(Pennisetum clandestinum)*.
    Makarikari grass *(Panicum coloratum)*.
    Plicatulum *(Paspalum plicatulum)*.
    Rhodes grass *(Chloris gayana)*.
    Western wheatgrass *(Agropyron smithii)*.
    Wheatgrasses *(Agropyron spp)*.

20. Heavy grazing, see 12.

21. Heavy treading, see 13.

22. Fire, see 9.

23. High altitude, see 11.

24. Wide range of temperatures, see 15.

25. Wide range of rainfall, see 18.

26. Very tolerant of low fertility, see 6a.

27. Pioneer conditions.
    Crested wheatgrass *(Agropyron cristatum)*.
    Green panic *(Panicum maximum* var trichoglume)*.
    Lovegrasses *(Eragrostis species)*.
    Mat grass *(Axonopus affinis)*.
    Molasses grass *(Melinis minutiflora)*.
    Plicatulum *(Paspalum plicatulum)*.
    Rhodes grass *(Chloris gayana)*.
    Weeping Lovegrass *(Eragrostis curvula)*.
    Western wheatgrass *(Agropyron smithii)*.
    Wheatgrasses *(Agropyron spp)*.

28. Flooding tolerant.
    African Star grass.*(Cynodon plectostachyus)*.
    Bahia grass.*(Paspalum notatum)*.
    Carpet grass *(Axonopus affinis)*.
    Centipede grass *(Eremochloa ophiuroide)*.
    Makarikari grass, *(Panicum coloratum)* var. makarikariense.
    Phalariscanary grass Harding grass. *(Phalaris tuberosa)*.
    Plicatulum *(Paspalum plicatulum)*.
    Vetiver grass. *(Vetiveria zizanioides)*.
    Napier Grass, *(Pennisetum purpureum)*.

29. Grows on Waterlogged soil.
    Plicatulum *(Paspalum plicatulum)*.
    Vetiver grass. *(Vetiveria zizanioides)*.

30. Grows on badly drained soil.
    Canary grass see Phalaris.

Makarikari grass *(Panicum coloratum).*
Phalariscanary grass Harding grass.
Setaria *(Setaria sphacelata).*

AP5.3d Section four. Climatic conditions

31.   Wide range of temperature tolerant.
Blue grama *(Bouteloua gracilis).*
Buffalograss.*(Buchloe dactyloides).*
Molasses grass *(Melinis minutiflora).*
Setaria *(Setaria sphacelata).*
Vetiver grass. *(Vetiveria zizanioides).*
Weeping Lovegrass *(Eragrostis curvula).*

32.   More temperate.
Carpet grass *(Axonopus affinis).*
Green panic *(Panicum maximum* var trichoglume).
Guinea grass *(Panicum maximum).*
Kikiyu grass *(Pennisetum clandestinum).*
Phalariscanary grass Harding grass.

33.   More tropical.
African Star grass.*(Cynodon plectostachyus).*
Bahia grass.*(Paspalum notatum).*
Birdwood grass *(Cenchrus setigerus).*
Buffel grass *(Cenchrus ciliaris).*
Guinea grass *(panicum maximum).*
Mat grass *(Axonopus affinis).*
Molasses grass *(Melinis minutiflora).*
Napier Grass *(Pennisetum purpureum).*
Plicatulum *(Paspalum plicatulum).*
Setaria *(Setaria sphacelata).*
Signal grass *(Brachiaria decumbens).*

34.   Flooding occurs.
African Star grass. *(Cynodon plectostachyus).*
Carpet grass *(Axonopus affinis).*
Centipede grass *(Eremochloa ophiuroide).*
Makarikari grass, *(Panicum coloratum)* var. makarikariense.
Phalariscanary grassHarding grass. *(Phalaris tuberosa).*
Plicatulum *(Paspalum plicatulum).*
Vetiver grass. *(Vetiveria zizanioides).*
Napier Grass *(Pennisetum purpureum).*

35.   Shade.
Centipede grass *(Eremochloa ophiuroides).*
Green panic *(Panicum maximum* var trichoglume).
Guinea grass *(Panicum maximum).*

36.   Grows with a wide range of rainfall, see 18.
Bahia grass.*(Paspalum notatum).*
Green panic *(Panicum maximum* var trichoglume)
Kikiyu grass *(Pennisetum clandestinum).*
Vetiver grass. *(Vetiveria zizanioides).*
Wheatgrasses *(Agropyron spp).*

AP5.3e Section five. Particularly for erosion control.

37. Good for erosion control.
    Bahia grass.*(Paspalum notatum)*.
    Blue grama *(Bouteloua gracilis)*.
    Buffalograss.*(Buchloe dactyloides)*.
    Buffel grass *(Cenchrus ciliaris)*.
    Kikiyu grass *(Pennisetum clandestinum)*.
    Lovegrasses.*(Eragrostis species)*.
    Molasses grass *(Melinis minutiflora)*.
    Napier grass *(Pennisetum purpureum)*.
    Rhodes grass *(Chloris gayana)*.
    Vetiver grass. *(Vetiveria zizanioides)*.
    Western wheatgrass *(Agropyron smithii)*.

38. Good for protecting soil.
    Bahia grass.*(Paspalum notatum)*.
    Buffalograss.*(Buchloe dactyloides)*.
    Lovegrasses *(Eragrostis species)*.
    Molasses grass *(Melinis minutiflora)*.
    Western wheatgrass *(Agropyron smithii)*.

39. Good for binding the soil.
    African Star grass.*(Cynodon plectostachyus)*.
    Bahia grass.*(Paspalum notatum)*.
    Buffel grass *(Cenchrus ciliaris)*.
    Carpet grass *(Axonopus affinis)*.
    Centipede grass *(Eremochloa ophiuroides)*.
    Eragrostis obtusiflora. See lovegrasses.
    Guinea grass *(Panicum maximum)*.
    Kikiyu grass *(Pennisetum clandestinum)*.
    Makarikari grass *(Panicum coloratum)*.
    Molasses grass *(Melinis minutiflora)*.
    Napier Grass *(Pennisetum purpureum)*.
    Rhodes grass *(Chloris gayana)*.
    Western wheatgrass *(Agropyron smithii)*.

40. Sod forming.
    Bahia grass.*(Paspalum notatum)*.
    Buffalograss.*(Buchloe dactyloides)*.
    Carpet grass *(Axonopus affinis)*.
    Kikiyu grass *(Pennisetum clandestinum)*.
    Molasses grass *(Melinis minutiflora)*.
    Paspalum grass *(Paspalum dilatatum)*.
    Western wheatgrass *(Agropyron smithii)*.

41. Good for waterways.
    African Star grass.*(Cynodon plectostachyus)*.
    Bahia grass.*(Paspalum notatum)*.
    Kikiyu grass *(Pennisetum clandestinum)*.
    Molasses grass *(Melinis minutiflora)*.
    Napier Grass *(Pennisetum purpureum)*.
    Phalariscanary grass Harding grass.
    Western wheatgrass *(Agropyron smithii)*.

42. Good for contour strips.
    Bahia grass.*(Paspalum notatum)*.

Molasses grass *(Melinis minutiflora)*.
Phalariscanary grass Harding grass.
Setaria *(Setaria sphacelata)*.
Vetiver grass *(Vetiveria zizanioides)*.
Western wheatgrass *(Agropyron smithii)*.

43. Vigourous grower.
    Birdwood grass *(Cenchrus setigerus)*.
    Buffel grass *(Cenchrus ciliaris)*.
    Centipede grass *(Eremochloa ophiuroides)*.
    Green panic *(Panicum maximum var trichoglume)*
    Guinea grass *(Panicum maximum)*.
    Napier Grass *(Pennisetum purpureum)*.
    Rhodes grass *(Chloris gayana)*.
    Setaria *(Setaria sphacelata)*.
    Western wheatgrass *(Agropyron smithii)*.

44. Quick growth after rain.
    Birdwood grass *(Cenchrus setigerus)*.
    Buffel grass *(Cenchrus ciliaris)*.
    Guinea grass *(Panicum maximum)*.
    Paspalum grass *(Paspalum dilatatum)*.

45. Long growing period.
    Buffel grass *(Cenchrus ciliaris)*.
    Crested wheatgrass *(Agropyron cristatum)*.
    Kikiyu grass *(Pennisetum clandestinum)*.
    Molasses grass *(Melinis minutiflora)*.
    Phalariscanary grass Harding grass.
    Plicatulum *(Paspalum plicatulum)*.
    Setaria *(Setaria sphacelata)*.
    Wheatgrasses *(Agropyron spp)*.
    Western wheatgrass *(Agropyron smithii)*.

46. Long life.
    Phalariscanary grass Harding grass.

47. Ease of establishment.
    Bahia grass *(Paspalum notatum)*.
    Kikiyu grass *(Pennisetum clandestinum)*.
    Molasses grass *(Melinis minutiflora)*.
    Rhodes grass *(Chloris gayana)*.

48. Wide adaptation.
    Blue grama *(Bouteloua gracilis)*.
    Buffalograss *(Buchloe dactyloides)*.
    Molasses grass *(Melinis minutiflora)*.
    Setaria *(Setaria sphacelata)*.
    Vetiver grass *(Vetiveria zizanioides)*.
    Western wheatgrass *(Agropyron smithii)*.
    Wheatgrasses *(Agropyron spp)*.

49. Fire resistant.
    Buffel grass *(Cenchrus ciliaris)*.
    Green panic *(Panicum maximum* var trichoglume)*.
    Guinea grass *(Panicum maximum)*.
    Rhodes grass *(Chloris gayana)*.

50.   Good pioneer crop.
      Crested wheatgrass *(Agropyron cristatum).*
      Green panic *(Panicum maximum var trichoglume).*
      Lovegrasses *(Eragrostis species).*
      Mat grass *(Axonopus affinis).*
      Molasses grass *(Melinis minutiflora).*
      Plicatulum *(Paspalum plicatulum).*
      Rhodes grass *(Chloris gayana).*
      Weeping Lovegrass *(Eragrostis curvula).*
      Western wheatgrass *(Agropyron smithii).*
      Wheatgrasses *(Agropyron spp).*

51.   Can tolerate heavy treading.
      Bahia grass *(Paspalum notatum).*
      Buffalograss *(Buchloe dactyloides).*
      Green panic *(Panicum maximum* var trichoglume)
      Kikiyu grass *(Pennisetum clandestinum).*
      Paspalum grass *(Paspalum dilatatum).*

52.   Poor soil tolerant.
      Bahia grass *(Paspalum notatum).*
      Birdwood grass *(Cenchrus setigerus).*
      Buffel grass *(Cenchrus ciliaris).*
      Centipede grass *(Eremochloa ophiuroides).*
      Lovegrasses *(Eragrostis species).*
      Mat grass *(Axonopus affinis).*
      Molasses grass *(Melinis minutiflora).*
      Plicatulum *(Paspalum plicatulum).*

52a   Very tolerant of low fertility.
      Bahia grass *(Paspalum notatum).*
      Lovegrasses *(Eragrostis species).*
      Mat grass *(Axonopus affinis).*
      Signal grass *(Brachiaria decumbens).*
      Weeping Lovegrass *(Eragrostis curvula).*
      Eragrostis obtusiflora. See lovegrasses.

53.   Tolerant of flooding.
      African Star grass *(Cynodon plectostachyus).*
      Bahia grass *(Paspalum notatum).*
      Carpet grass *(Axonopus affinis).*
      Centipede grass *(Eremochloa ophiuroide).*
      Makarikari grass, *(Panicum coloratum)* var. makarikariense.
      Phalariscanary grass Harding grass *(Phalaris tuberosa).*
      Plicatulum *(Paspalum plicatulum).*
      Vetiver grass *(Vetiveria zizanioides).*
      Napier Grass *(Pennisetum purpureum).*

AP5.3f Section 6. Values related to grazingfodder.

54.   Good for grazing and or fodder.
      Bahia grass.*(Paspalum notatum).*
      Birdwood grass *(Cenchrus setigerus).*
      Blue grama *(Bouteloua gracilis).*
      Buffalograss *(Buchloe dactyloides).*
      Crested wheatgrass *(Agropyron cristatum).*

Eragrostis obtusiflora. See lovegrasses.
Guinea grass *(Panicum maximum)*.
Napier Grass *(Pennisetum purpureum)*.
Phalariscanary grass Harding grass.
Plicatulum *(Paspalum plicatulum)*.
Setaria *(Setaria sphacelata)*.
Signal grass *(Brachiaria decumbens)*.
Western wheatgrass *(Agropyron smithii)*.
Wheatgrasses *(Agropyron spp)*.

55. Produces a lot of fodder.
Buffel grass *(Cenchrus ciliaris)*.
Crested wheatgrass *(Agropyron cristatum)*.
Eragrostis obtusiflora. See lovegrasses.
Guinea grass *(Panicum maximum)*.
Kikiyu grass *(Pennisetum clandestinum)*.
Lovegrasses *(Eragrostis species)*.
Napier Grass *(Pennisetum purpureum)*.
Plicatulum *(Paspalum plicatulum)*.
Setaria *(Setaria sphacelata)*.
Signal grass *(Brachiaria decumbens)*.
Weeping Lovegrass *(Eragrostis curvula)*.
Western wheatgrass *(Agropyron smithii)*.
Wheatgrasses *(Agropyron spp)*.

56. Good for summer grazingfodder.
Crested wheatgrass *(Agropyron cristatum)*.
Guinea grass *(Panicum maximum)*.
Makarikari grass *(Panicum coloratum)*.
Molasses grass *(Melinis minutiflora)*.
Napier Grass *(Pennisetum purpureum)*.
Paspalum grass *(Paspalum dilatatum)*.
Plicatulum *(Paspalum plicatulum)*.

57. Good for winter grazing.
Buffel grass *(Cenchrus ciliaris)*.
Guinea grass *(Panicum maximum)*.
Makarikari grass *(Panicum coloratum)*.
Phalariscanary grass Harding grass.

58. Can be grazed.
Bahia grass *(Paspalum notatum)*.
Buffalograss *(Buchloe dactyloides)*.
Buffel grass *(Cenchrus ciliaris)*.
African Star grass *(Cynodon plectostachyus)*.
Crested wheatgrass *(Agropyron cristatum)*.
Kikiyu grass *(Pennisetum clandestinum)*.
Makarikari grass *(Panicum coloratum)*.
Paspalum grass *(Paspalum dilatatum)*.
Rhodes grass *(Chloris gayana)*.
Signal grass *(Brachiaria decumbens)*.
Western wheatgrass *(Agropyron smithii)*.
Wheatgrasses *(Agropyron spp)*.

59. Heavy grazing tolerant.
Bahia grass *(Paspalum notatum)*.

Buffalograss *(Buchloe dactyloides)*.
Buffel grass *(Cenchrus ciliaris)*.
Crested wheatgrass *(Agropyron cristatum)*.
Kikiyu grass *(Pennisetum clandestinum)*.
Makarikari grass *(Panicum coloratum)*.
Paspalum grass *(Paspalum dilatatum)*.
Rhodes grass *(Chloris gayana)*.
Signal grass *(Brachiaria decumbens)*.

60. Combines well with legumes.
Birdwood grass *(Cenchrus setigerus)*.
Green panic *(Panicum maximum* var trichoglume)
Guinea grass *(Panicum maximum)*.
Plicatulum *(Paspalum plicatulum)*.
Rhodes grass *(Chloris gayana)*.
Setaria *(Setaria sphacelata)*.

61. Is not a good combination with legumes.
Carpet grass *(Axonopus affinis)*.

62. Vigourous grower.
Birdwood grass *(Cenchrus setigerus)*.
Buffel grass *(Cenchrus ciliaris)*.
Centipede grass *(Eremochloa ophiuroides)*.
Green panic *(Panicum maximum* var trichoglume)
Guinea grass *(Panicum maximum)*.
Napier Grass *(Pennisetum purpureum)*.
Rhodes grass *(Chloris gayana)*.
Setaria *(Setaria sphacelata)*.
Western wheatgrass *(Agropyron smithii)*.

63. Quick growth after rain.
Birdwood grass *(Cenchrus setigerus)*.
Buffalograss *(Buchloe dactyloides)*.
Guinea grass *(Panicum maximum)*.
Paspalum grass *(Paspalum dilatatum)*.

64. Responds well to winter rain.
Birdwood grass *(Cenchrus setigerus)*.
Green panic *(Panicum maximum* var trichoglume)

65. Long growing period.
Buffel grass *(Cenchrus ciliaris)*.
Crested wheatgrass *(Agropyron cristatum)*.
Kikiyu grass *(Pennisetum clandestinum)*.
Molasses grass *(Melinis minutiflora)*.
Phalariscanary grassHarding grass.
Plicatulum *(Paspalum plicatulum)*.
Setaria *(Setaria sphacelata)*.

66. Long lived.
Phalariscanary grass Harding grass.

67. Shade tolerant.
Centipede grass *(Eremochloa ophiuroides)*.
Green panic *(Panicum maximum* var trichoglume)
Guinea grass *(Panicum maximum)*.

68. Produces good stand-over, (foggageuncut hay.)
    Plicatulum *(Paspalum plicatulum).*

69. Grows with a wide range of rainfall.
    Bahia grass *(Paspalum notatum).*
    Crested wheatgrass *(Agropyron cristatum).*
    Green panic *(Panicum maximum* var trichoglume)
    Kikiyu grass *(Pennisetum clandestinum).*
    Vetiver grass *(Vetiveria zizanioides).*
    Western wheatgrass *(Agropyron smithii).*
    Wheatgrasses *(Agropyron spp).*

70. Has early spring growth.
    Crested wheatgrass *(Agropyron cristatum).*
    Green panic *(Panicum maximum* var trichoglume)
    Molasses grass *(Melinis minutiflora).*
    Phalariscanary grass Harding grass.
    Plicatulum *(Paspalum plicatulum).*
    Vetiver grass *(Vetiveria zizanioides).*
    Weeping Lovegrass *(Eragrostis curvula).*
    Western wheatgrass *(Agropyron smithii).*
    Wheatgrasses *(Agropyron spp).*

71. Gives worthwhile winter feed.
    Makarikari grass *(Panicum coloratum).*
    Phalariscanary grassHarding grass.

72. Good for contour strips.
    Bahia grass *(Paspalum notatum).*
    Molasses grass *(Melinis minutiflora).*
    Setaria *(Setaria sphacelata).*
    Vetiver grass *(Vetiveria zizanioides).*
    Western wheatgrass *(Agropyron smithii).*

73. Encourages rats.
    Napier Grass *(Pennisetum purpureum).*

74. Can exhaust the soil.
    Guinea grass *(Panicum maximum).*
    Napier Grass *(Pennisetum purpureum).*

AP5.3g Section seven. Growing conditions.

75. Grows on a wide range of soils.
    African Star grass *(Cynodon plectostachyus).*
    Blue grama *(Bouteloua gracilis).*
    Buffalograss *(Buchloe dactyloides).*
    Buffel grass *(Cenchrus ciliaris).*
    Centipede grass *(Eremochloa ophiuroides).*
    Crested wheatgrass *(Agropyron cristatum).*
    Guinea grass *(Panicum maximum).*
    Makarikari grass *(Panicum coloratum).*
    Molasses grass *(Melinis minutiflora).*
    Napier Grass *(Pennisetum purpureum).*
    Phalariscanary grassHarding grass.
    Rhodes grass *(Chloris gayana).*
    Setaria *(Setaria sphacelata).*

Vetiver grass *(Vetiveria zizanioides)*.
Weeping Lovegrass *(Eragrostis curvula)*.
Western wheatgrass *(Agropyron smithii)*.
Wheatgrasses *(Agropyron spp)*.

76. Grows on acid soil.
    Carpet grass *(Axonopus affinis)*.
    Centipede grass *(Eremochloa ophiuroides)*.
    Guinea grass *(Panicum maximum)*.
    Molasses grass *(Melinis minutiflora)*.

77. Acid soil tolerant, see 0.

78. Alkaline soil tolerant, see 1.

79. Wide range of Ph tolerant, see 14.

80. Does not tolerate salinity.
    Carpet grass *(Axonopus affinis)*.
    Centipede grass *(Eremochloa ophiuroides)*.

81. Grows on light soils.
    Bahia grass *(Paspalum notatum)*.
    Birdwood grass *(Cenchrus setigerus)*.
    Carpet grass *(Axonopus affinis)*.
    Mesquite grasses *(Hilaria)*.
    Weeping Lovegrass *(Eragrostis curvula)*.

82. Grows on heavy soils.
    Makarikari grass *(Panicum coloratum)*.
    Paspalum grass *(Paspalum dilatatum)*.
    Western wheatgrass *(Agropyron smithii)*.

83. Grows on badly drained soil.
    Carpet grass *(Axonopus affinis)*.
    Makarikari grass *(Panicum coloratum)*.
    Phalariscanary grassHarding grass.
    Setaria *(Setaria sphacelata)*.

84. Grows on waterlogged soil.
    Plicatulum *(Paspalum plicatulum)*.
    Vetiver grass *(Vetiveria zizanioides)*.

85. Flooding tolerant, see 4.

86. Sensitive to waterlogging.
    Buffel grass *(Cenchrus ciliaris)*.
    Molasses grass *(Melinis minutiflora)*.
    Napier Grass *(Pennisetum purpureum)*.
    Rhodes grass *(Chloris gayana)*.

87. Does best in moist conditions.
    Carpet grass *(Axonopus affinis)*.
    Centipede grass *(Eremochloa ophiuroides)*.
    Guinea grass *(Panicum maximum)*.
    Kikiyu grass *(Pennisetum clandestinum)*.
    Napier Grass *(Pennisetum purpureum)*.

88. Prefers well drained land.
    Bahia grass *(Paspalum notatum)*.
    Kikiyu grass *(Pennisetum clandestinum)*.

89.   Grows with a wide range of rainfall, see 18.

90.   Heat tolerant, see 5.

91.   Exceptionally drought resistant.
      Bahia grass *(Paspalum notatum)*.
      African Star grass *(Cynodon plectostachyus)*.
      Buffel grass *(Cenchrus ciliaris)*.
      Crested wheatgrass *(Agropyron cristatum)*.
      Green panic *(Panicum maximum* var trichoglume)
      Guinea grass *(Panicum maximum)*.
      Makarikari grass *(Panicum coloratum)*.
      Phalariscanary grassHarding grass.

92.   Cold tolerant, see 2.

93.   Survives frost.
      Blue grama *(Bouteloua gracilis)*.
      Buffalograss *(Buchloe dactyloides)*.
      Centipede grass *(Eremochloa ophiuroides)*.
      Crested wheatgrass *(Agropyron cristatum)*.
      Green panic *(Panicum maximum* var trichoglume)
      Guinea grass *(Panicum maximum)*.
      Makarikari grass *(Panicum coloratum)*.
      Weeping lovegrass *(Eragrostis curvula)*.
      Western wheatgrass *(Agropyron smithii)*.
      Wheatgrasses *(Agropyron* spp*)*.
      Kikiyu grass *(Pennisetum clandestinum)*.

94.   Can stand light frost.
      Buffel grass *(Cenchrus ciliaris)*.
      Molasses grass *(Melinis minutiflora)*.
      Phalariscanary grass Harding grass.
      Plicatulum *(Paspalum plicatulum)*.
      Rhodes grass *(Chloris gayana)*.
      Setaria *(Setaria sphacelata)*.
      Signal grass *(Brachiaria decumbens)*.

95.   Suffers from frost.
      Plicatulum *(Paspalum plicatulum)*.

96.   Grows with a wide range of temperature, see 15.

97.   More temperate, see 32.

98.   More tropical, see 33.

99.   Very tolerant of low fertility.
      Bahia grass *(Paspalum notatum)*.
      Eragrostis obtusiflora. See lovegrasses.
      Lovegrasses *(Eragrostis species)*.
      Mat grass *(Axonopus affinis)*.
      Weeping lovegrass *(Eragrostis curvula)*.

100.   Wide adaptation.
       Blue grama *(Bouteloua gracilis)*.
       Buffalograss *(Buchloe dactyloides)*.
       Crested wheatgrass *(Agropyron cristatum)*.
       Molasses grass *(Melinis minutiflora)*.

Setaria *(Setaria sphacelata).*
Vetiver grass *(Vetiveria zizanioides).*
Western wheatgrass *(Agropyron smithii).*
Wheatgrasses *(Agropyron spp).*

101. Ease of establishment.
Bahia grass *(Paspalum notatum).*
Kikiyu grass *(Pennisetum clandestinum).*
Molasses grass *(Melinis minutiflora).*
Rhodes grass *(Chloris gayana).*

102. Survives fire.
Buffel grass *(Cenchrus ciliaris).*
Green panic *(Panicum maximum* var trichoglume)
Guinea grass *(Panicum maximum).*
Rhodes grass *(Chloris gayana).*

103. Can grow at high altitudes.
Bahia grass *(Paspalum notatum).*
Crested wheatgrass *(Agropyron cristatum).*
Kikiyu grass *(Pennisetum clandestinum).*
Western wheatgrass *(Agropyron smithii).*
Wheatgrasses *(Agropyron spp).*

104. Salt tolerant, see 7.

105. Shade tolerant see 8.

106. Can tolerate heavy treading, see AP5.U

AP5.4 Description of the Grasses

*****African Star grass.** *(Cynodon plectostachyus).*

"A perennial grass with creeping stems which root well. It is native to East Africa. It has been successfully used in waterways on a wide range of soils, from sands to black earths. A high cyanide content has been recorded, but the grass is still used successfully under grazing" [63.]

As it is related to couch grass or wicken *(Cynodon datylon)* which is sometimes also called Star grass, Wire grass or Bermuda grass, it can be propagated by seeds or cuttings.

Care must be taken in introducing this grass as *Cynodon datylon* is a serious weed, particularly in irrigated plantation crops.

*****Bahia grass** *(Paspalum notatum).* *"Natural habitat light well drained soil; medium to high rainfall. Sea level to 5,000 ft (1,525 m). Distribution Mexico to Argentina, East and West Africa, Florida, (Taiwan.) Habit of growth, low growing, stiff dense cover from short shallow rhizomes; up to 18" (45 cm). Palatability, rather low being improved by breeding. Nutritive value, moderately good. Persistence. Very persistent under grazing and in time of drought. Planting method, root division for soil conservation work; seed of improved varieties for pasturage. Remarks, slow growing and low yielding. Excellent for soil conservation and extensive beef production. It can also be used for grass bunds and for monsoon waterways"* [67.]

An interesting discovery is described as follows "Recent work has shown that more significant nitrogen fixation may result from the activity of bacteria occurring in the rhizospheres of many tropical grasses and cereals, where they are able to derive energy from substances present in the root exudates. For example, an unimproved cultivar of the grass *Paspalum notatum*, occurring extensively on sandy soils in Brazil—, specifically stimulates the nitrogen-fixing bacterium *Azotobacter paspali* which becomes permanently established on the root surface in association with a thick layer of mucigel developed on parts of the root system." The chapter goes on to say that most of the important tropical grasses and cereals have now been shown to possess considerable nitrogenase activity in their rhizospheres". [18 p71.]

"This is a very dense, sod forming grass which is quite productive under low levels of fertility. It seldom grows taller than 40-50 cm (16-20") and is a good grass for use on airports adjacent to runways and for special soil conservation purposes. Likewise it is recommended for steep, gravelly or rocky hillsides where soils are poor and droughty (easily suffer from drought.) and where pangola grass may not provide a dense enough cover."[64]

"Its freedom from ergot (a fungus disease) is an advantage over common Paspalum"[63]

*Birdwood grass *(Cenchrus setigerus).This bunch type grass is related to Buffel grass, and has similar characteristics. It thrives best on alluvial soils but is more adaptable to soil types, and does better than Buffel grass in poor country with less than 15" (38 cm) of rain.*

A native of India and Eastern Africa it was in a consignment of drought resisting grass seeds sent by a Lord Birdwood to Australia, hence the name.

"An exceptionally drought resistant grass, it has been grown in areas of annual rainfall as low as 8" (20 cm). B. grass is a vigorous grower and is noted for its quick response to light rains and its plentiful seeding habit." [62])

"Seed can be produced a month from germination; hence it can grow and set seed in areas of very limited rainfall. B.Grass is found in 10" rainfall country. It seeds very freely indeed. B. has a rather short season of growth, and does not respond well to winter rain. It does best on sandy or at least free draining soils. It is quite palatable (tasty) and is sought (liked) by sheep. It will withstand heavy stocking it combines well with Townsville Lucerne." [63])

For sowing and management see Buffel grass.

*Blue grama *(Bouteloua gracilis).* "Blue grama is a low growing, long lived, native perennial that grows throughout the Great Plains.

(The vertically central section of the USA. author.)

The Northern Great Plains. Cool temperate with irregular and generally deficient rainfall. Generally termed in the east dry-sub-humid, in the central and western part semi-arid. Soils vary from thick black surface soils rich in organic matter in the most humid parts, to thin brown soils in the driest parts which are mainly used for grazing; and intermediate soils which come between the first two. About 75% of the precipitation comes in the six months from April to the end of September. Nearly half in May June and July.

The Southern Great Plains. "The climate is semi-arid to humid. Winters are extremely cold in places and mild elsewhere. The summers are usually hot and dry. The growing season is 180 days in the northern part to 300 days in the southern part. The topography varies from nearly level to strongly rolling. Surface draining is well developed except in the part which is in Texas. The soils range from slightly acid to calcareous and from deep sands and clays to thin soils. The cultivated crops are peanuts corn, sorghum and cereals, cotton, grass and legume forages."[20])

The leaves are 3-6" (8-15 cm) long---. It is found on all soil types including alkaline soils, but is most abundant on the heavier rolling upland soils. Its capacity to resist drought permits it to occupy the drier sites throughout its range of adaptation. Growth begins fairly late in the season and depends on how much moisture is available. The forage is relished by all classes of livestock. Growth ceases during long droughts but begins again upon the return of favourable moisture and temperature. Because of its wide distribution, high quality, hardiness, and growth habits, it is one of our (The American) most important range species. Blue grama is readily established from seed; excellent stands have been obtained by broadcast or solid drill sowing.

Seed rate: The usual rate is 8-12 pounds 3.6-5.4 kg) per acre = 20-30 pounds (9.0-13.5 kg) per hectare.

How to sow: For best results, the seedbed should be well prepared, and be free of weeds, but the seedlings are relatively persistent and compete with weeds and other grasses if they are not grazed until they become well established---.

Because of its wide adaptation, ease of establishment, and economic value, blue grama is used extensively for conservation purposes. Although its erosion control properties are effective when blue grama is seeded alone, the general practice is to make plantings with mixtures of other adapted grasses. Most re-vegetation seedings have been made on rangeland and abandoned cropland.

As a general rule, seed should be used near its point of origin. Experiments have shown that plantings of North Dakota seed have not been productive when planted in the Southern great plains and conversely, plants from Texas grown seed do not make satisfactory growth in North Dakota. This matter of plant adaptation is important with many of the native grasses and has led to the general caution that locally grown seed should be used whenever it is possible to do so."[65]

I would say that for use in other countries it would be advisable to have a selection of blue grama seed to determine which was most suitable for your region. It could be that one of the selections was quite or even very suitable even though the others were not at all suitable..

*Buffalograss.(*Buchloe dactyloides*). "B.G. is a fine leaved, native (in the USA,) sod forming perennial. It is the dominant species on large areas of upland on the short grass region of the central part of the Great Plains (Authors note, of the USA. see description of the region with Blue grama).

Generally it grows 4-6" (12-15 cm) high and produces leaves less than 1/8 of an inch wide (3 mm) and 3-6" (7.5-15 cm) long. It spreads rapidly by surface runners, and forms a dense matted turf. During the growing season the foliage is greyish green, which turns to a light straw colour when the plants cease growth. Growth begins in late spring and continues through the summer. Livestock like its forage. Its palatability, prevalence, and adaptation to a wide range of soil and climatic conditions make it an important forage species of the Great Plains.

It withstands long, heavy grazing better than any other native grass of that region; on ranges consistently subjected to heavy grazing (regularly heavily grazed), it often survives as a nearly pure stand

Because of its excellent ground cover, aggressive spread under use, wide climatic adaptation, and relative ease of establishment, buffalograss is ideally suited for erosion control on range and pasture lands where the soil does not contain too much sand.

The seed normally has a low germination rate. Soaking and chilling, processing with a hammer mill, or some other such treatment of the burs is usually required to insure good stands from planting at moderate rates.

Seed rate: 4-8 pounds (1.8-3.6 kg) per acre = 10-20 pounds (4.5-9.0 kg) per hectare.

Planting by the use of sod pieces is effective. Sod pieces about 4" (10 cm) in diameter are placed at 3-4 ft (91-122 cm) intervals on a well conditioned seedbed. Usually this results in a complete sod cover by the end of the second growing season." [65.]

*Buffel grass (*Cenchrus ciliaris*). "Buffel grass is native to north tropical and South Africa, India and Indonesia---. This perennial grass is very variable in habit, but is most drought resistant and withstands very heavy grazing once it is established. It has a large, strong root system, and more of the roots are found at deep soil depths than is the case with plants like Rhodes grass or Green panic. It has swollen stem bases which also accumulate more carbohydrates than green panic. Consequently it not only survives drought and firing but comes away very rapidly after drought breaking rains. It is sown in regions from 12-35" (30-89 cm) annual rainfall---.

Buffel grass makes most of its growth in summer, and is less cold tolerant than plants like green panic. Although it flowers early it is an unusual plant in that it continues to shoot strongly during flowering. It is adaptable in its soil requirements, preferring lighter textured soil but still performing well on self mulching soils. However it is sensitive to waterlogging, although the Tarewinnabar and to a lesser extent Biloela varieties recover better from short duration floods---. On high phosphate soils natural spread occurs" [63.]

"Buffel grass is adapted to areas of predominantly summer rainfall and makes its greatest growth during the hottest months of the year. Nevertheless, it does respond to winter rains when temperatures are not excessively low. It suffers from frost but quickly recovers during warm spells. It compares favourably with other species over a wide range of average annual rainfall. On the one hand it has grown and spread in areas where rainfall is as low as 12 inches. On the other hand, it can produce large quantities of valuable fodder in wet coastal areas of the tropics and subtropics---. Buffel grass seed undergoes a period of dormancy on reaching maturity. The extent and duration of this dormancy varies, but within a few weeks of harvest the germination rate is frequently only 2-3% and may not be much more that 50% after a whole year." [62.] The variety best suited for soil conservation seems to be Tarewinnabar, it differs from the other varieties in having bright green leaves, the others having a bluish tinge. This variety performs well over a wide range of conditions in Queensland Australia. It grows best in 25-30"(63-76 cm) of rainfall areas. It is a high producer with growth extending over a long season. Its early spring growth is better and it has outstanding seedling vigour, with rhizomes developing earlier than in the other varieties, this should mean better spreading and so soil protecting and holding ability.

"Where to sow: Buffel grass does well over a wide range of conditions, in rainfall areas from 12-35" (30-89 cm), depending on the variety used---. Although they generally prefer lighter soils, and need less fertility than Green Panic, they require reasonable fertility for good growth. Buffel grass will not tolerate waterlogging,

and although it does reasonably on self mulching soils, green panic or Makarikari are preferred for the heavy black clay soils.

When to sow: Sowing should be done when rain is assured, particularly if the seed is sown dry, as a number of days of moist conditions is needed to germinate the seed. Late autumn sowings are dangerous, as buffel seedlings are very susceptible to frost.

How to sow: It should only be sown 1/2 " (1.25 cm) deep. It responds best to a prepared moist seedbed, and rolling is a big help to consolidate the seed bed and press them firmly into available soil moisture. Seed tends to be fluffy and mixing it with sieved sawdust helps it to flow evenly.

Sowing rate: This varies widely according to the conditions. In the lower rainfall areas where thick stands are not desired, 1-2 pound (0.454-0.9 kg) per acre = 2.5-5.0 pounds (1.135-2.27 kg) per hectare is enough. In better rainfall areas, 3-5 pounds (1.36-2.27 kg) per acre = 7.5-12.5 pounds (3.4-5.67 kg) per hectare should be used.

Management: Buffel grass should be given every chance to establish itself in its first year, and build up its root system and food storage supplies, so that it can then withstand hard conditions and grazing. This usually means leaving it alone in its first year. Without a chance to develop these protective measures, it soon falls under pressure. Every four or five years it should be treated very lightly for a season to allow it to seed and restore its reserves." [61]

*Canary grass (see Phalaris).

*Carpet grass (*Axonopus affinis*). *see also mat grass.* "C. grass grows in highly acid to slightly acid soils of shallow to average depth and of fine sand to clay loam. It requires a great deal of moisture and tolerates swampy conditions. It is benefited slightly by the application of nitrogen---. It is a perennial, warm season grass, particularly adapted to low lying sands. It does not tolerate salinity" [20]

"It is a native of Central America and the West Indies---. It now grows in the Tropics of both hemispheres. A perennial creeping grass, it makes a dense sod and is distinguished by its compressed, two edged creeping stems which root at each joint, and by its blunt leaf tips. Because its sod is dense and its habit of growth is aggressive, legumes are maintained with difficulty in a pasture where carpet grass is used.

Seed rate: 5-10 pounds (2.27-4.54 kg) per acre = 12.5-25 pounds (5.67-11.35 kg) per hectare. Seed can be sown on a well prepared seedbed or broadcast on burned over open areas in timberland, (woodland). It is spread quite easily by grazing animals and by natural reseeding. Seeding is best done in spring, early summer or even midsummer---. Because it has no underground stems it has never become a pest in cultivated fields." [65]

*Centipede grass (*Eremochloa ophiuroides*). "Centipede grass grows in highly acid to neutral soils and is not tolerant of salinity. It needs shallow to deep sandy or gravelly loam to loam, and moist to very moist conditions. It grows at low nutrient levels, and is shade tolerant". [20]

"Centipede grass is a native of south eastern Asia---. A low growing perennial it spreads by stolons. It is adapted to a wide range of soils---. It will grow on clay soils and poor sand if enough moisture and plant foods are available for it to get started. It has withstood temperatures of 12° F.---. Its vigorous growth habit and ability to produce some seed makes it difficult to eradicate (destroy)

It is therefore not recommended in places where it might become a pest." [65]

*Crested wheatgrass. See wheatgrass.

*Elephant grass. see Napier grass.

*Green panic (*Panicum maximum var trichoglume*). *It is sometimes called Petrie green panic after the farmer who selected it.* This grass is a close relative of Guinea grass, it is an example of farmer selection and improvement. A farmer in Queensland, Australia noticed plants growing and spreading quickly in blue panic, grown from seed from India, though it originally comes from Africa. He was impressed by its growth and so selected and encouraged it. It has become the major sown grass in Queensland, north eastern Australia. It has a high requirement of nitrogen so should be grown with legumes. It also responds very well to phosphate so growing with desmodiums which seem able to concentrate phosphate and/or introducing and encouraging suitable mycorrhiza may prove worthwhile.

"It is a tall tufted perennial, although closely related to Guinea grass it is more palatable and drought resistant it is even more drought resistant than Rhodes grass, although it will not give as rapid a cover as Rhodes. However it is a free seeder and a thin stand will soon thicken up if allowed to seed. Green P. makes its main growth in the warmer months, and responds quickly to any rain. It will be cut by frost, but is not easily killed by frost, and gives a quick and valuable green pick after rain in the winter. It is probably the most tolerant of shade of the productive grasses, and its capacity to grow well when shaded by trees and Lantana (a wild plant in Australia), and to compete strongly with weeds, make it particularly valuable in many situations.

("The root system is fine and richly branched. More of the root system is concentrated in the upper layers cf the soil than in plants like buffel or Rhodes grasses. This may help to account for the rapid growth response to light showers displayed by green panic. Despite the root concentration in the surface, green panic shows first class drought resistance and survives well in situations where Rhodes grass dies out completely. It may be grown in areas down to 22" (56 cm) annual rainfall, and is also used in wet coastal or tableland areas receiving in excess of 70" (78 cm) annual rainfall. However, it is not as vigorous a plant as common guinea grass in high rainfall areas. Green panic top growth is frosted off, but the plant is not easily killed by frost. It has a capacity for making short bursts of growth when warm days are experienced after winter rain, and is superior to Rhodes grass in this respect---. Green P. will tolerate burning---. Green P. is efficient in that more of the surplus carbohydrates accumulates in the upper part of the plant, where it can be eaten. Proportionately less of the carbohydrate manufactured accumulates in the roots and stem bases than is the case with buffel grass. However, in comparing its tolerance of abuse with buffel grass, we are setting our standards of hardiness very high---. Green panic would not occupy the place it does today if it did not have resilient features"[63].

"Easy to manage it is taken readily by stock even when five feet (1.52 m) high and seeding, as it does not become woody when mature. It can readily be stood over in the autumn for winter grazing, particularly as the green pick (a flush of fresh growth after rain) can also be expected in the winter. It also makes good hay. Altogether this is one of the most useful grasses in use, although in the higher rainfall areas, the more productive grasses, such as setaria may take its place.

Where to sow: Green P. can be grown successfully in most areas with 22" (56 cm's) of rain and more, provided it is not in poorly drained situations. Where the rainfall reaches 60-70" (52-78 cm's) however, guinea grass is more productive.

It is best on the more fertile soils, but will compete readily with native pasture on the poorer soils. It is difficult to establish on the wide cracking heavy black clay soils of the Darling Downs, Queensland, Australia, and is not really happy on these soils. In other dark soils however it does extremely well---In the higher rainfall areas green panic has been a major component of mixtures using the newer legumes, and combines well with most legumes.

When to sow: The best times are mid to late summer, when moisture is assured or earlier when there is rain. However it has been successfully established in the autumn, using oats as a cover crop, or planted in Maize stands with the final hoeing.

How to sow: Best drilled into a prepared seedbed, into firm, moist soil, no deeper than 1" (2.5 cm) or preferably shallower. It can also be sown on scrub burns as a mixture with Rhodes grass or Buffel grass. The seed should be lightly covered if possible.

Sowing rate: From 0.5 pound (0.227 kg) per acre = 1.25 pounds (0.56 kg) per hectare in dry scrub areas in mixtures, to 3-6 pounds (1.36-2.7 kg) per acre = 7.5-15 pounds (6.75 kg) per hectare in the better areas, depending on soil preparation and seed quality. With legumes a sowing rate of 2 pounds (0.9 kg) per acre = 5 pounds (2.27 kg) per hectare, is generally satisfactory.

("Lucerne at 1-2 pounds (0.454-0.9 kg) per acre = 2.5-5.0 pounds (1.35-2.7 kg) per hectare may be added to the grass seed. In the sub tropics where frosts are not severe, siratro or townsville lucerne may be included at 2 pounds (0.9 kg) per acre = 5 pounds (2.27 kg) per hectare. In tropical high rainfall areas Glycine is a suitable combination and may be included at a similar rate."[62].)

Manure: Green P. responds remarkably to fertiliser, and use of phosphate will give markedly increased yields in most areas. It is also demanding of a good nitrogen supply for its best performance and should always be sown with a legume where a suitable legume can be used. In temperate conditions, it combines very readily with lucerne and phase bean. In the subtropical and medium rainfall coastal areas siratro, desmodium's and glycine's can be used, and in the wet tropics, centro, puero and stylo (the author would think greenleaf desmodium also) combine very well with it.

Management: One of the problems of management is the high palatability of green panic, as under heavy grazing, the cattle concentrate more on the Panic than on Rhodes or Buffel. However it is very resilient and even after being knocked about badly by grazing, can soon recover. It is best spelled at intervals with some form of rotational grazing to allow it to recover. New stands (plantings/sowings) should be given every chance to get well established before being heavily grazed, and if the stand is thin, it should be allowed to seed, as natural regeneration from seedlings is very good, and it will soon thicken up. Heavy grazing should be avoided just before winter, as a certain amount of foliage helps a great deal in frost protection" 61.

*Guinea grass *(panicum maximum.)* This did not grow as well as Setaria in Nepal on poor red and yellow lateritic soils at 4,700 ft.(1432.56 m) or in Taiwan on red soil and shaly soil at 200 ft. (61 m).

"Guinea grass is a native of Africa which has been introduced to many tropical countries---.("it has been cultivated widely in South America, the West Indies and South East Asia"63.) It forms tussocks up to 3 ft across and grows to several feet in height; it is a coarse vigorous perennial with leafy growth arising from the centre as well as from the outside of the tussock---.

Once established, Guinea grass is a hardy plant and, having a deep dense and fibrous root system, is able to withstand prolonged periods of dry weather. It will not tolerate heavy frosts, but recovers satisfactorily from occasional light frosts with the return of warmer weather. It is best suited to warm moist conditions---.

("It requires a minimum of 35"(89 cm) of rain but prefers much more" 61.)

("Its production is also reduced by cool temperatures more than some other pasture varieties, e.g. Setaria." 63.)

G. grass is tolerant of shading from other plants and this enables it to grow under such weeds as Lantana. When a sufficient bulk of grass has become established it is possible to set fire to it and so destroy the weed. The grass then recovers quite rapidly and replaces the lantana.

("It can be swamped by vigorous legume growth and still grow through to maintain a good grass/legume balance. It continues to make vigorous growth during the sustained cloudy conditions often experienced in the tropics in mid summer through to early autumn.---. It will not tolerate poorly drained conditions as well as para grass or para grasses relatives---. Mixed leguminous pastures containing Guinea grass have persisted well for over 20 years on the wet tropical coast of North Queensland, Australia, and maintained good productivity---. It combines well with Centrosema, and this mixture has maintained stable soil fertility conditions on rain forest soils in North Queensland. On the other hand, guinea grass growing alone has in 20 years caused a serious run down in soil nitrogen and soil organic matter. The need to grow guinea grass with a legume is obvious.63.)

G. grass will grow on a wide range of soils, but tends to produce poor stands on the more infertile types. it is particularly well suited to sloping land cleared in rain-forest country, where it will support high rates of stocking. It will tolerate acid conditions provided drainage is adequate.

The pasture may be established by planting rooted pieces of mature stools at intervals of 2 feet(60 cm's) in rows 4 feet (120 cm) apart. Sowing of seed is best carried out in spring or early summer so that the plants have a chance to establish themselves before the onset of high temperatures.

Sowing rate: 3-5 pounds (1.36 - 2.27 kg) per acre = 7.5-12.5 pounds = (3.4-5.67 kg) per hectare.

G.grass combines well with perennial summer growing legumes.(It should also combine well with Greenleaf desmodium, author) Glycine, for example, may be sown at 2 pounds (0.90 kg) per acre = 5 pounds (2.27 kg) per hectare with the grass at the recommended rate. ("care should be taken not to cover the seed deeply-1/4"(6 mm) is quite adequate." 63.)

"Guinea grass is very palatable and has a high feeding value, especially if it is not allowed to become too coarse. On the other hand it should not be subject to continuous heavy grazing to below 9-12"(23-30 cm). To maintain a good stand, G. grass should be allowed to re-seed itself at least once every two years." 62. Improved strains of Guinea grass are Coloniao grass and Hamil grass.

*Harding grass see Phalaris.

*Kikiyu grass *(Pennisetum clandestinum.)* "Kikuyu grass is a summer growing perennial originating in Kenya. Under suitable conditions it spreads rapidly over and through the ground by means of running stems (stolons or rhizomes). Both the surface runners and the underground stems root freely at the nodes, anchoring the plants firmly in the ground and forming a dense turf which withstands heavy trampling by livestock. The stems carry a large quantity of leaf and the stems themselves are very succulent. Under favourable conditions, K. grass makes a dense growth often as much as 2 ft (60 cm) in height. Although it is adapted to a wide range of climatic conditions, it does best in well drained, subtropical or temperate areas with a good rainfall. Country suitable for paspalum will therefore usually grow good kikuyu---.

Vigorously growing kikuyu will resist the invasion of weeds, such as narrow leaf carpet grass, more effectively than paspalum."[3]

"It is a very nutritious and palatable pasture grass and will withstand very close grazing. It is very drought resistant and produces a very dense sod which seldom permits invasion by weeds.

Trial plantings of this grass have been made for road stabilization in the mountains at elevations ranging from 1,000 m (3280 ft) - 2,700 m (8,856 ft). It has been very effective in binding and holding the soil on some very steep cuts and fills. Plantings for pasture purposes in this same area have been very productive and Kikuyu is recommended as the best pasture grass here."(Taiwan)[64].

"Kikuyu is especially suited to upland areas, and has a more even growth rhythm than Paspalum or Rhodes Grass. It is more cold tolerant and makes good autumn growth if nutrition is adequate. It will not be productive in areas having less than about 35" (89 cm) of annual rainfall, unless it is used for stabilising and reclaiming drainage lines which receive additional run-off water---.

Kikuyu is established from runners, cuttings or turves, either in the early spring or late summer period ---. K. is an excellent grass for erosion control, and is also very palatable and nutritious. The greatest difficulty in management comes from the tight sod which is formed. Firstly this mat of rhizomes and decomposing organic matter locks up nutrients and can cause the sward to become moribund (at the point of death.) Secondly, it makes the maintenance of a good legume/grass balance difficult.

Greenleaf desmodium, Silverleaf desmodium and Glycine may prove to be better companions for Kikuyu than clovers. The Desmodiums are probably better able to extract phosphate on the red loams commonly growing Kikuyu."[63]

"Planting is often done by hand in drills 3 feet (1 m) apart and at 3 feet (1 m) intervals within the rows. The operation may be accelerated (speeded up) on good soils by one man ploughing shallowly, followed by other dropping the plant material in every third or fourth row.

Kikuyu is useful for stabilising gullies, grassing waterways, and general soil conservation work. On the other hand, it is a tenacious weed of cultivated land and therefore it is unwise to establish it on land intended for cropping at a later date.

Winter legumes rarely combine well with kikuyu. Of the summer legumes glycine performs outstandingly and may be broadcast at 2 pounds (0.9 kg) per acre, = 5 pounds (2.27 kg) per hectare at planting time.

K. pastures usually grow slowly in their first year and therefore weed control measures should be employed to encourage their development and they should be stocked lightly.

Once established K. will withstand heavy grazing and trampling. After a few years indeed, it tends to grow rank and unproductive and should be opened up to maintain the quality and vigour of the sward."[62]

*Kus, see Vetiver grass.

*Lovegrasses. *(Eragrostis species.)* "Several species are recognized for their capacity to produce an abundance of seed and forage on soils of low fertility; hence they are used to provide vegetative cover on eroding sites. The best known species is *E. abysinica*, called Tef in Ethiopia, where it is cultivated as a cereal. It is cultivated also in India and Australia but usually as a forage plant.

E. obtusiflora. ranges from New Mexico and Arizona southward into Mexico, it is a hardy and rigid perennial that spreads by rhizomes and thrives on alkaline soil. It furnishes a great deal of forage where it grows naturally."[65]

*Weeping lovegrass *(Eragrostis curvula)*. "W.L. is the most widely adapted of the lovegrasses. Especially to dry sandy soils, and is used in the dry regions of the South and South West of the USA."[20]. "It came from the mountainous part of Tanganyika (now Tanzania), this grass is particularly adapted to the soils and climate of the Southern Great Plains of the USA.

(The conditions in that region are "The climate is semi-arid to humid. Winters are extremely cold in places and mild elsewhere. The summers are usually hot and dry. The growing season is 180 days in the northern part to 300 days in the southern part. The topography varies from nearly level to strongly rolling. Surface draining is well developed except in the part in Texas. The soils range from slightly acid to calcareous and from deep sands and clays to thin soils. The cultivated crops are peanuts corn, sorghum and cereals, cotton, grass and legume forages.)

Weeping lovegrass plants produce seed stalks 2-5 ft (61-152 cm) tall and numerous slender, curving basal leaves 10-20" (25-51 cm) long; the heavy forage is eaten readily by cattle in early spring but sparingly in summer. The normal presence of green shoots in the bases of the plants often induces cattle to eat old growth in the winter.

The plants resist summer heat and drought and survive temperatures as low as 11° F. if the soil has adequate moisture at the time of the first killing frost. Weeping lovegrass is easily established from seed. Its vigourous young seedlings quickly make an effective ground cover."[65]

Seed rate: 1-3 pounds (0.454-1.36 kg) per acre = 22.5-7.5 pounds (1.135-3.4 kg) per hectare.

*Makarikari grass, *(Panicum coloratum) var. makarikariense.* "This range of grasses, characterised by bluish leaves with a white midrib, came from the flood plains of Bechuanaland (now known as Botswana, author.) in Africa. They vary from erect, bunchy grasses of the bambatsi type, to lower growing grasses which spread by surface stems that root at the nodes and which are know as the makarikari types. Their strong, fibrous, and deep, root system makes them very drought resistant, yet they can stand very heavy clay soils and waterlogged conditions. Coupled with their frost resistance these features earn them a major place in many inland areas of Queensland and northern New South Wales, Australia. Although they make their major growth in the summer, they have the valuable capacity to sprout from the base following winter rains, and give a good bulk of fair quality winter feed. Altogether, they are hardy grasses which show great promise in areas where heavy soils, low rainfall and frost limit the performance of many other grasses.

(It seems from several descriptions that the variety Pollock is the one best suited for erosion control or for grassed waterways. Author.)

Pollock is a more stoloniferous type, which is a very poor seed producer. The stoloniferous types seem more adapted to areas subject to flooding. In view of the poor seed yield, this type is often planted from cuttings. ("it has better seedling vigour, ground cover and leafiness, but poorer seed production. On the Darling downs, Queensland, Australia it out yields green panic and Rhodes grass."[63])

(Note :- "the low germination capacity of Makarikari grass may be partly overcome by keeping the seed for at least a year, hammer milling to break the hard covering of the seed, and restricting sowings to periods when the soil is very moist."[62])

Where to sow: These grasses do best in the 20-25" (50-63 cm) rainfall areas, although in Africa they grow in regions down to 15" (38 cm) rainfall. They are fairly versatile as to soil types, and although preferring lighter soils, they persist on heavy black clay soils under waterlogged conditions.

When to sow: From spring to Autumn when rain is expected. Where there is likely to be a summer weed problem, autumn sowing may give better results, as the seedlings are fairly slow to establish and in the early stages are vulnerable to weed competition.

How to sow: Sow at about 0.5" (1.3 cm) deep.

Sowing rate: About 2 pounds (0.90 kg) per acre = 5 pounds (2.25 Kg) but may be a little more or less according to sowing conditions. (Planting by rooted cuttings can also be made.)

Management: The stand should be given every chance in the first year to become well established, and to thicken up, so it should be grazed lightly. Once well established, these grasses will stand very heavy grazing."[61]

*Mat grass *(Axonopus affinis) see also Carpet grass.* "A creeping grass native to Central America and the West Indies. It is very tolerant of low fertility conditions, and dominates considerable areas of Queensland Australia coastal and tableland pastures to the great disadvantage to the dairymen. (This suggests that care must be taken not to introduce it where it may spread in an undesirable way to pastures, it does not seem to be a likely problem in arable fields; author) It may be controlled by ploughing, seeding to other species and raising the fertility level by an appropriate legume/fertiliser combination. There is an export market for seed for use in lawns." [63]

*Mesquite grasses *(Hilaria).* "Include several species that grow in neutral to moderately alkaline soils and are tolerant of moderate salinity. They grow in shallow to deep sandy loams to clay loam in very dry or dry conditions. They are perennials and are adapted to conditions of the south west USA.

Galleta (*H. jamesii*) will tolerate drier sites." [20]

Seed rate: 4-8 pounds (1.8-3.6 kg) = 10-20 pounds (4.54-9.0 kg) per hectare.

*Molasses grass *(Melinis minutiflora).* Very good for covering and holding steep banks, earth walls of terraces, and even banks of shaly subsoil at 4,700 ft in Nepal. It makes a thick cushiony mat. It gets its name from the sweet sticky substance secreted by its leaves. It is highly digestible with a TDN= Total Digestible

Nutrients level of 20.2%

"This spreading perennial, which forms loose tussocks, is native to tropical Africa but has become naturalised in many parts of the tropics---.

It forms a fairly dense, straggling sward, smothering other plants with the dense mat produced.

The leaves are covered thickly with hairs, which exude a sticky secretion. The leaves contain a volatile oil which gives the grass a strong and distinctive odour. This does not cause tainting of milk or meat.

---. It has a long season of vegetative growth.

Molasses grass is not grown commercially in areas receiving much less that a 40 inch (1 m) annual rainfall. It is quite seriously affected by frost and (in Queensland, north eastern Australia) makes its best growth expression in warm coastal areas.

It is a pioneering species which will grow well on poor sandy soils and acid yellow clay loams as well as the red loam scrub soils on which it was first introduced. It will not persist on areas subject to periodic flooding, or to seasonal high water table levels.

It is sensitive to fire, but produces a very hot fire which carries well. Some crown survival or seeding regeneration will occur after firing, but the burning of a pure stand can lead to serious weed invasion and erosion. It is therefore sown most commonly with a more perennial, hardy grass such as guinea grass, which will form the eventual basis of the pasture.

Molasses grass is most valuable in initial land development, as it helps smother the initial weeds and carries the subsequent cleaning up fire.

As the growing points are carried rather high and the grass is not specially deep rooted, it is sensitive to over grazing and may be thinned out if grazing is too severe. It should not be grazed below 6-9" (15-23 cm); a rest during flowering is advantageous for subsequent stand regeneration.(Best to allow recovery to 12-18"(30-45 cm).

The seed is light and tends to bunch when being broadcast. A carrier such as sieved sawdust (or sand) is advisable to obtain even distribution in the field.

It is usually sown in spring or summer on cultivated land or in the ashes of a scrub burn.

Sowing rates: 2-4 pounds". [63] When sown in a mixture 1 pound (0.454 kg) per acre = 2.5 pounds(1.135 kg) per hectare.

"Molasses grass grows prolifically during the summer and autumn. It cannot tolerate severe frost but under light frosts will die back and later recover when growing conditions return. It has considerable drought resistance but grows best under conditions of high rainfall and good drainage---.

Molasses grass can be grown on a wide variety of soil types, but best results require adequate supplies of phosphate and nitrogen---.

(Note the author found it easy to grow from cuttings by splitting up clumps.) Molasses grass may be sown in mixtures containing other species such as Guinea grass or Para grass. In such cases the object is to achieve a quick ground cover with the molasses grass in order to suppress weeds while the other slower growing species become established. These mixtures are sometimes used following the clearing of timbered land. Molasses grass burns readily and produces a very hot fire. Once such a stand has produced a good cover it is often the practice to set fire to it in order to burn the remaining stumps and forest trash. The more fire resistant Guinea or Para grass is then able to dominate the pasture.

Molasses grass is also suitable for sowing in combination with perennial legumes such as Glycine or Siratro (The author found it grows well with Greenleaf Desmodium). Such pastures have a high feed value and can support heavy stocking rates.

It is essential to protect molasses grass from fire and grazing by livestock during the early stages of establishment.

Molasses grass is suitable for protecting gullies and exposed earth banks from soil erosion and provides a satisfactory cover for waterways established as part of soil conservation schemes"[62]

*Napier Grass, *(Pennisetum purpureum.)* The authors experience of this grass has not been good. It does produce well but at the expense of the soil fertility. On the poorer soils it exhausts the fertility and dies back in about 2-3 years, and as few third world farmers are going to plant grass on very fertile soils it is not appropriate.

"It rapidly exhausts the plant foods available in poor soils" [3]

It can of course be grown at the edge of fields and in contour strips but it will pull both moisture and fertility from the neighbouring crops. A further problem is that rats love it, its roots are sweet and they make their burrows among them. Not content with the N. grass they forage out into the neighbouring crops whether sweet potatoes, maize, peanuts, rice, etc. Their burrows can undermine banks and contour strips and lead to landslides. Furthermore although the grass does produce a lot of bulk the base of the stems become very woody and fibrous. The digestibility of the part eaten by stock (quite a lot is left !), is only 12.1% Total Digestible Nutrients compared with say Paspalum (15.9%), Sudan grass (14.4 - 18.6%), Molasses grass (20.2%).

It does continue to grow well in situations such as monsoon waterways and gullies where fresh nutrients are washed down each year. There it can be useful providing it is not near crops which may be eaten by rats living among the roots.

" This tall, bunch type perennial is native to tropical Africa, where it occurs mainly in areas receiving over a 40 inch (102 cm) rainfall---.

It is fairly tolerant of soil conditions but does not withstand flooding or waterlogging well. Deep loam or scrub soils give best growth. Being a very vigorous grower, it has a heavy demand for soil nutrients.

(As it does not set much viable seed, propagation is by cuttings) Vegetative planting should be carried out in the summer on well prepared land (when the soil is moist and more rain is expected). The stem pieces are best cut from hard stems about 6 months old, each piece possessing four or five nodes or stem joints. The furrows are run out 3-4"(8-10 cm) deep, the cuttings laid horizontally, and these are then covered with soil---.

If cuttings are spaced correctly- on 3 ft (1 metre) squares--- legumes planted between the rows can dominate the sward. Calapo has dominated Napier grass on scrub soils in the high rainfall tropics, and excellent mixtures of Napier grass/Centrosema, and Napier grass/Glycine have also been obtained. In a drier climate Lucerne has competed effectively with Napier grass." [63.]

"It is best sown in the summer when soil conditions are moist and further rain is expected.
How to sow? The seedbed must be well prepared and the furrows several inches deep."[61.]
Dallis grass, see paspalum.

*Paspalum grass/Dallis grass *(Paspalum dilatatum).* Looks rather like Signal grass, but in the authors experience is not so useful.

"It is a hardy perennial species which grows in clumps of varying size and has a knotted base of extremely short rhizomes. These underground shoots become entwined in time to form a dense mat or sod, and being inaccessible to stock enable the plants to withstand heavy grazing and trampling.

Long broad leaves are produced in abundance from the crown and, sparingly along the flowering stalk. Being set very close to the ground, these basal leaves cannot be grazed right back by stock, particularly cattle, and consequently the plants recover rapidly when the stock are removed. They are also responsible for rapid recovery following periods of drought---.

The seed heads are widely attacked each year by a fungus causing ergot, and some loss of production by dairy cattle can occur when the animals consume too many affected heads.(Ergot can cause poisoning and in some cases cause abortion- author.)

Paspalum does best on fertile, heavy soils. It does not do well on light soils in the absence of fertilizers.

Paspalum is essentially a summer growing plant. Best results are obtained when the annual precipitation exceeds 50"(127 cm), although satisfactory results can be obtained in the 30-40" zone.

Although it will not persist in drier areas, paspalum can survive protracted rainless periods. While somewhat susceptible to severe frost, it suffers only retarded growth from light ones."[62]

"*Seed rate 6-10 pounds (2.72-4.54 kg) per acre,= 15-25 pounds (6.75-11.35 kg)per hectare.* " [63]

*Phalaris/canary grass/Harding grass. *(Phalaris tuberosa).* "Phalaris is a perennial grass of great importance in southern (cooler) areas of Australia on account of its long growing period, its persistence, and the high feed value and palatability of its leaves in the young stages---. It is a deep rooted grass growing in leafy open clumps. The leaves have a characteristic blue green colour and are long and broad across the bases and hairless. The stems are bulbous towards the base and grow outwards from the crown for a few inches before becoming vertical. The flowering heads are spike like panicles, like cats tails in appearance, and are held erect. They may vary in length from 2-6" (5-15 cm). The flowering stems grow to a height of between 2.5-5,0 feet (76-152 cm). Once these stems are formed, however, the plant becomes very coarse and loses much of its feed value and palatability.

P. has been grown in southern areas of Australia with a rainfall as low as 17" (43 cm) and has remarkable persistence where autumn rains are unreliable or unsatisfactory. Under these conditions the plant persists mainly by sending out fresh shoots from the crown. On the other hand, it regenerates satisfactorily from seed where the rainfall regime is suitable. It can withstand hot dry summers, and long periods of drought as well as wet cold winters. It will even survive inundation (being flooded) for quite long periods.

P. grows all the year round except for the hottest period, and therefore is a useful species for most winter rainfall areas. It does not produce as much herbage per acre as perennial ryegrass or cocksfoot, but even where those species may be grown phalaris has a useful function in filling seasonal gaps in pasture production; it produces well throughout winter and spring, but the important feature is its ability to come away earlier in the autumn than most other grasses and to produce a greater amount of feed before the onset of winter---.

Phalaris is adaptable to a wide range of soil types but grows best in deep rich black soils. ("It can even grow well in poorly drained clay of medium or greater depth and tolerates moderate salinity".[20.]) It has a wide tolerance of pH values.

Establishment: P. establishes itself slowly and cannot tolerate competition from weeds or other plants during its early stages. On this account and because of the small size of the seed, it is important to sow onto a clean fine, and firm seed-bed. Initial grazings should be lenient (light) and of short duration (length), but once the root systems are well developed and the plants growing vigorously it is possible to graze heavily.

Sowing rate. 2-4 pounds (0.9-1.8 kg) per acre = 5-10 pounds (2.27-4.5 kg) per hectare. Sown in the spring. Because of its poor competition during establishment it is not usually sown with other grasses, but it may be sown with companion legumes. White and red clovers are suitable in the higher rainfall areas, and subterranean clover or various medics including lucerne in the drier regions. Maximum production, feed value and palatability are maintained by heavy grazing and mowing (cutting) once the pasture has become established.

Note: in some areas where phalaris is dominant in the pasture, a toxin (poison) is produced in the young phalaris leaves which arise following the autumn break. If consumed in sufficient quantity the toxin causes a debilitating (weakening) nervous disease in sheep and cattle known as phalaris staggers. The disease may be avoided by ensuring that the grass is well diluted (mixed) with other species in the sward, or by dosing the livestock with cobalt.

*Plicatulum *(Paspalum plicatulum).* "This tufty perennial grass is a native of central and South America. A very hardy grass, it is capable of high yields under favourable conditions, but has the advantage that it can establish and do reasonably well under fairly low fertility. This makes it a useful pioneer grass, particularly to include in a pasture mixture, so that it can `hold the fort' until nitrogen builds up under legume stands and allows more productive grasses to take over.

It has good drought resistance, but can take waterlogged conditions on heavy soils. Basically it is a summer and autumn grower, it does however come away well in the spring. As it does not seed early it has a very much longer productive season than Paspalum dilatatum. It does suffer from frost, but retains quite good nutritive value even after frosting---. The open nature of Plicatulum also makes it compatible with a wide range of legumes.

(Rodd's bay variety is probably better for soil conservation as it is a little more productive and is two or three weeks earlier in the spring. Frost causes much loss of nutritive value in the Rodd's Bay variety but not in Hartley. Both suffer less than other grasses such as Paspalum and Rhodes grass.)

Where to sow: In warm to hot climates where the rainfall is at least 30" (76 cm). It may be useful in the poorer drained situations. P. tolerates a wide range of soils, and has the ability to boost production where soil fertility is too low for more productive grasses like the Setaria's.

When to sow: From late spring to late summer, when soil moisture is adequate and follow up rain is expected.

How to sow: Best drilled or broadcast into a prepared firm seed bed, no deeper than one inch (2.5 cm).

Sowing rates: 3-4 pounds (1.36-1.8 kg) per acre = 7.5-10 pounds (3.4-4.5 kg) per hectare. Less when sown with other grasses.

Although it can perform under low fertility conditions. P. responds very well to good fertility, and is usually grown with legumes such as glycine, siratro and desmodiums.

Management:

In the early stages should aim at giving the legume in the mixture every chance to get well established before heavy grazing pressure is used. Pliculatulum makes valuable standover winter feed, even when frosted.

*Rhodes grass (*Chloris gayana*). Rhodes grass is a "tough and vigorous running perennial grass, from southern (author's note, other sources add central and eastern) Africa, which has been introduced to many of the tropical areas of the world---. R. Grass grows to a height of between 2.5 and 4 ft (76.-112 cm). It spreads by means of long surface runners, which root and send up leafy shoots at the nodes, the leaves are long and slender, tapering to a point, and are bright green in colour."[62.]

"It has played a major role in development of sub-coastal scrub land in northern New South Wales and Queensland, Australia. It is still used extensively in developing Brigalow (a kind of Acacia tree) country further inland.

Its ability to form a rapid cover of tufts from its running stems, which root down at the nodes, makes it a good smotherer of weeds in newly cleared scrubland and extremely useful in controlling regrowth of brigalow suckers. Rhodes grass is also quite tolerant of fire, and is often used to carry a fire for weed and regrowth control. The running habit makes it a good spreader, and thin stands will thicken up if lightly treated. Once well established, it will stand heavy grazing. Only moderately resistant to frost, its main growth is in summer. Although its strong fibrous root system gives it reasonable drought resistance, it cannot stand drought as well as green panic or buffel grass---. This root system, and the runner, make it very useful for erosion control work, and it is often used in rotation with tobacco for control of Nematodes.

Newer varieties of R. Grass have been developed, among them Katambora. A late flowering, fine leafed variety, which has proved extremely effective for erosion control work because of the dense mat it forms. (However it is not as drought resistant as R. grass.)

Where to sow: R. grass needs a rainfall of about 25-50"(63-127 cm) ("With the greater part of it falling in the summer months"[63.]) Where conditions are very wet its growth is poor.("Under very wet conditions R. grass becomes spindly in habit and yellow in colour and is out produced by Guinea grass and Napier grass" [63.])

It will establish and grow well on a range of soils, from sandy loams to heavy grey and brown soils, provided the drainage is not poor. Its relative salt tolerance compared with most other grass makes it popular in such areas as the Brigalow. (Brigalow is an acacia tree which grows along a region parallel to the east coast of Queensland, Australia.)

When to sow: it can be sown in spring where rains are fairly reliable, but it is usually sown in the summer when rain is more assured. In mild regions sowing can be as late as early autumn.

How to sow: The seed is very fine and so should be sown very shallow. Rolling is a big help where seed is broadcast. Often dry sawdust is used as a carrier to give more even distribution.

Sowing rates: When sown in a mixture with other grasses 0.5 pound (0.227 kg) per acre = 1.25 pounds (0.56 kg) per hectare. When sown mixed with legumes, 2-3 pounds (0.9-1.36 kg) per acre = 5-7.5 pounds (2.25-3.4 kg) per hectare. When sown by itself 3-6 pounds (1.36-2.7 kg) per acre = 7.5-15 pounds (3.4-6.75 kg) per hectare, depending on land preparation. It combines well with lucerne and phasey bean in the drier areas, and in more favoured regions Siratro and Centrosema have proved good partners to R. grass."[61.]

"It is advisable to allow a newly established pasture to set seed before stock are admitted for the first time. This will encourage formation of a dense sward and increase the life of the stand." [63.]

*Setaria (*Setaria sphacelata*). The author has been very pleased with Setaria grass in both Taiwan and Nepal; the experience in both countries was on acid soil. While it did not produce as much bulk as Napier grass the amount of digestible material will be as much as or greater; the reason being that Napier has a high percentage of tough, woody, and stalk pith material. Setaria keeps on producing long after Napier grass has exhausted the land and died back; it will grow on less fertile land than Napier, even on steep banks of fresh dug shaly subsoil; it starts growth earlier and goes on later than Napier. Setaria "Has perhaps the greatest range of adaptability of all tropical grasses ---. Setaria is found over most of the African continental plateau, but tends to be concentrated in areas where the annual rainfall exceeds 25 inches (63.5 cm)" [62.] Produces a lot of good fodder, not woody like Napier "The Setarias both Nandi (from Tanganyika- now Tanzania) and Kazungula (from Natal province of South Africa) are extremely vigorous, highly productive and versatile grasses. A striking feature of Setaria is the length of its growing period. The Setarias are well accepted by cattle and have relatively high digestibility. They combine readily with pasture legumes. Of the two, Kazungula is more vigorous and easier to establish, making it the one to use in more difficult situations and rougher country. It tends however to be rather coarser and stalkier than Nandi, but is very readily acceptable to stock.---. Setarias show fair drought resistance but can also tolerate fairly poorly drained conditions, and can stand prolonged periods of flooding without serious effects.(It can stand light frosts.)

Setaria needs at least 25"(63.5 cm) of rainfall for good production and doesn't reach it's peak at less than 35"(89 cm). Broadly speaking it will grow anywhere that paspalum grows, and it is much more productive and versatile. It will grow on almost any soil, from Wallum sands to heavy clays, subject to adequate plant food being available. Most situations suit it, except really swampy conditions.

Sow into a firm moist seedbed, one inch deep(2.5 cm) at the most. When broadcast sown, only very light covering is necessary. Rolling after sowing helps a great deal where soils do not cake. Avoid fluffy seedbeds.

Sowing rate: Nandi 1-2 pounds(0.454-0.9 kg) per acre = 2.5-5 pounds(1.135-2.25 kg) per hectare. Depending on soil preparation and quality of seed.

Kazungulu 0.5-1 pound(0.227-0.454 kg) per acre = 0.56-1.135 pounds per hectare depending on soil preparation and quality of seed." [61.] Once it is established from seed it produces lots of tillers which all make excellent planting material.

*Signal grass (*Brachiaria decumbens*). This grass does well on a hard infertile site in Nepal, even on shaly subsoil from 6 feet (2 m) deep; where weeds did not grow readily and other grasses would not grow. The altitude was 4,700 ft, (1,433 m) the rainfall about 60"(152 cm) almost all of which falls between the end June and the end of September. It can stand light frost.

"This grass is native to the open grasslands of Uganda, where it is favourably regarded---.

In North Queensland (north eastern Australia) It has always been recognized as a high producing grass---.

It is a creeping perennial, a relative of para (*Brachiaria mutico)* and Ruzi (*Brachiaria ruziziensis*). Signal grass is a good descriptive name, as the branches of the seed head are attached at right angles to the stalk (however Paspalum is similar in this respect, author.) It is on the wet tropical coast that signal grass has given it's best growth. It has produced significantly more dry matter than other grasses tested, 33,000 pounds (14,983 kg) per acre =82,500 pounds (37,455 kg) per hectare, being recorded in one year---. It is recommended for high rainfall frost free areas, and responds well under heavy stocking.

Seeding rate: 4-6 pounds (1.8-2.7 kg) per acre = 10-15 pounds (4.5-6.75 kg) per hectare." [63.]

It can also be propagated by cuttings.

*South African pigeon grass, see Setaria.

*Vetiver grass (*Vetiveria zizanioides*) "This is a tall perennial grass found wild in the drier tracts of western India. Its spongy, much branched fine rootlets contain a fragrant oil, which is a perfume by itself. The dry aromatic roots are made into curtains, mats, fans, etc to emit scented cool aroma when moistened. The oil is used as a valuable fixative for blending perfumes and cosmetics---. ("When manured and managed well- the yield is 3-5 tonnes of fresh roots yielding about 20 kg of oil per hectare. = 8 kg per acre." [40.]). "When planted correctly *V. zizanioides* will quickly form a dense permanent hedge. It has a strong fibrous root system that penetrates and binds the soil to a depth of up to 3 meters and can withstand the effects of tunnelling and cracking. It is perennial and requires minimal maintenance. It is practically sterile, and because it produces no stolons or rhizomes it will not become a weed. Its crown is below the surface, which protects the plant against fire and overgrazing. Its sharp leaves and aromatic roots repel rodents, snakes and similar pests. Its leaves and roots have demonstrated a resistance to most diseases. Once established it is generally unpalatable to livestock. The young leaves however are palatable and can be used for fodder.

It is both a xerophyte and a hydrophyte, and once established it can withstand drought, flood, and long periods of waterlogging. It will not compete with the crop plants it is used to protect. Vetiver grass hedges have been shown to have no negative effect on and may in fact boost - the yield of neighbouring food crops. It is suspected to have associated nitrogen fixing mycorrhiza which would explain its green growth throughout the year. It is cheap and easy to establish as a hedge and to maintain - as well as to remove if is no longer wanted. It will grow in all types of soil, regardless of fertility, pH, or salinity. This includes sands, shales, gravels, and even soils with aluminium toxicity. It will grow in a wide range of climates. It is known to grow in areas with average annual rainfall between 200 mm (8") and 6,000 mm (19ft 8" =236") (Author's note, I doubted this figure but [67 p52.] records that during August 1841 Cherrapunji received 241" of rain,

150 inches of which fell in 5 days. The author was himself in Taiwan when 64 inches were recorded in a typhoon lasting about 24 hours.) And with temperatures ranging from -9- 45° Centigrade. It is a climax plant, and therefore even when all surrounding plants have been destroyed by drought, flood, pests, disease, fire, or other adversity, the vetiver will remain to protect the ground from the onslaught of the next rains.

(Note :- This quotation is from a World Bank publication and The World Bank is promoting the use of vetiver grass hard. They are recommending it for contour strips. An Indian agricultural journalist I questioned about the viability of vetiver grass contour strips told me that it is not suitable for this purpose. One of the reasons being that it is not good fodder and farmers usually need good fodder. They prefer something else. The other reason is that the main point of growing vetiver grass is the extremely valuable oil which is extracted from the roots. This roots is in high demand for aroma therapy and other uses. Consequently when well developed the plants are dug up and so the contour bund is broken, runoff water flows through the breaks and erosion results.)

Planting: Obtain from or grow in a nursery. Vetiver grass nurseries are easy to establish. Inlets to small dams or water holding tanks make the best nursery sites because water enroute to the dam or tank irrigates the vetiver grass, which in turn removes silt from the water. Large gullies protected with vetiver grass also make good informal nurseries. For best results, the vetiver root divisions or slips should be planted in a double or triple line to form parallel hedges across the stream-bed. The hedgerows should be about 30-40 centimetres apart.

To remove a clump of vetiver grass dig it out with a spade or fork. The root system is too massive and strong for the grass to be pulled out by hand. Next tear a handful of the grass, roots and all from the clump. The resulting piece , the slip is what gets planted in the field. Before transporting the slips from the nursery to the field, cut the tops off about 15-20 cm (6-8") above the base, and the roots 10 cm (4") below the base. This will improve the slips chance of survival after planting by reducing the transpiration level and thereby preventing them from drying out."[1]

*Weeping Lovegrass, see lovegrasses.

*Western wheatgrass, see wheatgrasses.

*Wheatgrasses *(Agropyron spp.)* These include many native (USA) and introduced species. "Wheatgrasses are hardy, drought resistant and versatile. They produce abundant forage that is acceptable to all classes of livestock. Most of them are perennial, with or without running rootstocks. Usually the culms (stems) grow erect. The flower spikes resemble wheat, hence the scientific name Agropyron, which is derived from agrios = wild, and pyros = wheat. In the U.S. wheatgrasses have great value in the northern great plains, the inter-mountain region, and the higher altitudes of the Rocky Mountain states.

Some wheatgrasses form sod; others grow in bunches. The sod forming species are particularly valuable for erosion control. Germination of the seed, which usually is produced in abundance, is rapid and the young seedlings may become established in competition with weeds and other grasses. This seedling vigour permits the sowing of wheatgrasses with a minimum of seedbed preparation. Often good stands have resulted when the seed was drilled in weeds or small grain stubble. **This ease of establishment and adaptation to many kinds of soils, moisture conditions, and extremes of climate make wheatgrass of first rank for use in plantings that are intended to protect the soil.** Early spring growth, with high production of lush forage at the season when most needed by overwintering livestock, is another good characteristic of these grasses. The wheatgrasses have been used extensively for revegetating depleted range and abandoned farm lands. They are unexcelled for this purpose in the areas for which they are adapted---. The species probably most familiar is A. repens, commonly called quackgrass, which invades our cultivated fields and gardens with such aggressive persistence that it has fully earned its place in the category of weeds. Its aggressiveness, however, has been put to good use in holding soil in conservation practices.

*Crested wheatgrass *(A. cristatum).* Is a hardy perennial bunchgrass which came originally from Russian Turkestan; it produces an abundance of both basal and stem leaves---. C.W. is well adapted to the cool dry areas of the northern great plains, the inter-mountain region, and the higher elevations of the rocky mountain states of the United States. It has a wide spreading, deeply penetrating root system. Partly for this reason it can survive cold and drought, withstand grazing, and compete with weeds and associated grasses. Crested W. usually begins growth in early spring. It ceases to grow during long hot dry periods in summer, but it again makes growth when moist cool weather returns. By producing forage in early spring and early autumn, when the normal growth of native grasses has not yet begun or has ended, this grass provides succulent grass when it is most needed. High palatability, good quality and good volume of forage, combined with hardiness, drought resistance, and adaptation to widely different soil types, ("It will grow in slightly alkaline soils" [20];) make crested wheatgrass one of the most valuable of forage grasses in this (the USA) country.

Hay of excellent quality is obtained by cutting the plants early, just after heading and before blooming.

C.W. seedlings resist drought and withstand competition from weeds and grasses if given sufficient protection during the first 2 years of their establishment.

When to sow: If sufficient moisture is present in early autumn, seeding can be started in September and continued until field operations are stopped by the weather. In warmer areas spring seeding is satisfactory.

How to sow: Seedbeds may be well prepared or the seed may be drilled or broadcast in small grain stubble, weeds or depleted native range cover. However, when the latter methods of seeding are used, one or more additional seasons of protection are often needed for the newly established seedlings to overcome the competition of associated vegetation and arrive at the stage of readiness for use by livestock. Hence, if it is urgent that the newly seeded area be available for use in the shortest possible time, it often pays to make the additional effort needed for preparing a clean, firm seedbed.

Seed rate: Sow at the rate of 4-8 pounds (1.8-3.6)per acre = 10-20 pounds (4.5-9.0 kg) per hectare.

*Western wheatgrass, *(Agropyron smithii.)* Western W. is a native, perennial, sod forming grass distributed generally throughout the United States except in the humid Southeastern states---. Plant growth is vigorous with seed heads at a height of 2-3 ft (60-90 cm). Although it is adapted to a wide range of soils (including alkaline soils) it seems to prefer the heavy soils characteristic of shallow lake beds or along intermittent swales and water courses that receive excess surface drainage water. Under these conditions it may be found in almost pure stands; it also occurs in nearly pure stands on abandoned cultivated fields where the original stand of wheatgrass was not entirely eliminated (destroyed) by cultivation.

Western wheatgrass has several characteristics that make it exceedingly valuable for use in revegetation and erosion control. Its hardiness and drought resistance and its capacity to spread rapidly by means of underground rhizomes are outstanding values for conservation. It is excellent for terrace waterways and contour strip plantings for erosion control.

The extent of its underground rhizomes and roots depends on availability of moisture and on the soil fertility. Ordinarily these underground plant parts make a profuse, dense growth, resulting in a tough, fibrous sod that effectively binds the soil and offers protection from erosion. Growth starts fairly early in the spring and continues until limited by shortage of moisture or by continued hot summer periods. Abundant forage is produced and is relished (much liked) by all classes of livestock until it becomes harsh and woody during the summer. Mature plants cure well into a palatable, nutritious forage that provides excellent winter grazing. Leafy high quality hay may also be produced if proper precautions are taken to cut the grass while it is still succulent (juicy) not after the leaves and stems have become harsh and woody. Yield of hay depends upon moisture, particularly that available during the early part of the growing season. Seed of the western wheatgrass has low germination immediately after harvest. Its dormancy can ordinarily be overcome by 6-12 months of dry storage.

How to sow or plant: Best results have been obtained when sowing on well prepared clean seedbeds. Because the young seedlings are small and inconspicuous, the new stand often seems disappointing; but with full protection from grazing until the second growing season, the stand improves rapidly in vigour and density. Western Wheatgrass seedlings are drought resistant. They compete fairly well with weeds and other grasses, although not so well as crested wheatgrass. The plants spread rapidly by means of underground rhizomes. Thus if seeding results in a relatively thin stand, the spread will soon provide the density of cover required. The sod forming nature of western W. also provides a means of vegetative propagation on desired areas. The usual procedure is to start the new planting by use of sod pieces three or four inches square. This is an effective method of establishing a dense sod cover for a diversion channel, terrace water outlet or contour strip.

Seed rate. Very good stands have been obtained from sowing or broadcasting seed at the rate of 6-12 pounds (2.7-5.4 kg) of clean seed per acre = 15-30 pounds (6.75-13.5 kg) per hectare.

Its growth characteristics, drought resistance, hardiness, and wide adaptation to soil and climatic con ditions, make western wheatgrass one of the best grasses for revegetation and general farm use." [65]

LEGUMES FOR SOIL CONSERVATION

AP5.5 What is the role of Legumes in Soil Conservation ?

There is the very important role of legumes as soil improvers and alternative crops in rotations and companion plants in mixed cropping. They all improve fertility, sustainability and soil texture and so contribute to better crop stands which helps to protect and hold the soil, and in that way reduce erosion.

What is being considered now is the role of legumes on land which is, or is in danger of, eroding badly. On steep banks, and pastures, in gullies, and on landside or road side areas; remembering that much soil erosion is caused when roads are built in hilly areas. This is both directly as the earth crust is disturbed and soil moved; and indirectly, as soil stones and rocks which are disturbed or exposed in the road building, fall or are washed down into gullies and streams. When this happens, the extra volume of material added to the rainwater raises stream and river levels higher than usual; and so fields along the river banks become part of the river, further increasing the volume of flood water; and so the amount of flooding and damage downstream. The use of legumes in erosion susceptible places is in some cases to hold and protect the soil directly; in other places to maintain and improve the fertility and texture of the soil; so that grasses or other plants which hold and protect the soil may grow in a sustainable way. Plants which if they grew on their own would exhaust the soil, as, though the fodder they produce was taken away by grazing or cutting, little if any of the manure produced was returned to the soil. If grasses alone are planted, in time the areas can become exhausted and the grasses unable to perform their soil holding and protecting role.

Growing legumes with the grasses generally also increases the total amount and quality of fodder, and so the value of the areas on which they grow. They are then less likely to be dug and planted with erosion increasing crops.

"In Kenya, mixtures of Elephant, Guinea or Guatemala grass with Desmodium intortum or *D. uncinatum* have given higher dry matter and crude protein yields than those obtained from the grasses alone. (Kenya National Agricultural Research Station, 1970.) In one trial the crude protein content of elephant grass was raised from 6.6% to 7.1% and that of Guatemala grass from 6.6% to 10% by growing the grasses with D.Uncinatum, the crude protein of the mixtures being respectively 9.8 and 14%." [18 p353.]

Legumes have the ability to attract special nitrogen fixing bacteria to their roots. These bacteria take nitrogen from the air and turn it into plant food for the legumes and other plants to use. When this occurs little knobbly lumps called nodules can be seen among the roots; so legumes improve the soil. Many grasses if grown alone will exhaust the soil but if legumes are grown with them this is much less likely to occur. Legumes also improve the diet of animals fed on the mixture as they are generally higher in protein than grasses.

Before describing the different legumes it is important to note that *in some cases* legumes will not grow well unless the seed or the soil is inoculated with the nitrogen fixing bacteria, without the bacteria they will not get free nitrogen from the air, and when the soil nitrogen is short may suffer from nitrogen starvation just like any other pasture plant. Even though the legumes may find nitrogen fixing bacteria in the soil, sometimes special introduced bacteria may be more efficient in producing the nitrogen fertiliser and so the legumes and the grass growing with them will grow even better. I do not think that is as much a problem in the tropical areas as the temperate areas but it should be considered. It may be worth asking the seed supplier if it may be needed and if he can supply some inoculant. Once the bacteria is in soil or plants root systems, there is no need to buy more. Some soil or root nodules from where the legume is growing can be taken to a new area to introduce the bacteria to the soil there; or, a slurry (thick soil soup) may be made of the soil with water and the seed mixed in the slurry just before being sown into the soil.

For inoculating lucerne the following procedure is used in Australia. "Take the quantity of lucerne seed you wish to plant, and add to it the required quantity of dry inoculant, estimating the quantity from the weight of seed to be treated. Mix the two thoroughly and then add sufficient water to make the seed just moist. The innoculant is a dark colour and should be spread evenly over the surface of the seed coat.

If the seeds tend to stick together, insufficient water has been added to the mixture. After treating, the seed should be spread out **IN THE SHADE** to dry, and should be planted as soon as practicable after treatment." [69]

It is important to be careful with the inoculant (the bacteria). Whether purchased, or from local soil or plant roots.

1. They (the bacteria) don't like to be out of the soil.

2. They don't like dry conditions out of the soil.

3. They don't like heat.

4. They don't like most agricultural chemicals or other chemicals. If they are mixed in a container which is not clean as it had been used for disinfectant, soap, fertiliser or other chemicals the bacteria will die.

5. They don't like direct sunlight so don't spread the inoculant in the sun.

6. The sooner they are in moist soil the more of the bacteria will live.

The bacteria are our friends, we should look after them.

If the plants do not look healthy and green dig some up and look to see if there are nodules (little rough shaped lumps on the roots). If there aren't any, there are none of the right bacteria. Either you didn't add them or you didn't care for them properly or they were spoilt on the way from the supplier to you.

When the bacteria are there in the soil there is no need to add them again. They will stay there ready to develop next time there are plants to use them.

"The differences between tropical and temperate grazing legumes.

---Under the hot, humid conditions of summer rainfall area and where the supply of plant nutrients is adequate, the rate of growth of grasses and weeds is very rapid---. In order to compete satisfactorily under these conditions legumes must also be capable of rapid, lush growth, and therefore they are usually hardy robust plants capable of growing to a height of several feet and of covering a considerable area of ground from the one main root. The main period of growth occurs during the summer months, but under warm winter temperatures and in the presence of adequate soil moisture some growth also takes place during the rest of the year.

Tropical legumes commonly develop a climbing habit in association with other plants, and this makes them particularly suitable for growing in association with the tall tropical grasses and for smothering weeds. Tropical legumes have large leaves and coarse stems---. Most of them send down roots from the nodes and seed profusely. Survival is usually good provided that care is taken to avoid overstocking.

"Establishment: This is usually carried out at the beginning of the wet season either by sowing in drills or by broadcasting following a burn (when the area has had the growth burnt off by accident or to get rid of weeds and rubbish.) If drilled seed is placed every 3 or 4 ft (91-122 cm's) along rows up to 6 feet (1.83 m's) apart---, at no stage should livestock or the mower (or person cutting grass) be allowed to reduce the herbage to ground level.

Pastures containing tropical legumes should last for many years providing management is directed towards maintaining a satisfactory balance between grasses and legumes and weeds are kept in check (not allowed to develop strongly)" [62.]

AP5.6 Qualities of pasture/fodder legumes which may be used for soil conservation.

0. Acid soil tolerant.
1. Alkaline soil tolerant.
2. Cold tolerant.
3. Drought tolerant.
4. Flooding tolerant.
5. Heat tolerant.
6. Poor soil tolerant.
7. Salt tolerant.
8. Shade tolerant.
9. Wide range of pH tolerant.
10. Wide range of temperatures tolerant.
11. Hardy.
12. More temperate.
13. More tropical.
14. Good for erosion control.
15. Good for protecting soil.

16. Good for binding soil.
17. Good for improving soil.
18. Good for waterways.
19. Good grazing/fodder.
20. Good for summer grazing.
21. Good for winter grazing.
22. Can be grazed.
23. Heavy grazing tolerant.
24. Combines with legumes.
25. Vigorous grower.
26. Quick growth after rain.
27. Responds well to winter rain.
28. Responds well to summer rain.
29. Long growing period.
30. May become a weed.
31. Grows on a wide range of soils.
32. Grows on light soil.
33. Grows on heavy soil.
34. Grows on badly drained soil.
35. Grows on waterlogged soil.
36. Long lived.
37. Sod forming.
39. Ease of establishment.
40. Wide adaptation.
41. Survives fire.
42. Sensitive to waterlogging.
43. Survives frost.
44. Produces a lot of fodder.
45. Good seedling vigour.
46. Grows on acid soil.
47. Does not tolerate salinity.
48. Not good with legumes.
49. Grows with a wide range of rainfall.
50. Does best in moist conditions.
51. Can exhaust the soil.
52. Prefers well drained land.
53. Can tolerate heavy treading.
54. Encourages rats.
55. Good pioneer crop.
56. Suffers from frost.
57. Produces good standover feed (uncut hay) for winter.
58. Can stand light frost.
59. Exceptionally drought resistant.
60. Very tolerant of low fertility.
61. Can grow at high altitudes.
62. Has early spring growth.
63. Gives worthwhile winter feed.
64. Good for contour strips.
65. Combines with grasses.
66. Not good in orchards tea gardens etc.
67. Not good with grasses.
68. Chokes out weeds.
69. Grows year round if not too cold.

70. Climber.
71. Roots at the nodes.
72. Does not grow on poor soils.
73. Good in orchards, tea gardens, etc.
74. Can tolerate some waterlogging.
75. Fixes much nitrogen.
76. Deep rooted.
77. Slow to establish.
78. Resistant to nematodes.
79. Resistant to weevils.
80. Easily established from seed.
81. Good cover crop.
82. Very palatable.(livestock like to eat it).
83. Very susceptible to damage by fire.
84. Grows in alkaline soil.
85. Does not do well in high rainfall areas.
86. Grows in low rainfall areas.

AP5.7 Index of the qualities of 35 Legumes which are suitable for Soil Conservation.
(See the list of qualities to understand reference numbers.)

*Stars indicate species suitable for winter rainfall areas.

Alfalfa. see Lucerne.
Archer Axillaris.(Dolichos axillaris).
55.65.68.66.3.59.43.69.70.11.29.19.20.21.5.52.25.39.10.31.56.
13.15.76.86.
Ball clover, (Trifolium glomeratum).
27.29.20.21.17.0.86.
Barrel Medic. (Medicago truncatula).
36.49.1.12.21.43.3.17.2.86.84.
Berseem clover.(Trifolium alexandrinum).
55.13.17.19.84.85.86.
Bicolour lespedeza.(Lespedeza bicolor).See lespedeza species.
Black medic (M. lupulina).
2.43.9.84.
Bush desmodium, see Desmodium distortum.
Centro, (Centrosema pubescens).
66.70.71.50.13.8.22.23.15.25.72.
Cluster clover see Ball clover.
Dalrymple vigna. (Vigna marina).
25.71.34.55.81.56.19.29.31.23.82.
Desmodium distortum.
13.
Egyptian clover see Berseem.
Glycine.(Glycine javanica).
70.66.68.76.58.10.13.27.72.44.25.52.42.17.77.
Greenleaf desmodium (Desmodium intortum). 65.0.17.19.14.16.15.13.44.3.59.20.22.25.26.28.29.31.32.33.46.
60.73.71.56.74.75.23.76.79.82.6.61.
Intortum clover, see Greenleaf desmodium.
Kudzu bean. See Kudzu species.
Kudzu species.
Kudzu species Group 1.
Should be called Wet Tropics Kudzu,(Pueraria phaseoloides), now called Kudzu, Tropical Kudzu, Taiwan

Kudzu or, in Australia- Puero. It is more suited to the wet tropics, it is more tolerant of waterlogging, is more climbing than Dryer Climate Kudzu, and it does not develop roots at the nodes. It is of more use as a cover crop than for holding land on landslides, etc, which would otherwise erode. It grows very easily from seed. It does not root freely at the nodes.

13.50.74.70.81.39.20.15.57.56.8.0.55.25.82.68.83.46.62.31.18.

57.30.

Kudzu species Group 2.

Should be called Dryer Climate Kudzu,(Pueraria hirsuta). It is now called Kudzu, Kudzu bean, or Kudzu vine. It originated in Japan and is not suited to the wet tropics but suited to drier areas. When established it is drought resistant. It likes good drainage and does not do well if the drainage is poor. It likes a lighter soil and low to moderate rainfall. It can tolerate cooler winters better than tropical kudzu and it is a very useful plant for protecting and binding soil against erosion. It roots freely from the nodes thus establishing linked plants, it does not produce many seeds and so is normally propagated by rooted crowns, and according to one source rooted node cuttings.

61.71.76.14.15.16.17.5.59.19.82.29.31.6.32.42.62.36.86.

Kudzu vine. (Pueraria hirsuta). See Kudzu species.

Kudzu. (Pueraria lobata). See Kudzu species.

Kudzu. (Pueraria thumbergiana). See Kudzu species.

Kudzu. See Kudzu species.

Leichhardt Biflorus (Dolichos uniflorus, formerly D. biflorus).

64.8.19.65.26.3.59.44.57.75.13.5.31.42.6.52.63.20.21.86.57.

Lespedeza.sp.

0.32.6.60.29.22.

*Lucerne./Alfalfa in America. (Medicago sativa).

31.14.17.3.2.10.40.59.52.7.29.36.65.43.6.86.84.76.

Miles Lotononis bainesii).

13.50.43.2.29.31.0.46.32.33.65.82.19.74.52.59.78.79.35.4.18.

Puero. (Pueraria phaseoloides). See Kudzu species.

Rhodesian Kudzu. (Glycine Javonica.L). See Kudzu species.

Silverleaf desmodium.(Desmodium uncinatum).

0.36.13.14.15.16.17.19.20.22.23.26.28.31.32.33.44.46.56.60.65.71.73.74.75.82.6.61.76.

Siratro. (Phaseolus atropurpureus).

9.3.59.68.6.55.78.79.5.13.22.20.56.42.65.23.80.75.71.30.15.16.17.31.39.

*Strawberry clover. (Trifolium Fragiferum).

1.7.33.34.84.

*Subterranean clover. (Trifolium subteraneaum.)

17.10.13.86.

Taiwan Kudzu. (Phaseoloides Tonkinensis). See Kudzu species.

Tropical Kudzu. (Pueraria phaseoloides). See Kudzu species.

AP5.8 Selecting Pasture/Fodder Legumes which may be used for Soil Conservation and which Meet Specific Conditions.

The following table aids in selecting appropriate legumes for specific situations.

First look at the section index and select which is the most important factor for the situation being considered. After looking through the section and finding which species may be suitable, check in other appropriate sections to narrow down to the few which are likely to fit the situation. Those species should then be grown with the best local species to determine which would be suitable for the purposes required.

5.8a SECTION ONE. LEGUMES WHICH ARE PARTICULARLY TOLERANT OF/OR CAN SURVIVE CERTAIN CONDITIONS. Note:- There are other legumes which are adaptable to a wide range of conditions, see section two.

5.8B SECTION TWO. LEGUMES WHICH ARE TOLERANT OF A WIDE RANGE OF CONDITIONS.

5.8c SECTION THREE. LEGUMES SUITED TO DIFFICULT CONDITIONS.

5.8d SECTION FOUR. CLIMATIC CONDITIONS.
5.8e SECTION FIVE. PARTICULARLY FOR EROSION CONTROL.
5.8f SECTION SIX. VALUES RELATED TO GRAZING.
5.8g SECTION SEVEN. GROWING CONDITIONS.
5.8h SECTION EIGHT. SPECIAL FEATURES.

* **Stars indicate species suitable for winter rainfall areas.**

AP5.8a Section One. Legumes which are particularly tolerant of/or can survive certain conditions.

1. Legumes which are tolerant of acid soil.
    *Ball Clover. *(Trifolium glomeratum)*.
    Greenleaf Desmodium. *(Desmodium intortum)*.
    Kudzu species. Group 1.
    Miles Lotononis. *(Lotononis bainesii)*.
    Silverleaf Desmodium. *(Desmodium uncinatum)*.
2. Alkaline soil tolerant.
    *Barrel Medic. *(Medicago truncatula)*.
    *Strawberry Clover. *(Trifolium fragiferum)*.
3. Cold tolerant.
    *Barrel Medic. *(Medicago truncatula)*.
    *Lucerne / Alfalfa in America. *(Medicago sativa)*.
    Miles Lotononis. *(Lotononis bainesii)*.
4. Survives frost.
    Archer Axillaris. *(Dolichos axillaris)*.
    *Barrel Medic. *(Medicago truncatula)*.
    *Black Medic. *(M. lupulina)*.
    *Lucerne / Alfalfa in America. *(Medicago sativa)*.
    Miles Lotononis. *(Lotononis bainesii)*.
5. Can stand light frost.
    Glycine. *(Glycine javanica)*.
    Greenleaf Desmodium. *(Desmodium intortum)*.
    Silverleaf Desmodium. *(Desmodium uncinatum)*.
6. Drought tolerant.
    Archer Axillaris. *(Dolichos axillaris)*.
    *Barrel Medic. *(Medicago truncatula)*.
    Greenleaf Desmodium. *(Desmodium intortum)*.
    Leichhardt Biflorus. *(Dolichos uniflorus, formerly D. biflorus)*.
    *Lucerne / Alfalfa in America. *(Medicago sativa)*.
    Siratro. *(Phaseolus atropurpureus)*.
7. Exceptionally drought resistant.
    Archer Axillaris. *(Dolichos axillaris)*.
    Greenleaf Desmodium. *(Desmodium intortum)*.
    Kudzu species. Group 2.
    Leichhardt Biflorus. (Dolichos uniflorus, formerly *D. biflorus*).
    *Lucerne / Alfalfa in America. *(Medicago sativa)*.
    Miles Lotononis. *(Lotononis bainesii)*.
    Siratro. *(Phaseolus atropurpureus)*.
8. Heat tolerant.
    Archer Axillaris. *(Dolichos axillaris)*.
    Kudzu species. Group 2.
    Leichhardt Biflorus. (*Dolichos uniflorus,* formerly *D. biflorus*).
    Siratro. *(Phaseolus atropurpureus)*.
9. Can tolerate flooding.
    Miles Lotononis. *(Lotononis bainesii)*.

9a. Grows on waterlogged soil.

  Miles Lotononis. *(Lotononis bainesii).*

9b. Can tolerate some waterlogging.

  Greenleaf Desmodium. *(Desmodium intortum).*

  Kudzu species. Group 1.

  Miles Lotononis. *(Lotononis bainesii).*

  Silverleaf Desmodium. *(Desmodium uncinatum).*

9c. Grows on badly drained soil.

  Dalrymple Vigna. *(Vigna marina).*

  *Strawberry Clover. *(Trifolium fragiferum).*

10. Poor soil tolerant.

  Greenleaf Desmodium. *(Desmodium intortum).*

  Kudzu species Group 2/.

  Leichhardt Biflorus. *(Dolichos uniflorus,* formerly *D. biflorus).*

  Lespedeza.sp.

  *Lucerne/Alfalfa in America. *(Medicago sativa).*

  Silverleaf Desmodium. *(Desmodium uncinatum).*

  Siratro. *(Phaseolus atropurpureus).*

11. Very tolerant of low fertility.

  Greenleaf Desmodium. (Desmodium intortum).

  Lespedeza.sp.

  Silverleaf Desmodium. *(Desmodium uncinatum).*

12. Can grow at high altitudes. Over 5,000 ft (1,524 m).

  Greenleaf Desmodium. *(Desmodium intortum).*

  Kudzu species Group 2/.

  Silverleaf Desmodium. *(Desmodium uncinatum).*

13. Can tolerate heavy treading.

  None noted.

14. Can tolerate heavy grazing when established.

  Centro. *(Centrosema pubescens).*

  Dalrymple Vigna. *(Vigna marina).*

  Greenleaf Desmodium. *(Desmodium intortum).*

  Silverleaf Desmodium. *(Desmodium uncinatum).*

  Siratro. *(Phaseolus atropurpureus).*

15. Resistant to Nematodes *(eelworms).*

  Miles Lotononis. *(Lotononis bainesii).*

  Siratro. *(Phaseolus atropurpureus).*

16. Resistant to weevils.

  Greenleaf Desmodium. *(Desmodium intortum).*

  Miles Lotononis. *(Lotononis bainesii).*

  Siratro. *(Phaseolus atropurpureus).*

17. Salt tolerant.

  *Lucerne/Alfalfa in America. *(Medicago sativa).*

  *Strawberry Clover. *(Trifolium fragiferum).*

18. Shade tolerant.

  Centro. *(Centrosema pubescens).*

  Kudzu species Group 1.

  Leichhardt Biflorus. *(Dolichos uniflorus,* formerly *D. biflorus).*

AP5.8b Section Two. Legumes which are tolerant of a wide range of conditions.

19. Comparatively wide range of pH. tolerance.

  *Black Medic. *(Medicago lupulina).*

Siratro. *(Phaseolus atropurpureus).*
20. Wide range of temperatures tolerant.
    *Archer Axillaris. *(Dolichos axillaris).*
    Glycine. *(Glycine javanica).*
    *Lucerne/Alfalfa in America. *(Medicago sativa).*
    *Subterranean Clover. *(Trifolium subterraneum.)*
21.· Wide range of soils tolerant.
    *Archer Axillaris. *(Dolichos axillaris).*
    Dalrymple Vigna. *(Vigna marina).*
    Greenleaf Desmodium. *(Desmodium intortum).*
    Kudzu species Group 1/.
    Kudzu species Group 2/.
    Leichhardt Biflorus. *(Dolichos uniflorus,* formerly *D. biflorus).*
    *Lucerne/Alfalfa in America. *(Medicago sativa).*
    Miles Lotononis. *(Lotononis bainesii).*
    Silverleaf Desmodium.*(Desmodium uncinatum).*
    Siratro. *(Phaseolus atropurpureus).*
22. Wide adaptation.
    *Lucerne/Alfalfa in America. *(Medicago sativa).*
23. Wide range of rainfall tolerant.
    *Barrel Medic. *(Medicago truncatula).*

AP5.8c Section Three. Legumes suited to difficult conditions.

24. Hardy (can survive hard conditions), plants.
    *Archer Axillaris. *(Dolichos axillaris).*
    Heavy grazing see 14/.
    Heavy treading see 13/.
25. Fire none noted as able to survive fire.
    High altitude see 12/.
    Wide range of temperatures see 20/.
    Low fertility see 10/. & 11/.
26. Pioneer conditions.
    *Archer Axillaris. *(Dolichos axillaris).*
    *Berseem Clover. *(Trifolium alexandrinum).*
    Dalrymple Vigna. *(Vigna marina).*
    Kudzu species Group 1.
    Siratro. *(Phaseolus atropurpureus).*
    Heat see 8.
    Cold see 3.
    Frost see 4. & 5.
    Drought see 6. & 7.
    Acid soils see 1.
    Alkaline soils see 2.
    Saline (salty) see 17.
    Flooding see 9.
    Waterlogging see 9a. and 9b.
27. Grows on badly drained soil, see 9c.
    Shade see 18.

AP5.8d Section Four. Climatic conditions.

Wide range of temperatures tolerant see 20.
28. More temperate.

      *Barrel Medic. *(Medicago truncatula).*
29. More tropical.
      *Archer Axillaris. *(Dolichos axillaris).*
      *Berseem Clover. *(Trifolium alexandrinum).*
      Centro. *(Centrosema pubescens).*
      Desmodium Distortum.
      Glycine. *(Glycine javanica).*
      Greenleaf Desmodium. *(Desmodium intortum).*
      Kudzu species Group 1.
      Leichhardt Biflorus. *(Dolichos uniflorus,* formerly *D. biflorus).*
      Miles Lotononis. *(Lotononis bainesii).*
      Silverleaf Desmodium. *(Desmodium uncinatum).*
      Siratro. *(Phaseolus atropurpureus).*
      *Subterranean Clover. *(Trifolium subterraneum).*
      Can grow in shade see 18.
      Can grow where there is a wide range in the annual rainfall, see 23.
30. Year round growth if the winter is not too cold.
      *Archer Axillaris. *(Dolichos axillaris).*
31. Does not do well in higher rainfall areas.
      *Berseem Clover. *(Trifolium alexandrinum).*
32. Grows in low rainfall areas.
      *Archer Axillaris. *(Dolichos axillaris).*
      *Ball clover. *(Trifolium glomeratum).*
      *Barrel Medic. *(Medicago truncatula).*
      *Berseem Clover. *(Trifolium alexandrinum).*
      Kudzu species Group 2.
      Leichhardt Biflorus. *(Dolichos uniflorus,* formerly *D.biflorus).*
      *Lucerne/Alfalfa in America. *(Medicago sativa).*
      *Subterranean Clover. *(Trifolium subterraneum).*

AP5.8e Section five. Particularly for Erosion control.

33. Good for erosion control.
      Greenleaf Desmodium. *(Desmodium intortum).*
      Kudzu species Group 2.
      *Lucerne/Alfalfa in America. *(Medicago sativa).*
      Silverleaf Desmodium. *(Desmodium uncinatum).*
34. Good for protecting soil.
      *Archer Axillaris. *(Dolichos axillaris).*
      Centro. *(Centrosema pubescens).*
      Greenleaf Desmodium. *(Desmodium intortum).*
      Kudzu species Group 1.
      Kudzu species Group 2.
      Silverleaf Desmodium. *(Desmodium uncinatum).*
      Siratro. *(Phaseolus atropurpureus).*
35. Good for binding soil.
      Greenleaf Desmodium. *(Desmodium intortum).*
      Kudzu species Group 2.
      Silverleaf Desmodium. *(Desmodium uncinatum).*
      Siratro. *(Phaseolus atropurpureus).*
36. Good for improving soil.
      Ball clover. *(Trifolium glomeratum).*
      *Barrel Medic. *(Medicago truncatula).*
      *Berseem Clover. *(Trifolium alexandrinum).*

Glycine. *(Glycine javanica).*
Greenleaf Desmodium. *(Desmodium intortum).*
Kudzu species Group 2.
*Lucerne/Alfalfa in America. *(Medicago sativa).*
Silverleaf Desmodium. *(Desmodium uncinatum).*
Siratro. *(Phaseolus atropurpureus).*
*Subterranean Clover. *(Trifolium subteraneaum).*

37. Good for rainy season waterways, particularly on slopes.
Kudzu species Group 1.
Miles Lotononis. *(Lotononis bainesii).*

38. Good for contour strips.
Leichhardt Biflorus. *(Dolichos uniflorus,* formerly *D. biflorus).*

39. Vigorous grower.
*Archer Axillaris. *(Dolichos axillaris).*
Centro. *(Centrosema pubescens).*
Dalrymple Vigna. *(Vigna marina).*
Glycine. *(Glycine javanica).*
Greenleaf Desmodium. *(Desmodium intortum).*
Kudzu species Group 1.

40. Quick growth after rain.
Greenleaf Desmodium. *(Desmodium intortum).*
Leichhardt Biflorus. *(Dolichos uniflorus,* formerly *D. biflorus).*
Silverleaf Desmodium. *(Desmodium uncinatum).*

41. Long growing period.
*Archer Axillaris. *(Dolichos axillaris).*
*Ball Clover. *(Trifolium glomeratum).*
Dalrymple Vigna. *(Vigna marina).*
Greenleaf Desmodium. *(Desmodium intortum).*
Kudzu species Group 2.
Lespedeza sp.
*Lucerne/Alfalfa in America. *(Medicago sativa).*
Miles Lotononis. *(Lotononis bainesii).*

42. Long life.
*Barrel Medic. *(Medicago truncatula).*
Kudzu species Group 2.
*Lucerne/Alfalfa in America. *(Medicago sativa).*
Silverleaf Desmodium. *(Desmodium uncinatum).*

43. Ease of establishment.
*Archer Axillaris. *(Dolichos axillaris).*
Kudzu species Group 1.
Siratro. *(Phaseolus atropurpureus).*

44. Wide adaptation, see 22.

45. Good pioneer crop, see 26.
Can tolerate heavy treading. None noted.

AP5.8f Section Six. Values Related to Grazing

46. Can be grazed.
Centro. *(Centrosema pubescens).*
Greenleaf Desmodium. *(Desmodium intortum).*
Lespedeza sp.
Silverleaf Desmodium. *(Desmodium uncinatum).*
Siratro. *(Phaseolus atropurpureus).*

47. Good for grazing and/or fodder.
   *Archer Axillaris. *(Dolichos axillaris).*
   *Berseem Clover. *(Trifolium alexandrinum).*
   Dalrymple Vigna. *(Vigna marina).*
   Greenleaf Desmodium. *(Desmodium intortum).*
   Kudzu species Group 2.
   Leichhardt Biflorus. *(Dolichos uniflorus,* formerly *D.biflorus).*
   Miles Lotononis. *(Lotononis bainesii).*
   Silverleaf Desmodium. *(Desmodium uncinatum).*

48. Produces a lot of fodder.
   Glycine. *(Glycine javanica).*
   Greenleaf Desmodium. *(Desmodium intortum).*
   Leichhardt Biflorus. *(Dolichos uniflorus,* formerly *biflorus).*
   Silverleaf Desmodium. *(Desmodium uncinatum).*

49. Good for summer grazing/fodder.
   *Archers Axillaris. *(Dolichos axillaris).*
   *Ball Clover. *(Trifolium glomeratum).*
   Greenleaf Desmodium. *(Desmodium intortum).* Kudzu species Group 1.
   Leichhardt Biflorus. *(Dolichos uniflorus,* formerly *D.biflorus).*
   *Lucerne/Alfalfa in America. *(Medicago sativa).*
   Silverleaf Desmodium. *(Desmodium uncinatum).*
   Siratro. *(Phaseolus atropurpureus).*

50. Good for winter grazing/fodder.
   *Archer Axillaris. *(Dolichos axillaris).*
   *Ball Clover. *(Trifolium glomeratum).*
   *Barrel Medic. *(Medicago truncatula).*
   Leichhardt Biflorus. *(Dolichos uniflorus,* formerly *D biflorus).*
   Heavy grazing tolerant, see 14.

51. Combines well with grasses.
   *Archer Axillaris. *(Dolichos axillaris).*
   Greenleaf Desmodium. *(Desmodium intortum).*
   Leichhardt Biflorus. *(Dolichos uniflorus,* formerly *D biflorus).*
   *Lucerne/Alfalfa in America. *(Medicago sativa).*
   Miles Lotononis. *(Lotononis bainesii).*
   Silverleaf Desmodium. *(Desmodium uncinatum).*
   Siratro. *(Phaseolus atropurpureus).*

53. Very palatable *(tasty to animals).*
   Dalrymple Vigna. *(Vigna marina).*
   Greenleaf Desmodium. *(Desmodium intortum).*
   Kudzu species Group 1.
   Kudzu species Group 2.
   Miles Lotononis. *(Lotononis bainesii).*
   Silverleaf Desmodium. *(Desmodium uncinatum).*
   Vigorous grower, see 39.
   Quick growth after rain, see 40.
   Year round growth if weather is not too cold, see 30.
   Poor soil tolerant, see 10.
   Very tolerant of low fertility, see 11.
   Tolerant of flooding, see 9.
   Year round growth if not too cold, see 30.

54. Easily established from seed.
   Siratro. *(Phaseolus atropurpureus).*
   Grows in low rainfall areas, see 32.

55. Produces good standover.
      *Archer Axillaris. *(Dolichos axillaris).*
      *Ball clover. *(Trifolium glomeratum).*
      *Barrel Medic. *(Medicago truncatula).*
      *Berseem clover. *(Trifolium alexandrinum).*
      Kudzu species Group 2.
      Leichhardt Biflorus. *(Dolichos uniflorus,* formerly *D.biflorus).*
      *Lucerne/Alfalfa in America. *(Medicago sativa).*
      *Subterranean clover. *(Trifolium subteraneaum).*
      Shade tolerant, see 18.
      Grows with a wide range of rainfall, see 23.
57. Has early spring growth.
      Kudzu species group 1.
      Kudzu species Group 2.
58. Gives worthwhile winter feed.
      *Archer Axillaris. *(Dolichos axillaris).*
      *Ball clover. *(Trifolium glomeratum).*
      *Barrel Medic. *(Medicago truncatula).*
      *Berseem clover. *(Trifolium alexandrinum).*
      Kudzu species Group 2.
      Leichhardt Biflorus. *(Dolichos uniflorus,* formerly *D.Biflorus).*
      *Lucerne/Alfalfa in America. *(Medicago sativa).*
      *Subterranean clover. *(Trifolium subteraneaum).*
57. Gives worthwhile winter feed.
      Archer Axillaris. *(Dolichos axillaris).*
      *Ball clover *(Trifolium glomeratum).*
      *Barrel Medic. *(Medicago truncatula).*
      *Berseem clover, *(Trifolium alexandrinum).*
      Kudzu species. Group 2.
      Leichhardt Biflorus *(Dolichos uniflorus,* formerly *D. biflorus).*
      *Lucerne./Alfalfa in America. *(Medicago sativa).*
      *Subterranean clover. *(Trifolium subterraneum).*

AP5.8g Section Seven. Growing conditions.

   Grows on a wide range of soils, see 21.
   Comparatively wide range of pH tolerance, see 19.
   Poor soil and poor fertility tolerant, see 10. and 11.
58. Fixes much nitrogen.
      Greenleaf Desmodium *(Desmodium intortum).*
      Leichhardt Biflorus *(Dolichos uniflorus,* formerly *D. biflorus).*
      Silverleaf Desmodium. *(Desmodium uncinatum).*
      Siratro. *(Phaseolus atropurpureus).*
      Acid soil tolerant, see 1.
59. Grows on acid soil.
      Greenleaf Desmodium *(Desmodium intortum).*
      Kudzu species Group 1.
      Miles Lotononis. *(Lotononis bainesii).*
      Silverleaf Desmodium. *(Desmodium uncinatum).*
      Alkaline soil tolerant, see 2.
60. Grows in alkaline soil.
      *Barrel Medic. *(Medicago truncatula).*
      *Berseem clover. *(Trifolium alexandrinum).*

    *Black medic *(M. lupulina).*
    *Lucerne./Alfalfa in America. *(Medicago sativa).*
    *Strawberry clover. *(Trifolium fragiferum).*
Can grow at high altitudes, (over 5,000 ft = 1,524 m) see 12.
Wide range of temperatures tolerant, see 20.
FOR OTHER TOLERANCES SEE SECTIONS ONE AND TWO.
More temperate, see 28.
More tropical, see 29.
Hardy, see 24.

61. Grows on light soils.
    Greenleaf Desmodium *(Desmodium intortum).*
    Kudzu species Group 2.
    Lespedeza.sp.
    Miles Lotononis. *(Lotononis bainesii).*
    Silverleaf Desmodium. *(Desmodium uncinatum).*

62. Grows on heavy soils.
    Greenleaf Desmodium *(Desmodium intortum).*
    Miles Lotononis. *(Lotononis bainesii).*
    Silverleaf Desmodium. *(Desmodium uncinatum).*
    *Strawberry clover. *(Trifolium fragiferum).*
    Grows on badly drained soil, see 9c.
    Can tolerate some waterlogging, see 9b.
    Grows on waterlogged soil, see 9a.
    Can tolerate flooding, see 9.

63. Does best in moist conditions.
    Centro, *(Centrosema pubescens).*
    Kudzu species Group 1.
    Miles Lotononis. *(Lotononis bainesii).*

64. Prefers well drained land.
    *Archer Axillaris. *(Dolichos axillaris).*
    Glycine. *(Glycine javanica).*
    Leichhardt Biflorus *(Dolichos uniflorus,* formerly *D. biflorus).*
    *Lucerne./Alfalfa in America. *(Medicago sativa).*
    Miles Lotononis. *(Lotononis bainesii).*
    Does not do well in higher rainfall areas, see 31.

65. Sensitive to waterlogging.
    Glycine. *(Glycine javanica).*
    Kudzu species Group 2.
    Leichhardt Biflorus *(Dolichos uniflorus,* formerly *D. biflorus).*
    Siratro. *(Phaseolus atropurpureus).*
    Grows in low rainfall areas, see 32.
    Quick growth after rain, see 40.

66. Responds well to winter rain.
    *Ball clover, *(Trifolium glomeratum).*
    Glycine. *(Glycine javanica).*

67. Responds well to summer rain.
    Greenleaf Desmodium *(Desmodium intortum).*
    Silverleaf Desmodium. *(Desmodium uncinatum).*
    Long growing period, see 41.

68. Suffers from frost.
    Dalrymple Vigna. *(Vigna marina).*
    Greenleaf Desmodium *(Desmodium intortum).*
    Kudzu species Group 1.

Silverleaf Desmodium.*(Desmodium uncinatum).*
Siratro. *(Phaseolus atropurpureus).*
Has early spring growth, see 56.
Has year round growth if weather is not too cold, see 30.
69. Does not grow on poor soils.
Centro, *(Centrosema pubescens).*
Glycine. *(Glycine javanica).*
70. Very susceptible to damage by fire.
Kudzu species Group 1.
Grows on alkaline soil, see 60.

AP5.8h Section Eight. Special Features.

71. Chokes out weeds.
*Archer Axillaris. *(Dolichos axillaris).*
Glycine. *(Glycine javanica).*
Kudzu species Group 1.
Siratro. *(Phaseolus atropurpureus).*
72. May become a weed.
Kudzu species Group 1.
Siratro. *(Phaseolus atropurpureus).*
73. Good in orchards.
Greenleaf Desmodium *(Desmodium intortum).*
Silverleaf Desmodium *(Desmodium uncinatum).*
74. Not good in orchards.
*Archer Axillaris *(Dolichos axillaris).*
Centro *(Centrosema pubescens).*
Glycine. *(Glycine javanica).*
75. Climber.
*Archer Axillaris. *(Dolichos axillaris).*
Centro *(Centrosema pubescens).*
Glycine. *(Glycine javanica).*
Kudzu species Group 1.
76. Roots at the nodes.
Centro *(Centrosema pubescens).*
Dalrymple Vigna. *(Vigna marina).*
Greenleaf Desmodium *(Desmodium intortum).*
Kudzu species Group 2.
Silverleaf Desmodium. *(Desmodium uncinatum).*
Siratro. *(Phaseolus atropurpureus).*
77. Deep rooted.
*Archer Axillaris. *(Dolichos axillaris).*
Glycine. *(Glycine javanica).*
Greenleaf Desmodium *(Desmodium intortum).*
Kudzu species Group 2.
*Lucerne/Alfalfa in America. *(Medicago sativa).*
Silverleaf Desmodium. *(Desmodium uncinatum).*
Fixes much nitrogen, see 58.
78. Slow to establish.
Glycine. *(Glycine javanica).*
Easily established from seed, see 54.
Resistant to nematodes, see 15.
Resistant to weevils, see 16.

79. Good cover crop.

> Dalrymple Vigna. *(Vigna marina).*
> Kudzu species Group 1.
> Very susceptible to damage by fire, see 70.

AP5.9 A Description of Suitable Legumes

Note:-* indicates species suitable for areas with winter rainfall though they may grow in areas with mainly summer rainfall.

*Alfalfa. *See Lucerne.*

Archer Axillaris.(Dolichos axillaris).* The author has no personal experience of this plant. Its origin is in Kitale, Kenya.

"This perennial legume is capable of remarkable year round growth on frost free locations. Its twining (twisting along other things) climbing habit makes it an ideal competitor to weeds, and it combines very successfully with tall growing grasses. Its very strong root system is extremely good at extracting moisture from the soil, and this makes it one of the most drought resistant of the legumes, at least equivalent to Siratro in this respect.

Even when frosted, it has a very long productive season. It can come away in spring before the season breaks, while other plants may still be waiting for the rain, and keep growing when conditions are too dry for other legumes.

It is, however, quite susceptible to frost but will recover quickly---. ("The Archer variety--- grows well where frosts are not too severe" [63].

The leaves are extremely bitter to taste, and it was thought that there would be a palatability problem, but once accustomed to it, cattle take to it readily. Altogether, its general performance and its outstanding dry weather growth earns it a place in any pasture mixture in coastal areas 40" (102 cm) rainfall, and its drought resistance takes it further into the dry areas that most new legumes." [61] " Axillaris has made more growth under intermittently hot, dry conditions than any other tropical legumes, e.g. Greenleaf Desmodium.(Authors note:- I have no experience with this legume but Greenleaf Desmodium is itself amazingly drought resistant once established.) It is complementary to Siratro for well drained soils.

A special feature of axillaris is its relative freedom from pests and diseases which attack other legumes. Thus it has some tolerance of Amnemus weevil, and seems to be resistant to the powdery mildew which detracts from *Leichhardt uniflorus*, (Biflorus). It does suffer from little leaf virus---. It establishes readily. Inoculation with Rhizobium is an insurance measure, but benefits are only occasionally observed with this species." [63]

"Where to sow: A wide range of soil types are suitable for *Dolichos axillaris*, but it will thrive only in well drained situations. It has shown some promise in a 30 inch (76 cm) annual rainfall belt, and outstanding results in rainfall areas of 45 inches (114 cm) or better.

When to sow: anytime between spring and autumn, but early plantings are favoured, provided soil moisture conditions or anticipated rainfall are adequate to ensure a good strike. (Good germination rates.)

How to sow: As good a seedbed as possible is desirable---. Plant no deeper than one inch (2.5 cm).

Seed should be inoculated and lime pelleted with the correct culture strain before sowing. Pelleting is essential if seed is mixed with fertiliser for sowing.

Seed rate: if it is the only legume in the sward up to 3-4 pounds (1.36-1.8 kg) per acre = 7.5-10 pounds (3.4-4.5 kg) per hectare. However the general recommendation is for growing with other legumes such as glycines and desmodiums. In this case 1-2 pounds (0.45-0.9 kg) per acre = 2.5-5.0 pounds (1.135-2.27 kg) per hectare, will be sufficient to give the pasture mixture a degrees of drought resistance.

Management: As with all tropical legumes, grazing should be lenient (light) until the stand is well established. It can seed (produce seed) both during winter and summer, and it has shown an ability for seedlings to thicken up the stand (the sward). Cattle readily transport the seed in the manure. Rotational grazing to a height of six inches will encourage the highest possible productivity.

*Ball clover/*Cluster clover *(Trifolium glomeratum).* Ball clover is a valuable winter growing annual which occurs naturally in many areas and is particularly common in the New South Wales and Victorian (Australia) wheat belts---.

It produces a large quantity of green feed in the spring followed by prolific quantities of seed at the end of the season, which provide useful concentrated feed during the summer months.

Ball clover is a valuable species for improving stock carrying capacity and soil fertility in areas where annual rainfall is between 12 and 18 " (30-45 cm) and mainly of autumn winter spring incidence. Like most of the subterranean clovers it prefers soils which are neutral to slightly acid in reaction.---." [62.].

"It is not tolerant of salinity. Soils should be shallow or of average depth and fine sand to silt loam. It needs moist to heavy moisture conditions"[20.]

*Barrel Medic. *(Medicago truncatula).* "During the growing season the low growing medic plant sends out runners from a central crown and taproot system. In a sward there is a dense profusion of these runners which cover the ground by the end of winter. The leaves are borne on these runners---. The flowers are various shades of yellow and when they dry and fall off within a few days, the immature pods, whose shape depends on the species are revealed. These seed pods enlarge and harden and at the end of the growing season they fall to the ground. The seed pods remain on or in the ground until conditions permit germination.

The seed pod of Barrel medic is distinctive among the medics. It is hard and woody and consists of tightly wound coils which carry short straight spines. A large proportion of the seed lying in the paddock (field) at any time is "hard". This hard seed will not germinate even when adequate moisture is available because it is covered by a waterproof seed coat. "Soft" seed on the other hand is seed whose waterproof seed coat has been broken so that moisture can enter.---. The significance of this hard seed coat to the farmer is that once a reserve of medic seed has accumulated on an established pasture area, the medic may not need to be resown.(Even though the field may be used for other crops for a time.)" [71.]

"Barrel medic is adaptable to a wider range of rainfall conditions than most other medics. In Western Australia it is able to persist and survive in a growing season as short as four months. It is well adapted to soils having a high lime content and generally produces best on those with a moderately alkaline reaction. It is widely grown throughout the wheat growing areas of Australia." [62.]

"Barrel medic is best suited to a temperate type climate, but will grow in colder subtropical areas where reasonable cool season rains occur. The competition of summer growing species prevents barrel medic from reaching its full development, but it can provide useful feed in some years.

(The author finds this statement confusing as when he was in Australia he saw a related crop snail medic, growing in Queensland west of Brisbane, on droughty, rough, scrub land. The farmer was very happy with it and said that it germinated in late summer, it grew quickly and gave very good winter grazing. In the late spring it died back and was dormant till the late summer. The initial sowing had been carried out by broadcasting from an aircraft after the scrub had been cleared.)

It is counted more drought resistant than burr medic, and has good frost resistance.---. It is responsive to nutrients, and superphosphate is necessary on some soils for good growth of barrel medic."[63.]

"Owing to the extremely high proportion of hard seed which commonly occurs in seed samples of barrel medic it is always advisable to scarify home grown seeds before sowing." [62.]

(Scarify, to cut, scratch, or graze.) This is done to hard coated seeds to allow moisture to enter the hard waterproof coating which some seeds have to allow them to survive for long periods before germinating. Some times it is done by making a little cut in the seed coat, sometimes by cutting off a piece of the seed-coat. In other cases the seeds are put through a mill or ground between stones, or treated with acid to remove part of the waxy coating.)

In Australia Barrel medic is sometimes sown with wheat, oats, barley or linseed. This will be better for soil than wheat alone as the legume will improve the soil, it will cover the ground between the rows of wheat, and it will provide livestock feed. It is a more economically attractive way to establish pastures.

"Many failures to establish pastures in the wheat belt result from the use of too high a sowing rate of the wheat cover crop.

The wheat should be sown at no more than 25 pounds (11.35 kg) of seed per acre = 62.5 pounds (28.37 kg's) per hectare. Further reductions to 20 pounds of wheat seed has little effect on the wheat yield but has large effects on the final pasture pod yield at the end of the crop year. It is the yield of medic pod at the end of the first year that governs the density of pasture establishment in the following year.

Results of experiments — have shown that, on the average, there was no reduction in wheat yield by reducing the sowing rate from 45 to 25 pounds per acre. Increasing the pasture sowing rate from 2 pounds to 4 pounds (0.9 to 1.8 kg) per acre = 5 to 10 pounds (2.27 to 4.5 kg) per hectare, had no effect on the wheat yield.

Sowing should be performed when moisture levels are suitable (When the moisture level will continue to be adequate): Sowing during mid April to mid May (in the southern hemisphere, mid October to mid November in the northern hemisphere) is ideal if moisture conditions are satisfactory. This is because pasture germination and seedling establishment can take place before winter.

It is recommended that medic seed should be inoculated before sowing with improved strains of Rhizobium. This low cost procedure--- gives a much greater likelihood that the most efficient bacteria form the nitrogen nodules on the medic roots.---. The seed should be inoculated just before sowing. In addition, this seed may be lime pelleted. When the soil acidity in the surface 6 inches is less than Ph 6.0 then the inoculated seed should always be lime pelleted. The technique consists of mixing a weak adhesive, such as 1 pint of sugar solution = 1 imperial tablespoon (25 grams) of sugar in 1 pint (0.568 litre) water; and peat inoculum with every 30 pounds (13.62 kg) of barrel medic seed. The resultant mix is fairly dry if the correct quantities of seed and solution are used. If further drying is necessary, the seed can be laid out on hessian. But it must be protected from direct sunlight in a cool place as the rhizobia are killed by the ultraviolet rays in sunlight. The inoculated seed should be sown as soon as possible, if it is not sown within 2 days then it should be re-inoculated. After a packet of inoculum is opened the inoculant should be used within two days, if not it should be resealed.

Lime pelleting. The technique of lime pelleting consists of mixing pasture seed coated with inoculum and an adhesive with fine lime to produce a firm pellet in which the seed is enclosed. The lime pellet protects the inoculum from the effects of soil and fertiliser acidity and improves inoculant survival. It also keeps a large number of rhizobia near the seed.

The procedure of lime pelleting 30 pounds (13.62 kg) medic is as follows: 3.5 oz (99 g) of coarse ground gum acacia are dissolved in 8 fluid oz (227.28 ml) of fresh water to make 0.5 pint (0.284 litre) of sticker. The gum acacia can be quickly dissolved if the mixture is heated. When the solution is cool the inoculant is added and thoroughly mixed. The sticker and inoculant are then poured over the 30 pounds (13.62 kg) medic seed and the seed is mixed for 1-1.5 minutes until it is evenly coated.

7.5 pounds (3.4 kg) of fine plasters lime is then added and mixed for 1-1.5 minutes until the seed is evenly pelleted. A clean cement mixer is a convenient method of mixing seed, sticker, and lime---. The pelleted seed can be sown immediately, but it can be stored up to one week in a cool place.

Farmers should ensure that the inoculum is the correct strain for medic seed, (the same inoculum can be used for all the medics and lucerne) that it is stored in the refrigerator until used, and that it is not older that the expiry date marked on the packet.

When sowing medics without a cover crop:

When moisture conditions are favourable, 3-4 pounds (1.36-1.8 kg) medic seed per acre = 7.5-10 pounds (3.4-4.5 kg) per hectare can be sown onto a seedbed. Superphospate fertiliser at 0.5-1 cwt (25-50 kg) per acre = 1.25-2.5 cwt (62.5-125 kg) per hectare should also be sown with the pasture." [72]

*Berseem clover/*Egyptian clover, *(Trifolium alexandrinum).* "It is a good pioneering legume. It has an erect habit of growth, which makes it suitable for cutting. It is sown in the autumn and is best treated as an annual crop.---. It is the least winter hardy of all clovers" [20].

"a cool-season annual legume with white flowers, is a native of Egypt, Israel, Syria and Cyprus. In Egypt, berseem clover is the most important soil building legume in rotation with irrigated cotton in the Nile valley.

In India, it is useful as a green manure crop, a winter fodder, and for reclaiming alkaline soil. It produces from four to seven cuttings a year, with a total production of approximately 40 tons of green fodder per acre = 100 tons per hectare. It is an important crop in the Punjab and in western Uttar Pradesh. It is also grown successfully in Bihar, Madhya Pradesh, Madras and Mysore---. It has not been a success in the high rainfall areas of Assam, West Bengal, Orissa and Kerala. In most areas in India, Berseem clover responds to applications of superphosphate fertiliser. Without superphospate fertiliser the yield was 8 tons per acre; with an application of 80 pounds of superphospate per acre the yield was 21 tons.

Seed rates: 15-20 pounds (6.8-9.0 kg) per acre = 37.5-50 pounds (17-22.5 kg) per hectare." [26]

Note : Another source Ref 70 gives the seed rate at 4-6 kg per hectare = 1.6-2.4 kg per acre.

Bicolour lespedeza.*(Lespedeza bicolor).* "Is a perennial woody species, used mostly for erosion control and bird feed. It is adapted best to loam and clay soils." [20 p653]

The author does not have personal experience of this plant and has not been able to find any more information than the above. It is presumed to be a summer rainfall legume.

*Black medic *(M. lupulina).* "Is the most winter hardy of the listed species. It is a winter annual in the South (USA) and a summer annual in the north. It is less exacting (more adaptable) in its calcium requirement than the other species.

Bush Desmodium, see Desmodium distortum.

Centro, *(Centrosema pubescens).* The authors experience in Taiwan on poor to medium acid soils was that Centro was much inferior to both Greenleaf and Silverleaf Desmodium. Both in its production of fodder and in its characteristics. Both of the Desmodiums were better at rooting into terrace banks and so holding and protecting the soil. They were also better at covering bare spaces including rocks, utilising the space and giving shade cover to reduce heat and evaporation as air passed over hot areas. They were better in orchards, as the Centro would climb up into fruit trees making a tangled mess and an unhealthy micro climate which encouraged pests and diseases. The Desmodiums did sometimes climb into trees but only occasionally. They were also more resistant to drought than Centro. However under different conditions Centro may be more suitable.

"This creeping, twining perennial legume is native to South America. The leaves are trifoliate---. The stems are long and vigorous growing, rooting moderately at the nodes and capable of extending fourteen feet (4.27 m) in a summer. A dense mat of foliage is produced. Centro is a very late flowering species. The pea type flowers are mauve and very showy; they give rise to dark brown pods about 5" (13 cm) in length and containing up to 20 brownish seeds.

Centro flourishes best in areas receiving over 50" (1.27 m) of annual rainfall. It has a good root system and will withstand a long dry season, but has never become well adapted in areas receiving less than about 40" (1 m) of rainfall. It is essentially a plant for warm lands; in south eastern coastal Queensland, Australia, it will only flourish on the warmer sides of the hillsides and makes a poor showing in the valley bottoms.

Centro will climb on anything it encounters. It can be seen growing 50 ft (15.24 m), and its use as a cover crop in plantations has been restricted because of its contamination of and smothering action on tea hedges etc.

When growing with pasture grasses it is not readily shaded out, if soil nutrient conditions are satisfactory. It is moderately palatable, but withstands heavy grazing---. Some failures have been attributed to nodulation difficulties. Centro is not as fast at germination as puero or siratro but quicker than glycine. The build up of nodules is sometimes delayed but not as much as in Glycine." [63.]

"In pure stands centro forms a dense mat about 1.5 ft (46 cm) deep.---. Fertile soils are desirable for optimum growth---.

Seed rate: The usual rate is 3-4 pounds (1.36-1.8 kg) per acre = 7.5-10 pounds (3.4-4.5 kg) per hectare. When sown in bare fallow in drills 6 ft (1.83 m) apart a good cover may be expected in 6 months". [62.]

*Cluster clover, *see ball clover.*

Dalrymple Vigna. *(Vigna marina).* "D.V. is a vigorous, leafy perennial legume with twining runners which readily (easily and quickly) root down into the soil---.

("One of its chief virtues is its ability to grow vigorously and nodulate freely on poorly drained soils"[68.])

Outstanding features of this plant are its rapid early growth in its first year, and its ability to grow in poorly drained situations where other legumes fail. Its rapid growth makes it a useful pioneer crop, or as a cover crop for slower growing legumes, particularly as its performance seems to decline after its first season, and it may be only a short lived perennial, needing reseeding to persist. Vigna is very sensitive to frost, but recovers rapidly with warm, moist conditions. It is also very attractive to insects although damage is not usually serious. It is highly palatable (tasty) to stock, which rapidly spread the seeds. Early spring and late autumn growth make it a valuable contributor to the grazing cycle.

Where to sow: D.V. needs a rainfall of 40"(102 cm) and more for good performance.---. It grows well on a wide range of soil types, particularly on friable (easily crumbled) soils which allow its runners to root down easily.

When to sow: in late spring and summer , when moisture is adequate.

How to sow: drilled or broadcast into a firm moist seedbed, no deeper than 1 " (2.5 cm), and lightly covered. Seed should be inoculated and preferably lime pelleted.

Sowing rate: in a mixture of legumes 1-2 pounds,(0.454 -0.9 kg) per acre, = 2.5-5 pounds (1.35-2.27 kg) per hectare. Alone, 3-5 pounds (1.36-2.27 kg) per acre = 7.5-12.5 pounds (3.4-5.67 kg) per hectare.

Management: it is very attractive to stock, so overgrazing of young stands should be avoided. Once the stand is well established, with its runners pegged (rooted) firmly into the soil, it is much more resistant to grazing."[61.]

Desmodium distortum. This is a woody bush; in the western hills of Nepal at about 4,500 ft (1,371 m) it grows about 3 ft (1 m) tall on poor soil. It dies back in the winter (even though the coldest temperature would be 4° C.) and dry season and few plants survive, so it needs planting every year. Its production was not very good and so we gave up growing it after two years.

Dryer climate Kudzu. see Kudzu species group 2.

*Egyptian clover, *see Berseem.*

Glycine.*(Glycine javanica).Sometimes wrongly called Rhodesian Kudzu,* this plant is similar in form to Greenleaf Desmodium but is finer stemmed. In the authors experience it does not do as well on poor soils as Greenleaf or Silverleaf Desmodium. It has the bad habit of climbing up and choking fruit trees which seldom occurs with the Green and Silverleaf Desmodiums. It is not as good at rooting from the nodes and thus binding the soil of terrace banks or slopes.

"Glycine is widespread in the East Indies, Manchuria, tropical Asia, Ethiopia, tropical East Africa and parts of southern Africa---.("It likes something to climb, which makes it ideal for use with vigorous and tall growing grasses, like setaria, green panic and the guinea grasses, and it is extremely useful in fighting weeds." [61.])

Glycine is a deep rooted plant. In Kenya it has been observed to bring up soil phosphate from lower levels in the soil, and to cycle it into the upper layers where more shallow rooted grasses could take advantage of it.

(Author's note:- I believe a similar effect occurs with the Green and Silverleaf Desmodiums.) It has better drought tolerance than Centro or the Desmodiums. Although burnt off by severe frosts, it is more tolerant than many other tropical legumes and shades (is better than) siratro in this respect. It is a warm climate plant, but where centro is restricted in its range to the tropics, glycine is very useful in the cooler sub-tropics---.

In two successive winters at the Kairi research station on the Atherton Tableland in New South Wales, Australia, where frosts are mild, Tinaroo glycine put up a surprising performance. Its winter growth far exceeded that of lucerne---. This growth rate is sufficient to maintain one cow per acre (2.5 cows per hectare) ("Glycine is rather slower to establish than other legumes in the first season, but once established, is extremely vigourous and produces large quantities of high quality feed, capable of carrying two dairy cows per acre (five per hectare) for lengthy periods" [61.]) irrespective of any surplus summer growth which might be present. Its main growth is made in the summer, but its longer cool season growth can be used to advantage to maintain livestock production. Heavy frosting will destroy much of the feed value, but light frosts can be tolerated, and superior spring vigour will result from this kind of management. Glycine is thought to be not as shade tolerant as Puero, or Dolichos lablab.

Glycine may be grown in areas receiving more than 30" (76 cm) of annual rainfall; it is not suited to very high rainfall conditions above about 70" (178 cm)---.

Glycine is more demanding in its soil requirements than other tropical legumes--- it seems more at home on well drained red scrub loams. However given adequate nutrition it has grown well on a number of other soils, including yellow clays derived from shale, black self mulching soils and medium textured forest soils. It has a higher demand for nutrients (especially phosphorous, potash and lime) than many other species. It is not suited to very acid soils, and is intolerant of waterlogging. Glycine is not badly affected by many pests and diseases.

Glycine/green panic mixtures have been very effective in restoring the fertility of red loams --- , which had been run down and eroded by a long history of continuous maize cropping---.

A considerable number of lines of glycine have been introduced the three main varieties in commercial practice in Australia are Tinaroo, Clarence and Cooper.

Tinaroo—from Kenya not as good for summer vigour but none will produce as well in the autumn or winter. It is therefore best used in the more humid areas where the potential for a longer growing period may be exploited (used). It has soft, rather thin leaves, which are bright green in colour.

Clarence— from South Africa ("This is the earliest of the three to come away (start to grow) in the spring, and is commonly used in mixtures to provide early feed."[61.]) The leaflets are slightly more elongated and the veins more prominent than those of Tinaroo. The leaves are dark green in colour.

Cooper— from Tanganyika (Tanzania)("The mid season variety, which tends to be more hardy and drought resistant than the other two."[61.]) It has larger, coarser leaves and longer internodes (more distance along the stems between leaves) than Tinaroo. The leaves are often a dull green colour."[63.]

"Where to sow: although it has performed well in 35" (89 cm's) rainfall (with Cooper 30" (76 cm's) rainfall, it prefers 40-50" (102-127 cm's)rainfalls and higher. Glycine is no where near as versatile as siratro or the desmodiums as to soils, but on deep fertile, well drained soils, particularly volcanic and scrub and shale soils, it is extremely productive and hard to beat. It will not take poorly drained soils or low fertility.

When to sow: any time during spring and summer, when soil moisture is adequate, and when follow up rain is expected, but the earlier in this period the better.

Sowing rate: Drilled 4 pounds (1.8 kg), broadcast 4-6 pounds (1.8-2.27 kg) per acre = 10-15 pounds (4.5-6.75 kg) per hectare.

Fertiliser: Glycine will not tolerate low fertility and should be sown with 2-5 cwt (101-254 kg) per acre = 5-12.5 cwt (252-635 kg) per hectare of molbdenised superphosphate, with annual maintenance dressings of 1-2 cwt superphosphate.(51-101 kg) per acre = 2.5-5 cwt (127.5-252.5) per hectare.

Management: Glycine is slower than other legumes to establish, and should be given every chance to develop. Dolichos lab lab at 5-7 pounds (2.27-3.17 kg) per acre = 12.5-17.5 pounds (5.67-7.92 kg) per hectare. is a useful nurse crop to help to control weeds and give a bulk of feed in the early stages."[61.]

Greenleaf Desmodium *(Desmodium intortum),* also called Intortum Clover. In Australia, it and Silverleaf D. are called the Tick trefoil's.

Where the author first worked in Taiwan this species had been sown and had hardly grown. It was difficult to find any of it among the razor grass type of *Imperata cylindrica,* a weed grass which grew everywhere if it had a chance. We cut the grass down twice to weaken it prior to digging it up. Then the Greenleaf Desmodium which was also cut sprang into life and soon we had a three feet (1 m) high crop of high quality fodder which choked out the razor grass. We then found that if we planted as cuttings, lengths of stem about 9" (23 cm) long; at an angle of 45° with 2/3 in the soil and the base end butting against soil, we had much quicker and better development.

When the author went to Nepal, he asked about Greenleaf Desmodium, and at two research stations was told it was no use. I tried it and it seemed to perform like it had in Taiwan. I asked one grass nursery if I could have more planting material, they told me I could have it all. Later we had shown it was in fact very suitable, they asked for planting material back. I think they had been too gentle with it. If left after planting and not cut it is weak and straggly, if cut (though not too short) it becomes vigorous and produces much high quality feed palatable to all classes of farm livestock, cattle, sheep, goats, pigs and poultry, probably horses too.

We later supplied cuttings to several projects and farmers in Nepal with differing weather and soil conditions, and all were very pleased with it. We have not needed to inoculate with nitrogen fixing bacteria or apply lime to acid soils in either Taiwan or Nepal. It is said to be a balanced food for rabbits. We fed it as a basic ration for both New Zealand and Angora rabbits very successfully, with the occasional addition of sweet potatoes or carrots, and some soyabean cake for milking mothers.

Greenleaf Desmodium is very useful for soil conservation, it may be planted along the edge of terraces and banks; in the rains it grows out and hangs down. Where it touches the earth it sends out roots which grow into the soil and there is soon a matted growth holding and protecting the bank. As it develops it will grow out from the bank and can be cut for quality animal feed and returned as manure; or it can be spread over the terraces as a mulch or ploughed in as a green manure crop. When established it had good root nodule production in different environments, in both subtropical areas in Taiwan and more temperate areas in Nepal. When planted on poor soil it improves the soil. In one unusually dry year in Taiwan even the wild trees were showing stress and losing leaves, and on the ground the crops were dying, yet there were some bright green patches on the hillside which people noticed; they asked us what they were? they were the areas of Greenleaf Desmodium. We planted cuttings in stony ground round an old cement water tank 2.5 x 5 metres. Later when an old student came to visit he asked what we had done with the tank as it was gone. We showed him that it was still there, though it could not be seen as it was completely covered by the Desmodium.

Desmodium and Centro were planted between the trees in the citrus orchard. The Centro proved to be less productive and needed pulling out of the trees, with continual cutting it was weakened and eventually only the Desmodium survived. This we cut and put in piles between the trees, in shady areas when possible. This turned into compost which we applied around the drip line of the trees as a manure and mulch. We found that our citrus trees yielded almost as well as others which were heavily fertilised, at the same time the pest and disease levels were much reduced. I have became a great enthusiast for Greenleaf Desmodium.

Research in Taiwan comparing the yields of Pangola grass when on its own compared with legumes produced the following results:

Combination.	Yield kilograms per hectare			Percent legumes by dry weight	
	Green weight	Dry matter	Crude protein	1st cutting October'60	4th cutting June'61
Pangola grass alone	82,715	13,507	1,348	—	—
Pangola grass and *Desmodium intortum*	81,997	13,456	1,837	32.0	10.6
Pangola g. and *Centrosema pubescens*	70,721	12,518	1,340	49.0	1.5
Pangola g. and *Glycine javonica*[64]	68,150	12,325	1,656	12.0	0.9

"In Kenya, mixtures of Elephant, Guinea or Guatemala grass with *Desmodium intortum* or *D. uncinatum* have given higher dry matter and crude protein yields than those obtained from the grasses alone"[18 p353].

"This trailing perennial legume was introduced from Central America. I consider it represents one of the most hopeful prospects for coastal pasture improvement, and that it will exert as significant effect on coastal productions as any one species of the newer tropical legumes. Greenleaf Desmodium is a rather coarse spreading plant with rather thick stems which root down well. The leaves have fine hairs and bear a characteristic (special) reddish brown to purple flecking on the upper surface. The flower racemes are lilac (blue pink) to pink in colour, and produce pods which curve back to the flowering stem---.

Its growth rhythm is more even than that of centro; good spring and autumn vigour is displayed (shown). It is one of the first tropical legumes to grow in the spring and because it flowers about a month later in the autumn than silverleaf desmodium its autumn growth is also pleasing. Its autumn performance is comparable with that of the late flowering Tinaroo Glycine, and can be used similarly for autumn saved grazing. It is burnt back by frost---. It shows more drought tolerance than Silverleaf Desmodium.

Greenleaf d. is very versatile in its soil requirements. It possesses some tolerance of waterlogging and makes excellent growth on a variety of soils, from grey sands to yellow clays and red loams. It is tolerant of acid conditions and does not require heavy liming in areas where this would be necessary for white clover or lucerne growth. It is most responsive to phosphate, ("The desmodiums are probably able to extract phosphate better than white clover on red loam soils."[63 p28.]) and will fix large quantities of nitrogen. It has good nutritive (feeding) value and is well accepted by stock. ---it will withstand heavy stocking once it is well established. ---Weeds have succumbed to its smothering action.(They are quickly choked out). Greenleaf Desmodium will quickly build up a layer of decomposing duff on the soil surface and add to the moisture holding capacity and fertility of sandy soils. It combines well with Nandi Setaria, Guinea grass, Green Panic, Molasses grass, Plicatulum and other grasses.

Seeding vigour is satisfactory, but not as good as pioneer legumes such as Puero and Calopo.(Authors note:- I have not included Calopo as it is unpalatable to stock and it is short lived.) Light first season grazing is advocated."[63.]

"It produces long runners which will root down easily at the nodes, and scramble through and over grasses and weeds. Under proper management, the rooting stems form a close lattice work on the soil surface, and build up of organic matter in the top layers of soils is fast under this legume. Nitrogen fixation is very good.---. It has proved very useful in pure stands, in pasture mixtures, and for hay.(Author's note, if used for hay it should not be moved too much or the leaves will become loose.) Seed pods stick readily to stock aiding the spread of this useful legume.

Greenleaf is resistant to most diseases, although seedlings can succumb to (be affected by) disease in the early stages, and little leaf virus can be a problem in the wet tropics. Amnemus and white fringed weevil have proved troublesome in some stands.

Where to sow: Greenleaf requires a minimum of 35-40 inches of rain for good production, although its deep tap root gives it good drought resistance. It will grow successfully on most soil types, and it is fairly tolerant to poor drainage and flooding. Its ability to do well on light and sandy soils makes it very useful on light soil ridges.

When to sow: Depending on moisture conditions, from early October (early April in the northern hemisphere) to the end of February (end of August in the northern hemisphere), although good stands have been established as late as April (October in the northern hemisphere). Generally speaking early sowings have shown up best, providing that there has been adequate soil moisture for good establishment, as once established, greenleaf can withstand long dry periods.

How to sow: Desmodium responds well to good land preparation, and should be given as good a seedbed as is reasonable in the conditions. sowing should be into a firm moist seed bed no deeper than 1"(2.5 cm's) Seed should be inoculated and preferably lime pelleted. Lime pelleting is essential if seed is to be mixed with fertiliser for sowing.

(Author's note:- We have not needed to inoculate with nitrogen fixing bacteria or apply lime to acid soils in either Taiwan or Nepal.)

Only a light covering is needed and rolling will help establishment considerably on soils which do not cake (crust over when rolled or trodden.)

Sowing rate: 1-2 pounds (0.454-0.9 kg) per acre, = 2.5-5.0 pounds (1.135-2.27 kg's) per hectare. Lesser rates can be used when it is sown in mixtures such as with Siratro, Cooper and Tinaroo Glycine, where it provides earlier spring growth.

(Author's note. Planting:- Once a plot has been established from seed it should be cut back to six "(15 cm's); if left after planting and not cut it is weak and straggly, if cut it becomes vigorous and produces much high quality feed palatable to all classes of farm livestock. Greenleaf Desmodium produces long leafy shoots which can root from all the nodes, so this material can also be used for planting. The shoots are cut into lengths 7-10" (18-25 cm) long, depending on how many nodes there are. Sometimes the nodes are far apart and sometimes close together. When the nodes are close a shorter cutting is best. The cutting can be planted as it is although it is better to trim back the stalk to the nearest nodes, see below.

It is best to cut the upper end with a slant and the lower end straight across so that it is easy to know which is the lower end when planting.

Take a suitable tool such as a trowel, spade or mattock. Make a slit in the soil at an angle of about 45° = 100% slope. Place the cutting in the slot lower end downwards and with its stalk end butting against soil. About 2/3 in the soil and 1/3 above the soil. Press the soil down firmly and keep grazing animals away. The cuttings can be planted anytime of year in countries which are always warm and the soil is moist; however cuttings do not take as well soon after flowering as food reserves have been used up making flowers and seeds. In other countries, the cuttings should made and planted in the warmer months when the soil is moist, and will stay moist for a few weeks. The cuttings should be planted as soon as possible after being cut. While waiting they should be kept cool, damp, and out of the sun.

Cuttings can be prepared and made into bundles covered with sacking or paper, and sent to another part of the country. If it is very dry or the journey is long, they should be watered before the journey, and during the journey if they are drying out. They should not be too wet, but just damp, so that they do not dry and wither. They should be planted as soon as possible. If managed well they can probably survive for 3-5 days but should not be just left for a week as one consignment was. The people then found that the percentage which grew was not good and complained. We found that the cuttings had been neglected and left in a hot store for a week. Also that the workers had not been told which end of the cutting should go downwards. All the other consignments grew well.

Management: Treat lightly the first season to allow its runners to become well established and root down when they can. Once well established, however, it will stand heavy grazing.

(Author :- but grazing or cutting should leave at least 5" (13 cm) from the roots; then recovery will be quick and good.)

Should grass tend to smother the legume in the early stages, a quick grazing (or cutting) will help to open the sward and let light into the legume."[61]

The author has found that after about 5 years the stems of the mother plants become woody and the plants have less vigour. In a field situation this will probably not matter as the vines will have rooted and produced new plants. However when we have greenleaf planted along the edges of terraces to hang down over walls producing fodder and organic manure there will be less production. It may be worth planting some cuttings or pulling up some of the vine-like stalks and pressing them into the soil along the terrace edge. They will then root and make replacement plants.

Intortum clover, *see Greenleaf desmodium.*
Kudzu. see Kudzu species.
Kudzu bean. see Kudzu species.
Kudzu species.

The author has not had personal experience of these species. Kudzu has several names. I have been told that there are two kinds Kudzu and Tropical Kudzu and that while Tropical Kudzu has become a troublesome weed pest in some countries Kudzu does not cause such a problem. In researching the information for this book the author has found confusion within the 12 sources, with several names both common and scientific.

The best conclusion seems to be that there are in fact only two kinds, though they may be given different names in different places.

Group 1. Should be called *Wet Tropics Kudzu,(Pueraria phaseoloides);* now called Kudzu, Tropical Kudzu, or in Australia Puero. It is more suited to the wet tropics, it is more tolerant of waterlogging, it is more climbing than Kudzu and it does not develop roots at the nodes. It is of more use as a cover crop than for holding land on landslides etc which would otherwise erode. It grows very easily from seed. It does not root freely at the nodes. See Group 1/ below.

Group 2. Should be called *Dryer Climate Kudzu,(Pueraria hirsuta).* It is now called Kudzu, Kudzu bean, or Kudzu vine. It originated in Japan and is not suited to the wet tropics but suited to drier areas. When established it is drought resistant. It likes good drainage and does not do well if the drainage is poor. It likes a lighter soil and low to moderate rainfall. It can tolerate cooler winters better than tropical kudzu and it is a very useful plant for protecting and binding soil against erosion. It roots freely from the nodes thus establishing linked plants, it does not produce many seeds and so is normally propagated by rooted crowns, and according to one source rooted node cuttings. See group 2/ below.

Not in either group is-Rhodesian Kudzu. (Glycine javonica.L); as this is in fact Glycine, see Glycine.

Group 1. *Tropical Kudzu. (Pueraria phaseoloides). Puero. (Pueraria phaseoloides).*
Kudzu. (Pueraria lobata).   *Taiwan Kudzu. (Phaseoloides tonkinensis);*
SHOULD ALL BE CALLED TROPICAL KUDZU.

Tropical Kudzu. *(Pueraria phaseoloides).*
Professor Masefield writes that it is a **tropical** relative of Pueraria hirsuta. [4.]

"Tropical Kudzu (Pueraria phaseoloides) is better suited to southern Taiwan (the tropical part) at elevations below 500 meters (1,640 ft). In northern Taiwan it loses all of its leaves in the winter months and has difficulty recovering productiveness in the spring. In the south, it grows very dense and produces an abundance of seeds. It will undoubtedly find a place as a legume for limited grazing during the rainy season with the maximum growth held in reserve as carry over in the dry season. It may be grown in combination with Napier or Guinea grass."[64.]

Puero, (Pueraria phaseoloides)," a native of the East Indies is a vigorous climbing legume, often producing runners 15-20 feet (4.5-6.3 m's) in length. These with a number of secondary shoots, form a dense cover 24-30" (60-75 cm's) in depth within 9 months of sowing.---. It is somewhat susceptible to frost and also to damage from heavy grazing and although it soon forms a dense cover the amount of fodder available on this legume is restricted to the relatively thin cover of top leaf growth. Puero seed pods mature irregularly, thus making seed harvesting a difficult task. It would appear that puero will be of more value as a long term cover crop used to control weeds and to provide light grazing than as a permanent pasture legume in the wet tropics." [3.]

"Puero (Pueraria phaseoloides). ---is a summer grower which responds well to warm conditions; it is less cold tolerant than glycine. Puero will withstand shading. It requires a heavy rainfall for satisfactory development, and shows some tolerance of waterlogging. Puero is tolerant of acid conditions but responds to phosphatic fertilisers. It has been used on the wet tropical coast of north Queensland (Australia) with success as a pioneering legume.

Seedling vigour is superior to that of calapo and siratro which have good reputations in this respect. Puero is highly palatable, and is sown with legumes, like centro, which are more persistent under grazing.

Puero is a useful addition to mixtures sown on new land because of its rapid seedling growth, its capacity to smother weeds and its ability to provide early grazing---. As a permanent pasture component puero's place is in more doubt." [63.]

"Where to sow: Puero is a summer grower which demands warm and wet conditions, and its use is mainly limited to the wet tropical area, with at least 70" (178 cm) of rainfall. It has reasonable tolerance to waterlogging, and is tolerant of acid soils.

How to sow: On as good a seedbed as is reasonable. should be sown mixed with other legumes and grasses.

Sowing rates: 2-4 pounds (0.9-1.8 kg) per acre = 5-10 pounds (2.7-4.5 kg's) per hectare, as a component of a mixture.

Fertiliser: Responds well to superphosphate in most areas.

Maintenance: Care is needed in early management, so avoid overgrazing in the early stages. Puero is also susceptible to damage by fire." [61.]

This legume has a similar habit and geographical range to that of Centro. It is less resistant to heavy grazing, but it is a good pioneering species." [68.]

"Kudzu. *(Pueraria lobata)* grows in highly acid to neutral soils. It is not tolerant of salinity. Shallow to deep, gravelly loam and moist to heavy-moist conditions are suitable. It is benefited by the application of mineral nutrients when they are needed, although the plants can utilise nutrients from relatively unavailable sources. It is a perennial with a vine type of growth. It is adapted to the Southern states of the United States. It is not hardy---. It is good for waterways in places where it is adapted. If kudzu is used, the size of the waterway must allow for the bulky growth of the plant." [20.]

Taiwan Kudzu. (Phaseoloides tonkinensis); is only mentioned by one source, [10.] with no description given. It is not mentioned in a publication from Taiwan. It seems to be incorrect. Phaseoloides should be the last name. I would guess that it is in fact *Pueraria phaseoloides* which originates in the East Indies. Tonkinensis is probably referring to the Gulf of Tonkin in the East Indies where *Pueraria phaseoloides* originates.

Group 2. Kudzu vine. *(Pueraria hirsuta). Kudzu.*
Kudzu. (Pueraria thumbergiana). Kudzu bean.
SHOULD ALL BE CALLED 'KUDZU'

Kudzu vine. *(Pueraria hirsuta);* is described by the reliable Professor G.B. Masefield, as, "The Kudzu vine which has been much used in soil conservation work in America and introduced into some tropical highland areas".[4.]

Another source states "Kudzu Vine *(Pueraria hirsuta)* is a perennial legume, being extensively used in the southern part of the United States of America for soil conservation and pasture purposes.

In India it has been profitably utilized for reclamation of gullies, stabilization of soil along highways, and as a palatable and nutritious fodder. The plant sends out branches in all directions; sometimes the length of a shoot exceeds 50 ft (15 m). Roots may develop at each node, so that each node turns out to be an individual plant the next year. These roots and the dense cover which the plant forms, are responsible for preventing the washing away of the soil. Since the roots are fleshy and go quite deep, kudzu vine, once established, does not require any further irrigation. The leaves are shed during winter and add a large amount of organic mater to the soil, thus enriching it. With the approach of spring new growth commences and the plant remains beautifully green during the hot summer and autumn months. It is an excellent fodder and when introduced in crop rotations, the yields of the succeeding crops are greatly increased.

Kudzu can grow on all types of soils of medium to high fertility or even on poor lands, if adequately manured. Light soils and low to moderate rainfall are favourable for the growth of the plant. Poorly drained or waterlogged areas are not conducive to its healthy growth. In India the plant grows well in both plains and hills to an elevation of 5,000 ft (1,500 m). In the plains, where winter is milder, the leaves tend to remain green throughout the year. Where temperatures are likely to go below 50° F. (10° C), Kudzu sheds its leaves and remains dormant during the winter.

Kudzu produces seed in only small quantities, generally three years after planting, and hence it is normally propagated by vegetative means. Raising plants from stem cuttings does not give successful results. The best method is the planting of crowns during January and February. A crown is a swollen node with two or three fleshy roots and a few viable buds. Swollen nodes without fleshy roots are also difficult to establish. Younger crowns, one to two years old, are better to establish than older crowns. These crowns are dug from the field during December and January, when the plant remains dormant due to low temperatures. From an acre, approximately 15,000-20,000 = from a hectare 37,500-50,000 well developed crowns can be dug. These crowns can stand a transport period of over ten days and they show a good survival when planted in the field. The crowns can be planted directly on the desired site, [at 20 ft (6 m) by 20 ft (6 m) spacing] [in a nursery bed, or in earthen (or plastic) pots (about 12" (30 cm) deep, (planted out after the beginning of the monsoon at a 20 ft (6 m) by 20 ft (6 m) spacing), depending on the facilities available.

Planting directly on the desired site is advantageous, since the young plants are left to grow undisturbed from the very beginning. This method is, however, applicable only where proper care for watering and protection of the growing plants is possible.

In order to provide a good start to the growing plants, it is desirable to prepare pits, about two x two x two feet (60 x 60 x 60 cm) about 20 ft (6 m) in all directions, well in advance, and to fertilize them with an adequate quantity of fertilizer and farmyard manure. it is necessary to see that the manure is adequately decomposed or else it is liable to attract white ants and thus cause damage to the plants. While transplanting the young plants into the pits care should be taken that the block remains about 4 " (10 cm) above the soil surface to avoid stagnation by water. The soil should be thoroughly packed.(no big air spaces in the pits.) If the soil is liable to form a crust, it should be hoed from time to time. Complete protection from grazing and trampling, at least for one full season, should be provided.

If planted on sloping land, it may be necessary to make small ridges to protect the plants from being washed away by the direct impact of rushing water. Under favourable conditions, the plant produces a fairly dense cover in two or three years. Once established it does not need any irrigation."[26]

As the above two sources mention that the plant was introduced from America it is interesting to look at what an American publication has to say. It names the plant :-

Kudzu. *(Pueraria thumbergiana)*. The description agrees with the other two sources except for the scientific name. It does however mention that it was introduced to America from Japan in 1876. Hirsuta could be from a Japanese name so I suggest it is the same plant. Additional points made in the American source are as follows.

"On land rough or otherwise hard to work, kudzu will reduce erosion while furnishing an abundance of good feed. On tillable land kudzu increases soil fertility in long rotations, as shown by increased yields of crops following it in rotations- but kudzu is not a crop to be grown on poor land without fertilizer. Ample use of fertilizer at time of planting is necessary to assure a good stand.---. If dug before planting time, the crowns should be stored in a cool, dry storage room with the roots in moist (not wet) sphagnum moss or sand. In new plantings the plants are set in rows 20 - 30 feet (6-10 m) apart and spaced so as to use about 500 plants per acre = 1,250 plants per hectare.---. Row plantings of corn (maize) or other row crops can be grown between the kudzu rows during this period.

(Author: if this is to be done I would suggest a legume such as soyabeans or climbing lentils with the Maize, perhaps also an inter-row crop such as potatoes or sweet potatoes, pumpkins or cucumber.)

When once established the stand will last for many years if properly managed." [65]

Kudzu bean. description fits *Pueraria hirsuta* although the scientific name is not mentioned. It is described as a native of Japan. It also states -

"It is possible that it will prove of use in the development of tropical grazing areas and for erosion prevention in districts of heavy rainfall." There is a description of propagation by stem cuttings.

"Stem cuttings should preferably be rooted in damp soil before being set out in the field". [3]

Kudzu vine. *(Pueraria hirsuta);* see Kudzu species.

Leichhardt Biflorus *(Dolichos uniflorus, formerly D. biflorus).* An annual legume for the drier tropics which self re-seeds if allowed to. It could be very useful for strips, for pasture between trees, for improving the nutrition of stock to allow more stock to be kept and so more animal manure, and more income.

"It can be used in many ways, in combination with native grass or fodder crops, or as a pure stand. Probably its greatest value is as a protein rich supplement for dry season grazing.

Its ability to give rapid and early growth makes it ideally suited to areas with a clearly defined "Wet season" followed by dry conditions which limit the use of other legumes. Biflorus produces high yields of seed, which holds well in the pod even after the plant matures in early winter. This protein rich seed together with the plant material gives valuable standover feed into the dry months following, particularly when used as a supplement (added to) in conjunction with (together with) low protein feed like native pasture--- Good quality hay can also be made from biflorus.

It's drought resistance, ease of establishment, resistance to pests and disease, and ability to provide valuable protein and soil nitrogen in regions too dry for many other legumes, makes this versatile plant extremely valuable.

Where to sow: Ideally suited to the marked seasonal rainfall pattern of North Queensland, Australia, in areas of 25-40" (63-102 cm) rainfall. In higher rainfall areas, other legumes can be used, and rainfall after mid autumn will actually be damaging to the mature crop held as standover feed.

Fairly versatile as to soils, biflorus, thanks to its drought resistance, will even grow well on shallow and sandy soils; although yields are not as good as on the better soils. It will not stand waterlogging, and poorly drained soils must be avoided. So far it has not been tried extensively on the heavy soils, and it is expected that it will not do well on very alkaline soils.

When to sow: planting in the mid summer months is usually best. Drought tolerance allows it to be sown on storm rains early in the summer, and although plants can be established later, the plants get less chance to produce growth, owing to their early maturing nature.

How to sow: Biflorus is easily established on most soils suitable to its growth. Although it can be established on soils with loose surface texture simply by burning and broadcasting, some form of cultivation is advised on hard surfaced soils---. Aerial sowing should prove very useful for large-scale (or difficult to sow areas) establishment provided suitable soils are chosen.

Inoculation is unnecessary where native legumes occur in the area.

Sowing rates: Drilled after soil cultivation, 5-8 pounds (2.27-3.6 kg) per acre = 12.5-20 pounds (5.67-9 kg) per hectare. Broadcast after soil cultivation, 10-12 pounds (4.54-5.4 kg) per acre, = 25-30 pounds (11.35-13.5 kg) per hectare. Broadcast with no cultivation 12-15 pounds (5.4-6.8 kg) per acre = 30-37.5 pounds (13.5-17 kg) per hectare.

Fertiliser: Biflorus responds well to phosphate.

Management: Where biflorus is used as a legume in native pasture, its best value comes from not grazing during its growing season, to allow it to make maximum growth, and set large quantities of seed. After it matures in early winter, cattle can go on from mid winter to the following summer when native pasture is very low in protein. Once seed has set, cattle grazing the area will trample spilt seed into the ground for the next season, and spread it through their droppings. Pure stands of biflorus should be used to supplement rough feed rather than for straight grazing. The high protein content allows the cattle to make much better use of any crop stubble, straw or dry native pasture available." [61.]

Lespedeza.*sp.* Often called bush clovers. "Optimum pH range is 4.5-6.5" [67 p37.] "have revolutionised 20 million acres of acid sandy land in the southern and eastern parts of the USA.

The species used commercially are all of Asiatic origin. These are *Common lespedeza.(Lespedeza striata). Korean lespedeza (L.stipulacea). (and Sericea Lespedeza (L.cuneata), Sericea L.(and L. bicolor,* see below, author) are perennial, the others annual.---.

Lespedezas are especially useful on acid soils of low fertility. Sowing is carried out in early spring, either broadcast or in close drills, using 20-25 pounds (9-11.27 kg) of seed per acre = 50-62.5 pounds (22.5-28.17 kg) per hectare. In no case has inoculation of the seed been necessary in the USA" [65.]

Korean L. called Korean clover in Australia "it does best on sandy or loamy soils and, when seasonal conditions are suitable, it may, as the seed matures, make a useful contribution to grazing in late autumn and early winter." [3.]

Lotononis, *see Miles lotononis.*

*Lucerne./*Alfalfa in America. *(Medicago sativa).* Lucerne is sometimes classed as being a summer rainfall legume. Perhaps we should say that it is both. It does make most of its growth in summer, but in the sub tropics it will also grow through the winter. It grows in a wide range of conditions though it does not like acid soil unless it is limed. Some writers say that it cannot tolerate waterlogged conditions but this writer has seen it grow very well on a field which was almost pure sand hardly higher than the river which had deposited the sand. The water table must have been from 1-2 ft (30-60 cm) and at flood time the river would rise to the level of the field surface. Although it is not a plant which binds the soil surface as those with stolons or rhizomes do, it makes quite a dense cover when established, it improves the soil. The sandy field mentioned above grew nothing but a few sparse weeds until it was sown to lucerne with a nurse crop of oats. After two years it was sown to a grass ley which grew quite well. I imagine that the deep roots of the lucerne brought up plant foods which were beneath the sand. Although I have not seen it I have read that plant breeding of lucerne with closely related species Glutinosa lucerne *(Medicago glutinosa)* and rhizoma lucerne *(Medicago falcata × Medicago media)* has resulted in creeping lucerne. Although these are less productive and poorer seeders than Hunter river or traditional lucerne, they are valued for their creeping habit which should make them better for protecting and holding the soil. They are also reported to be better at tolerating close grazing, and in tolerating dry conditions better than normal lucerne.

It will grow in very dry conditions, and it will, due to its very deep root system continue to produce fresh growth in the winter in areas where other legumes find it too cool or where there is insufficient moisture for their winter growth.

"Lucerne is a perennial legume which originated in the Middle East. It was the first cultivated herbage plant, being sown by the Persians about 700 B.C. It is now grown all over the world from the arctic circle to the sub-tropics, and is also grown in some tropical regions.

It is an erect perennial, the young stems being square and the leaves and stems sparsely hairy" [63.]

"Its extraordinary adaptability and persistence have made possible its cultivation in all the main pastoral and arable regions of Australia---.

Under ideal environmental conditions it may produce annually as much as four tons of high protein hay of good palatability per acre (10 tons per hectare). At the other extreme, lucerne has compared favourably in terms of quantity and quality with other legumes under an annual rainfall as low as 9"(23 cm) in South Australia.

Lucerne is remarkably drought resistant on account of its deep root system and yet has a large capacity for production under irrigation and in good seasons.

Lucerne prefers soils of moderate fertility which are deep, well drained, and neutral to slight alkaline in reaction. It has a high degree of salt tolerance. It will grow on sands, loams or clays, where drainage is satisfactory, but will not tolerate waterlogged conditions for even short periods.

The main season of production extends from spring through summer to early autumn. Lucerne makes greater growth during the summer months than any other grazing legume available for use in the cooler sub tropics --- In locations where winter temperatures are mild it will continue to grow during the winter, but elsewhere growth is retarded from autumn to early spring.

The persistence of lucerne can vary over a wide range and is largely dependent on soil type and the standard of management. Pure stands under ideal conditions may last as long as 15 or twenty years, crops cut for hay are likely to persist longer than when grazed off." [62.]

"Some form of rotational grazing or regular spelling (resting) is absolutely necessary if lucerne is to be maintained. It is usually lost within a year of continuous grazing" [63.]

"*Sowing:* Lucerne grows relatively slowly in the early stages following sowing and hence is very susceptible to competition from weeds during establishment. It is therefore necessary to ensure that it is sown on as clean a seedbed as possible; to this end it is often wise to cultivate the ground several times before sowing in order to cause weeds seeds to germinate and then be destroyed. As lucerne is more cold tolerant than many of the weeds where it is grown it is often sown in the spring. Where weeds and pests are not very troublesome however, autumn is usually a better time for establishing a sward since moisture is less likely to be a limiting factor than in the spring, and in any case the plants have a better chance of becoming well established before the onset of dry summer conditions.

Seed rate: this varies according to annual rainfall. In high rainfall areas, or under irrigation, lucerne in pure stands may be sown at rates of 15-20 pounds (6.8-9.0 kg) per acre = 37.5-50 pounds (17-22.5 kg) per hectare. In the drier regions— lower rates may be used. For example, where annual rainfall is less than 15" (38 cm), rates of 2-3 pounds (0.9-1.36 kg) per acre = 5-7.5 pounds (2.27-3.4 kg) per hectare. are quite ample. ---. Lucerne may be sown in combination with a number of species of grasses---. It has been used successfully at 1-2 pounds (0.454-0.9 kg) with 4-5 pounds (1.8-2.27 kg) of buffel grass or the same quantity of green panic, per acre; = 2.5-5 pounds (1.35-2.27 kg) lucerne with 10-12.5 pounds (4.5-5.67 kg) of buffel or green panic per hectare. Another successful combination used is 8 pounds (3.6 kg) of lucerne with 6-8 pounds (2.7-3.6 kg) of Rhodes grass per acre = 20 pounds (9.0 kg) lucerne and 15-20 pounds (6.75-9.0 kg) Rhodes grass per hectare.

(Author: *Seed inoculation:* As stated earlier if lucerne, burr medic or other close relatives of lucerne have not been grown it will usually be beneficial to inoculate the soil by spreading some soil from a field which has grown lucerne or its close relatives on the top of the field to be sown. As this may not be practical, inoculant can be supplied from the seed merchant when the seed is bought. The instructions should be followed carefully. See also the beginning of this section. If inoculant is not available it may be possible to dig up the roots of plants already growing together with the soil around them. Crush the roots and mix them with the soil in a thick soup and coat the seeds with the soup before sowing. The seeds should then be sown into the soil quickly. Later plants can be taken from the resulting crops with their roots and replanted in other fields. The soil in which they grew can be spread on other areas before the seed is sown and so the most suitable bacteria can develop in the area. It may be that there will be suitable wild legumes which have useful bacteria living with them. When sowing the seeds of another legume the Leucaena bush or tree, in Nepal, I did not have enough inoculant for all the seed. I went about the area looking for wild legume bushes, taking soil from the root areas. I mixed it up and put a little with the soil in which the seeds were planted. Those seeds did just as well as the ones with the correct inoculant, though this will not happen in every case.

Management: Careful control of the grazing of lucerne is particularly important for both the plants themselves and the health of the stock.

Since individual stems die back following grazing and regeneration takes place in the form of new shoots arising from the crown, the plants should be allowed a chance to recuperate (recover). Under continuous grazing these shoots are destroyed as they emerge, food reserves are rapidly depleted, and the plants die. The most satisfactory technique is to divide the area into units and stock heavily so that the plants are eaten down quickly. The stock are then removed and the lucerne is allowed to recover before further grazing.

Lucerne is very liable to cause *bloat* (see glossary), particularly when the material is wet and the animals stomachs are empty or when they are granted unrestricted access to it before their digestive organs have become accustomed to it. (Author. This means if animals are allowed to eat as much as they like before they have become used to it they may get bloat. Give them other food before giving them the lucerne so that they cannot fill their stomachs with lucerne). Sheep goats cattle (and similar animals) should therefore be admitted with some care (not be allowed to have too much) for the first time to pure stands of lucerne or to pastures containing a high proportion of it. (Author. I would say not just for the first time but for the first 7-10 days to allow the animals to adjust to it.)

Suitable precautionary measures are, restricted daily periods of grazing in the early stages, and ensuring that stock have had something else to eat before they are admitted to the lucerne." [62]

Miles Lotononis. *(Lotonis bainesii).* Miles l. is a creeping perennial legume native to Southern Africa. In early evaluation trials in Queensland, Australia it was overlooked until successful nodulation was achieved with a further introduction of the right rhizobium (bacteria.) strain. Lotononis is suited to moist environments, and will grow well in most subtropical and tropical climates receiving more than 35" (89 cm) of annual rainfall. Its special virtue is its frost tolerance, it is the most frost resistant of all the commercial tropical legumes. This long season of growth represents a very real advance. It is tolerant of waterlogging, poor drainage and flooding. It will grow under very acid conditions. Lotononis responds well to improved nutrition---.It has performed satisfactorily on a range of soils. It grows well on sandy soils, but is also found on heavier textured soils. It may be that indifferent persistence on tight soils under grazing may be associated with the failure of its stems to root down well under these conditions.

Lotononis is one of the few legumes to combine well with Pangola grass, but it has also been grown with a variety of both tall and low growing grasses. It is quite the most palatable tropical legume available, and there is a danger that selective grazing will remove it. Lotononis provides high quality feed for stock; its crude protein digestibility is unusually high." [63.]

"*Where to sow:* Lotononis is at its best on light, sandy soils---. It has also done very well on light soil ridges, and to some extent on heavier soils. It doesn't mind a certain amount of waterlogging or flooding but persistence is better on well drained soils---. In the wet tropics it may be affected by fungus. Lotononis has difficulty in establishing in competition with carpet or mat grass.

When to sow: usually from mid summer onwards with the main sowing period in early autumn. it can also be sown early in spring if soil moisture is adequate.

How to sow: Lotononis has an extremely fine seed and should be sown as shallowly as possible, usually on the surface, and merely given the lightest of cover. It appreciates a good seedbed which must be firm, and not fluffy in the top layers. Rolling is very useful in firming the seed into the soil and moisture where soils do not cake with rolling. Dropping on the surface and rolling in has proved most effective.

It is essential that the seed be inoculated with the correct strain of bacteria before sowing. and pelleting must be used if the seed is to be mixed with chemical fertiliser for sowing.

Sowing rate: 1 pound (0.454 kg) per acre, = 2.5 pounds (1.135 kg) per hectare.

Fertilizer: Lotononis responds very well to fertilizer, and as it is usually sown on very sandy or light soils, severely lacking in plant foods, substantial dressings of fertilizer may be used. On the Wallum sands in Australia, which are extremely low in fertility, at least 6 cwt's [note a hundredweight or cwt is 112 pounds] (305 kg) per acre = 15 cwt's (762 kg) per hectare of superphosphate, 1 cwt (51 kg) per acre = 2.5 cwt's (127 kg) per hectare of muriate of potash, and molybdenum, copper, and zinc, are needed for best results. Up to 15 cwt's (762 kg) per acre = 45 cwt's (1905 kg) per hectare of lime is also used." [61.]

"*Management:* Since natural seeding regeneration (replacement) is rather rare, survival depends upon not grazing the plant right back to the crown continually. If the pasture is allowed to grow up, the stems do not root down into the soil so well, but climb over the associated (neighbouring) grass. Under these conditions very heavy intermittent grazing is probably more damaging than a more constant grazing pressure which is not too severe to prevent stems rooting. Lotononis has at times almost disappeared from pastures and then built up again in a subsequent (later) year. The reasons for this are not always clear.

"One of its drawbacks has been a certain amount of unpredictability in its performance, and for this reason it should be sown with another legume such as Greenleaf Desmodium. On the other hand during dry seasons it has shown remarkable drought resistance." [61.]

(Author :- What this means is that unlike most of the other legumes which do well with intermittent grazing such as strip grazing, fairly heavy grazing followed by a rest period; lotononis does better to have steady but not too heavy grazing. That way the runners do not get the chance to climb high and can be trodden down and then root into the ground, and so grow into replacement plants. If this is not done the parent plants will suffer from too heavy grazing and die back and it will take a while to produce new rooted rhizomes to bring back strong growth.)

It is very susceptible to little leaf virus but if the affected material is grazed off, the plants usually recover. It is resistant to nematodes and is not significantly troubled by other pests or diseases." [63.]

Puero, *(Pueraria phaseoloides),* see Kudzu species.

Rhodesian Kudzu. *(Glycine Javonica.L),* not in either Kudzu group as this is in fact Glycine, see Glycine.

Silverleaf Desmodium. *(Desmodium uncinatum).* In Australia it is called Tick trefoil together with Greenleaf D. This is very similar to Greenleaf but it does not yield as much fodder, it is more troubled by attack of a pest which eats holes in the leaves, but which hardly bothers Greenleaf D whether grown next to it or further away. It has a shorter growing season and is less vigorous and is less tolerant of a wide range of soils; and is less drought resistant than Greenleaf D. I have not found any reason to grow it if Greenleaf D can be grown. The five sources of information the author has consulted agree.

Siratro. *(Phaseolus atropurpureus).* "This rugged and versatile perennial legume was bred by Dr Mark Hutton of C.S.I.R.O. (Australia) from plants collected in Mexico.

"Its ease of establishment, drought resistance, ability to compete successfully with weeds, and tolerance of relatively low fertility conditions, make it an ideal pioneer legume, particularly where seed bed preparation is difficult. A major advantage is its resistance to the nematodes (eelworms) which attack lucerne and other medics, and to the amnemus and white fringed weevils which can prove troublesome with clovers and some of the newer legumes. On the other hand it is susceptible to bean fly, particularly if planted after midsummer." [61.]

The authors experience has been limited but in both Taiwan and Nepal, Siratro has been disappointing compared with Desmodiums and Glycine. It was a thinner plant which died out after a year or two not having produced much fodder. It could be that it was a poor strain or our management was not suitable for it though suitable for the Desmodiums; or that our climate was not suitable; too dry in the hottest part of the year, mostly cooler due to lack of sunshine in the rainy season.

"A moist, warm, subtropical or tropical climate is required for Siratro. It responds well to warm conditions and under very hot conditions its growth is less depressed than that of glycine. Its main growth period is summer and early autumn; it is a little slower than the desmodiums or glycine in the spring. It is burnt off by frost, and moderately severe frosts kill the plant back to the older crowns, thereby slowing down subsequent spring growth. It has also been noted that if a body of growth accumulated and some check to plant growth occurs (eg moisture stress) leaf fall is substantial. However, despite this Siratro is one of the hardiest tropical pasture legumes---. Its persistence under dry conditions and under steady grazing pressure is very good. It may be sown with confidence in areas receiving more than 30" (76 cm) of annual rainfall, provided drainage is not too impeded.It has persisted on softwood scrub-soils in areas drier than 30"---. It has shown promise in the Northern Territory of Australia, New Guinea, Fiji, the Philippines, Mexico and Southern Rhodesia (now Zimbabwe), to name a few.

Siratro is versatile in its soil requirements;---. It has a faculty for developing on quite shallow soils; it is a good hillside legume. Siratro is less demanding in its nutrient needs than glycine or lucerne, and more demanding than Townsville lucerne or stylo. However it is most responsive to improved nutrition.

Siratro is compatible with a wide range of grasses. It has been successfully grown with Rhodes grass, setaria paspalum, scrobic, plicatulum, green panic, guinea grass and many other grasses. The important thing is the splendid nitrogen contribution made by Siratro to its associated grass.

Siratro is resistant to grazing once well established. In experiments--- it was shown that heavy and frequent cutting reduced the growth of glycine, centro and siratro, but that this reduction was least in the case of siratro---.

Siratro is readily established from seed. The plant nodulates (develops nodules on its roots) well, and develops a good body of nodule tissue earlier than glycine or centro. Siratro is nodulated by a wide range of native rhizobium strains; inoculating seed is an insurance measure only---.

One of the special advantages of Siratro is its resistance to nematodes. Susceptibility to nematodes is one of the main weaknesses of phasey bean, a distant relative of siratro. Little leaf virus has been but (only) rarely recorded in Siratro and this disease does not have the importance it does in lotonis or silverleaf desmodium. Bean fly can attack young Siratro plants, and seed dusting is a successful preventative measure. The mature plants are fairly resistant to attack."[63.]

"Under good conditions, the plant sends out lengthy stolons which root readily at the nodes and produces a dense mat of herbage." [62.]

"It is a good spreader— it produces seed freely throughout its growing season and throws it considerable distances. Grazing cattle also help to spread the seed.

Where to sow: Siratro likes warm conditions and a rainfall of about 35"(89 cm) and more, although it is much more tolerant of dry conditions than most of the legumes. In the really wet areas of 100" (254 cm) and more it tends to suffer from disease. it will grow on practically any soil except under poorly drained conditions, and will tolerate low fertility. It is a logical choice for poorer or more difficult country where it is difficult to establish more productive legumes.

Overall, it merits a place in most sowing mixtures, particularly in the lower rainfall areas and where conditions are tough.

When to sow: Spring and summer, and the earlier in the season the better, when soil moisture is adequate. Late sowings increase the risk of bean fly attack.

How to sow: Siratro responds to good land preparation, and does best when sown shallow in a firm, moist seed bed. It has the great value however, of being able to establish where little or no seedbed preparation is possible, and is probably the easiest of the new legumes to establish.---. Inoculation is desirable, although usually not essential.

Sowing rates: 3-6 pounds (1.36-2.7 kg) per acre, = 7.5-15 pounds (3.4-6.75 kg) per hectare. According to the extent of land preparation.

Management: It should be given an opportunity to let its runners root down and set some seed and get a good foothold. Once established it will stand strong grazing, but, like most of the legumes, performs best with intermittent or rotational grazing."[61.]

"Satisfactory regeneration each summer depends on an adequate quantity of seed. It is therefore advisable to remove livestock from the stand at flowering time, at least once every two years, and keep them off until seed setting has finished."[62.]

*Strawberry clover . *(Trifolium Fragiferum)*. "Grows in neutral to slightly alkaline soil. It tolerates moderate salinity. It grows in shallow to deep, sandy loam to poorly drained clay. It needs moist to heavy moisture conditions. It tolerates flooding. It is a perennial and is adapted to the poorly drained, salty soils of the Western United States."[20]

*Subterranean clover. *(Trifolium subteraneaum.)* Named because the flower heads are developed on or near the surface of the soil and when the plant dies back the seed burrs are deposited on the soil or under it where the soil is soft enough for penetration to take place.

The author's experience of this legume has not been good; however, as there are many strains of this species it could be that the strain we were given to try, was an unsuitable one for poor acid hill soil in West Nepal, or the climate was too dry. It grew poorly and needed weeding. I will not give much information about sub. clover as due to the range within the species, much would have to be written. As it does not root at the nodes and so bind the soil, it would not seem particularly good for soil conservation purposes. It would seem it may be useful for soil improvement.

"A free seeding, self regenerating annual, sub clover has undoubtedly made by far the greatest contribution of any species to pasture improvement in Australia. In addition to being the most widely distributed of exotic (introduced) herbage plants, through its outstanding ability to raise the fertility of soils, it has made it possible for other legumes and grasses to thrive in conditions where they would often not have survived.---. It has a prostrate habit of growth and when left ungrazed sends out surface runners which do not root at the nodes---. It is particularly suited to the mediterranean type of climate,--- where hot dry summers are followed by generally mild temperatures and annual rainfall is between 20"and 30" (51-76 cm).With the newer varieties adapted to a wider range of conditions however, the use of subterranean clover is extending into rainfall areas which have much lower annual precipitation (rainfall). Indeed where other factors such as seepage and evaporation are not excessive, and where the proportion of the rainfall occurring in the cooler months is sufficiently high, it is possible to grow the variety Dwalganup satisfactorily under an annual precipitation as low as 12 "(30 cm) or even 10" (25 cm).

(There are several species, with differing characteristics. To find more about them I suggest interested readers send off for information to the firms listed at the end of this chapter).

One species of particular note is Yarloop.

Yarloop: This variety generally flowers about two weeks later than Dwalganup. It requires a minimum of six and a half months of effective rainfall and is suited to a wide variety of soil types. Unlike other subterranean clovers, it is adapted equally to well drained soils and waterlogged conditions and is a good substitute for strawberry clover in areas otherwise unsuitable for that species. Being an erect plant it stands up well to weed competition but tends to dominate other species in the sward if not controlled by suitable grazing management." [62.]

Taiwan Kudzu. see Kudzu species.
Tropical Kudzu. see Kudzu species.
Wet Tropics Kudzu,*(Pueraria phaseoloides)*, see Kudzu species group 1.

AP5.10 Readers who would like further information, or to obtain seeds or other propagating material for local trials may enquire of the following firms.

WRIGHT, STEVENSON & CO. LTD.
HEAD OFFICE, AUSTRALIA: 330 ST KILDA ROAD, MELBOURNE, 3004.
Postal address; GPO BOX 4339. MELBOURNE 3001 TELEX 31342.

TERRANOVA TROPICAL PASTURE SEEDS.
SEED PRODUCTION DIVISION OF ANDERSONS SEEDS LTD.
68 BOUNDARY STREET, BRISBANE. QUEENSLAND. AUSTRALIA.

ELDERS PASTURE SEEDS AND FERTILISERS INFORMATION SERVICE.
ELDER SMITH GOLDSMITH MORT LTD.
AUSTRALIAN HEAD OFFICE;
ELDER HOUSE 27-39 CURRIE STREET. ADELAIDE.
SOUTH AUSTRALIA. TELEX AA82211.

ELDERS-GM HONG KONG.
36TH FLOOR CONNAUGHT HOUSE, CONNAUGHT CENTRE. HONG KONG.

ELDERS-GM LONDON.
3 ST HELENS PLACE, BISHOPSGATE. LONDON EC3 TELEX 885608

QUEENSLAND AGRICULTURAL SEEDS PTY LTD
PO BOX 1052, TOOWOOMBA. QUEENSLAND. AUSTRALIA.
TEL 6176301000, TELEX 140571QASEES,
FAX 61-76-30 1005
Has inoculants for legumes if needed.
Catalogue available with a variety of tropical legumes,
Lucerns and tropical grasses.

THE INLAND AND FOREIGN TRADING COMPANY LTD.
BLOCK 79A INDUS ROAD #04-418/420,
SINGAPORE 0316
TEL 2782193,2722711. TELEX RS25254 IFTCO. FAX 2716118.
An extensive listing includes legumes,
Cover crop/pasture seeds, flowering shrubs.

KENYA SEED COMPANY LTD
CONTACT MR CHARLES NDAGWA MACHARIA.
ELGON DOWNS FARM RESEARCH CENTRE
PO BOX 13 ENDEBESS, KITALE,
KENYA.
TEL (0325) 20941/2/3, TELEX 38046.
Carries rhizobial inoculants
A wide range of legumes and probably also carries grass seeds.

T.S.L LIMITED
CONTACT G.E.KENT, MANAGER.
PO BOX 66043
KOPJE
HARARE. ZIMBABWE.
TEL HARARE 68685,
TELEX 26167TSllTD
Legumes available in quantities probably have tropical grasses also.

SETROPA LTD.
CONTACT JAN KOITER
PO BOX 203
1400 AEBUSSUM
THE NETHERLANDS.
TEL (31) 215258754,
TELEX 73255SETRO
FAX (31) 215265424.

An extensive seed catalogue is available that includes legumes and inoculants, grasses and many varieties of trees.

EMPRESA DE SEMILLES FORRAJERAS, SEFO
LA PAZ, BOLIVIA.
TEL 41975, FAX 375042.
Carries a wide range of fodder seeds.

MILTON B FLORES, CIDICCIO
APARTADO 3385
TEGUCIGALAPA, HONDURAS.
This organization conducts research on cover crops in the tropics as alternatives to fertilisers, as erosion control, and sources of food and fodder.

WATERWAYS

1. Classification of Vegetal Covers As to Degree of Retardance.

Retardance.	Cover	Condition.
A	• Weeping lovegrass	Excellent stand, tall (average 30")
	• Yellow bluestem Ischaemum	Excellent stand, tall (average 36")
B	• Kudzu	Very dense growth, uncut
	• Bermudagrass	Good stand, tall (average 12")
	• Native grass mixture (little bluestem, blue grama, and other long and short Midwest grasses)	Good stand, unmowed
	• Weeping lovegrass	Good stand, tall (average 24")
	• Lespedeza Sericea	Good stand, not woody, tall (average 19")
	• Lucerne	Good stand, uncut (average 11")
	• Weeping lovegrass	Good stand, mowed (average 13")
	• Kudzu	Dense growth uncut
	• Blue grama	Good stand, uncut (average 13")
C	• Crabgrass	Fair stand, uncut (10-14")
	• Bermudagrass	Good stand, mowed (average 6")
	• Common lespedeza	Good stand, uncut (average 11")
	• Grass/legume mixture summer (orchard grass, redtop, Italian	Good stand, uncut (6-8") rye-grass, and common lespedeza)
	• Centipedegrass	Very dense cover (average 6")
	• Kentucky Bluegrass	Good stand, headed (6-12")
D	• Bermudagrass	Good stand, cut to 2.5" height
	• Common lespedeza	Excellent stand, uncut (average 4.5")
	• Buffalograss	Good stand, uncut (3-6")
	• Good stand, uncut (4-5")	
	• Grass/legume mixture-autumn, spring (orchardgrass, redtop,	
	• Italian ryegrass, and common lespedeza	
	• Lespedeza sericea	Very good stand before cutting
E	• Bermudagrass	Good stand, cut to 1.5"height
	• Bermudagrass	Burned stubble.

Note: The covers classified have been tested in experimental channels. They were green and generally uniform.
FROM REFERENCE HANDBOOK OF CHANNEL DESIGN FOR SOIL AND WATER CONSERVATION. TP61
U.S.DEPARTMENT OF AGRICULTURE, SOIL CONSERVATION SERVICE.

2. Maximum lengths of channel terraces.

	Sandy soils		Clay soils	
	m	ft	m	ft
Normal maximum	250	900	400	1200
Absolute maximum	400	1200	450	1500
(If spacing is reduced)				

These distances all refer to the maximum length of flow. A channel which sheds run-off on both sides from a high point can have the maximum length on either side of the high point.

From:-Dept. of Conservation and Extension, Government of Rhodesia, Dept. In-service Manual.

Found in [10 p197].

PRACTICAL SURVEYING OF LAND FOR SOIL CONSERVATION

7.1 Introduction.
7.2 Methods of surveying land for conservation.
7.3 The Good News Level.

7.1 Introduction

When we look at an area of slope land and plan soil conservation we need to be able to mark out the lines of contour strips, terraces, or paddy fields, in level lines.

Sometimes we need to mark accurate sloping lines for drainage or for laying water pipes or ditches.

We also need to be able to decide and measure the height difference between contour lines or terraces according to the depth of reasonable soil.

Another important measure is how much slope a field has so that we can tell if it is too steep for some use but suitable for other use?

If we are making terrace walls of different materials what should the slope of the walls be?

When we make a terrace we should have a back slope so that surface water will run to the back of the terrace and not over the edge of the terrace, how to set the slope?

It is often assumed that to do such work we will have to get a surveyor or engineer to make the plans for us, that is not correct, farmers can do it all themselves whether they have been to able to go to school or not. Perhaps that sounds impossible but it is a fact. Farmers have made and used the "Good News Level" themselves, to do all those things. At first when they start to use it they may have difficulty, but as with most things with a little practice they become able to do it and feel very pleased to be able to do so. It is in fact better for farmers to do it than engineers or surveyors, as they understand their land better and can make adjustments to fit the special needs of their own situations.

7.2 Methods of surveying land for conservation

In my experience methods of surveying land for conservation fall into one of four categories. They are either 1) too theoretical, 2) require expensive equipment and special training, 3) simple but not practical or accurate. 4) simple, practical, and multipurpose–using the Good News Level.

1) *Theoretical surveying* was carried out on the land of the Yu Shan Agricultural Training Centre by a specialist surveyor; as happens too often with 'specialists' it was both very expensive and useless. It did not take account the depth of soil when setting the height between terraces. It did not take into account the fact that some sections of the planned terraces had deep soil while others had very shallow soil; so some should be taken from the deep soil to add to the poor areas. Some areas were very stony and so the stones were used for terrace walls, in both cases the planned level was reduced and so the resultant contours were far from level.

Another point was that there were areas of large rocks which were too big to break up, and so it was not practical to terrace, these areas were used for other uses, they could not be used as the survey planned,

The A Frame seems an appropriate and simple device. Three sticks fastened into an 'A' shape with a mark at the centre of the crossbar and a bobline hanging down and over the crossbar. Leg A is placed at the starting point, then leg B is moved over the ground until the bobline is over the centre mark showing that the frame is level. The A frame is then pivoted and leg A moved to find the next level position. At suitable places pegs are placed to **mark the contour lines.**

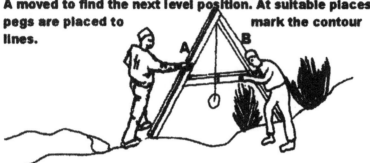

This process has to be repeated many times. On ploughed, rough or stoney ground progress is particularly slow and difficult. When it is windy the bobline swings and surveying is impossible

Fig. 1

all these factors interrupted the planned water flows etc.

The result was that the expensively planned, graded terraces, would be neither suitable nor carefully graded on completion. Excess water would have accumulated at the lower points and then overflowed and caused erosion, little would have gone to the carefully planned waterways.

The above points to the fact that it is best if farmers can do the surveying and planning themselves according to their own situation. This is possible, and in this chapter the whole process is covered in detail.

2) Using expensive complicated levels and surveying methods which neither the extension workers nor the farmers can use properly.

Sometimes NGO's purchase such equipment, it looks good but is seldom if ever used effectively.

3) Using apparently simple methods which are impractical.

As an example "A" frames are often advocated

At first sight the A frame seems practical and simple, if you are using it in a classroom or a sports field. Unfortunately most places where soil conservation is needed are very different to sports fields. See Fig. 1.

An A frame has two legs, one is placed at the starting position. The other leg is pivoted about the starting position until when it is placed on the ground the weighted string which hangs from the top of the two legs hangs over the centre mark on the 'A' crossbar in the middle of the frame. That shows that both the legs are at equal height. It sounds simple till you try it on a ploughed field for instance. Where do you put the legs? on top of the furrow slice ? at an average height whatever that is ? or on the bottom of the furrow ?. If on a rocky hillside or a hillside with bushes and clumps of grass where do you put the legs ?

The legs of the A frame are 3 or 4 ft (0.91 -1.22 m's) apart, about 1 metre. So to mark out 100 metres the legs have to be pivoted about till the right place is found 100 times. Each of the 100 times there can be a slight error. If it is a windy day (which it often is at the time of year late autumn and early spring when the fields are empty,) the weighted string will be swinging about and it will be hard to determine when the marker string is over the mark.

It may not seem too much difference, the error will not be very great but if there is a very slight error every two metres across a slope it will become an important error over 100 metres or even 50 metres.

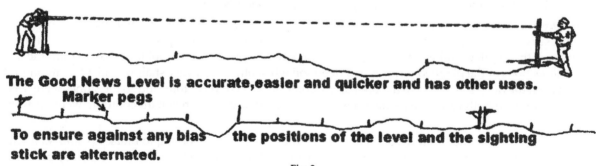

The Good News Level is accurate, easier and quicker and has other uses.
Marker pegs

To ensure against any bias   the positions of the level and the sighting stick are alternated.

Fig. 2

In Nepal a colleague of the authors did not think that there would be a problem, till he taught a group of farmers how to use an 'A' frame for marking out contour lines on a hillside. When they found that the contour lines they were demonstrating were coming together, they realised that it was not accurate enough, and so the next day he came for one of our Good News Levels described in this chapter. See Fig. 2.

The Water Pipe Level

The simplest form of this is a long clear plastic tube and two marked pegs. The ends of the pipe are fastened in an identical way onto the two marked pegs. Water is poured into the pipe till the pipe is almost full with the water finding its own level as the two pegs are placed side by side. The pegs are marked at the height of the water in each end of the tube, the level in both the pipes will of course be the same. One end of the tube is placed at the base position and the other taken to a position further along the hillside. The peg is moved up and down the slope until the position is found where the level of the water is the same as the water level at the base position. The place is of course level with the first position, the process is repeated along the contour. This seems a good idea, I have made and used one for marking out the foundations of buildings for which it was very good, however we found problems when using it for marking contours on a hillside. When we are moving about, particularly when carrying the pipe coiled up to a village, and then filling it on a hillside. In such situations it is easy to have air locks (air bubbles), these can be unnoticed and the level seeming to work. However, the problem is that the accuracy is affected, so that the contour lines plotted are not level, and the work was a waste or worse. If the lines dip in places, water will concentrate in such places and flow down causing erosion.

Spirit Level Levels

These come in three or more designs, one is extremely impractical. It is a frame like a large table with a spirit level on the middle. The table is carried to the hillside, and carried to the starting position. One leg of the table is placed at the base position, then the table is swivelled till with its feet on the ground the spirit level is level. The place of the feet of the table at the base + 1 is marked, and the table picked up and carried so that the base end of the table is placed on the base + 1 mark and the process repeated to mark the base + 2 position. It has the difficulties of carrying a table about on a hillside, it is cumbersome to use, it is not accurate, it is slow, it can only serve one purpose. It depends on an accurately made and robust frame and a good large spirit level, not often found on third world hill farms.

Another is with the spirit level hanging from a rope which is stretched between two poles, this cannot be very accurate, and will take many readings to cover a hillside, it will be affected by winds. As mentioned before the wind factor is important, as the slack time in the farming year is often in the autumn after the summer crops are harvested; or in the spring before the land is prepared for the crops of the rainy season, at these times windy weather is more frequent than is normal.

The third is probably the best of this kind, being simply a long board with a spirit level fastened on top of it, it is moved about in a similar way to the table or the "A" frame. It will not be suitable on a rocky or uneven hillside or where there is long grass. However it will be easier to manage than the other two.

THE ABOVE LEVELS CAN ONLY BE USED FOR MAKING LEVEL MARKS.

7.3 Good News Level

THIS LEVEL IS SIMPLE TO MAKE AND USE. It is more accurate, quicker and more practical to use than the other 'simple' levels.

It is "GOOD NEWS", as local people in the hilly areas can make it. With an hours practice illiterate people can use it. They can set out quickly, and accurately, level contour lines or graded contour lines; and can determine where to set the succeeding contour lines. They can also learn other surveys and calculations needed for soil and water conservation projects.

It can be used for many purposes

1). For making level contour lines on hillsides or on gentle slopes.

2). For measuring the percentage or grade of slope of hillsides, so that the land can be classified as to whether it should be for terraces, fruit tree orchard, forest etc. Symbols can be used where some people may not understand percentages.

3). For making graded (sloping) contour lines on hillsides, and for plotting roads, waterways, pipelines etc.

4). When working on irrigation or drinking water systems it can be used for measuring the difference in height, (the head) between a water source and various components of the system.

5). For setting the relative position of contour lines for terraces etc. above or below an existing terrace or contour line, according to the average depth of the soil.

6). For setting the angles appropriate for building different kinds of terrace walls.

7). For setting the angle of backslope of terraces.

The requirements are simple:-

Three pieces of wood.

One :- 147.5 cms (4'10") long, 3 cms (1 3/16") wide, 1.5 cms (9/16") thick.

One :- 113 cms (3'8 1/2") long, 4 cms (1 9/16") wide, 1.5 cms (9/16") thick.

One :- 67 cms (2'2 1/2") long, 3 cms (1 3/16") wide, 3 cms (1 3/16") thick.

The width and thickness of the wood is not critical, but old wood which will not warp or twist is best.

Three screw bolts and nuts and washers:- About 4 cms (1 1/2") long and 5 mm(1/4") thick.**One piece of flexible string or strong thread** 50 cms (19 1/2").

Fig. 3

One bob-weight, for example a piece of flat metal or heavy wood 10 cms (4") long 3 cms(1 1/4") wide with a hole in the centre at one end. The exact specifications are not important. It should be heavy enough so as not to swing easily in a wind, about 250 gms (1/2 lb).

One pole (the sighting stick) 1.6 metres or longer which has a mark on it which is the same height as the top of the sighting bar.

Fig. 4

To mark out level contour lines

The man holding the level aims the sighting bar at the stickman, and signals him to move the stick up or down the slope with the top of his hand level with the sight mark until the top of his hand is level with the sight bar. The stick man holds the sight stick upright and stands at the side of the stick with the top of his hand level with the equal height mark on the stick. It is best if his arm is in line with his hand so that it is easier for the level man to see when at a distance.

A boy or other man checks that the level is upright by observing if the bobline is over the centre mark, if it is not he moves the level till it is. Now the level man lowers his eyes and looks very carefully along the top of the sight bar like aiming a gun, so that his eyesight follows the sight bar exactly. He then signals fine adjustments to the stick man until the top of the stickmans hand is exactly in line with his sight. A marker peg is placed at the place where the bottom of the sighting stick is, and the stick man moves away to the next position.

As an extra check, rotate the positions with the level man moving away from the stick man, see Fig 2.

The Good News Level is not only for marking out level lines.

1. Marking out level control lines

See figure 8 No 1.

2. For measuring the slope of a hillside.

See figure 8 No 2 I have found that farmers very quickly understand how to do this. If need be symbols can be added to the sight bar to indicate the recommended use for each slope angle.

3. For marking out or measuring a % slope for a ditch or water pipe.

See figure 8 No 3. The basic idea is that a new mark is added to the sighting stick which is 1 metre above or below the equal height mark. The stick man goes a 100 metres across the field and then he moved up or down until when his hand is on the new mark his hand is in line with the sighting bar on the level. That position is one metre above or below the position of the level man so the angle of the line is of 1% slope. This could be for the slope of a ditch or water pipe. If the same marks were used for a 50 metres length the slope would be 2%.

4. *For measuring the head or vertical height difference between two positions such as for a drinking water system.*

See figure 8 No 4. This uses the same principle as for 3/. Except that a longer sighting stick is needed and another stick with a cross bar to take the place of the stickmans' hand. The positions are marked and the height is recorded. The level is moved to the stickmans' position, the stick man moves to the next position and the process is repeated. The height differences are added together to make the difference in height between the top and bottom positions.

Marking Out. It is best to use pegs with a white or yellow top. Paint is best but otherwise plastic, paper or rag will do.

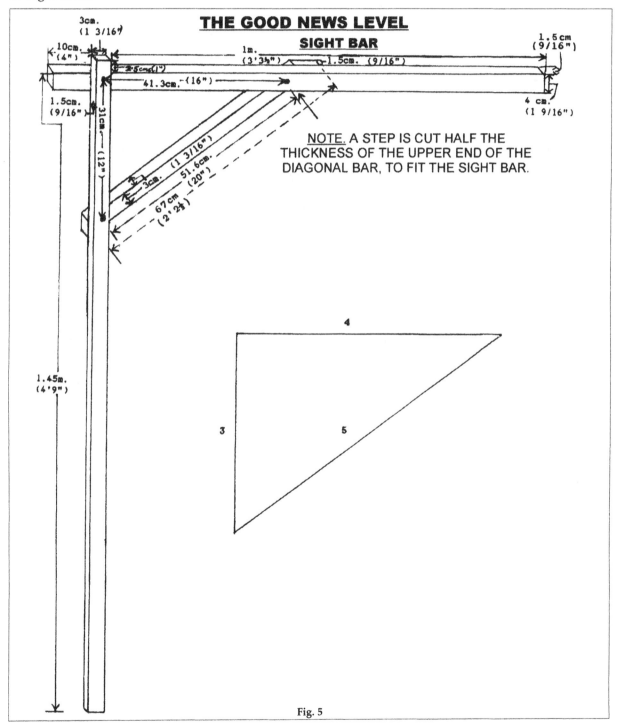

Fig. 5

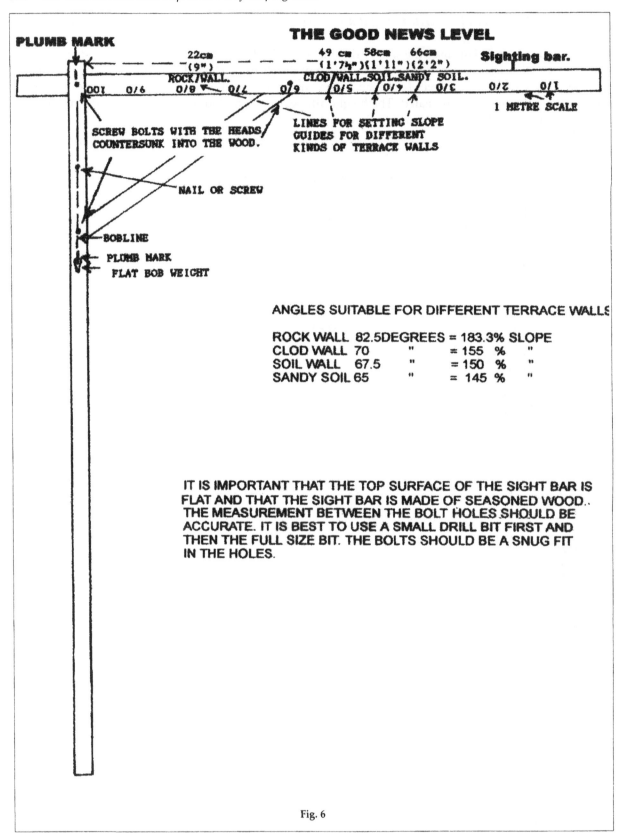

THE GOOD NEWS LEVEL

PLUMB MARK

22cm
(9")

49 cm 58cm 66cm
(1'7½")(1'11")(2'2")

Sighting bar.

ROCK/WALL.

CLOD/WALL.SOIL.SANDY SOIL.

1 METRE SCALE

SCREW BOLTS WITH THE HEADS
COUNTERSUNK INTO THE WOOD.

LINES FOR SETTING SLOPE
GUIDES FOR DIFFERENT
KINDS OF TERRACE WALLS

NAIL OR SCREW

BOBLINE

PLUMB MARK
FLAT BOB WEIGHT

ANGLES SUITABLE FOR DIFFERENT TERRACE WALLS

ROCK WALL 82.5DEGREES = 183.3% SLOPE
CLOD WALL 70   "   = 155 %   "
SOIL WALL 67.5   "   = 150 %   "
SANDY SOIL 65   "   = 145 %   "

IT IS IMPORTANT THAT THE TOP SURFACE OF THE SIGHT BAR IS
FLAT AND THAT THE SIGHT BAR IS MADE OF SEASONED WOOD..
THE MEASUREMENT BETWEEN THE BOLT HOLES SHOULD BE
ACCURATE. IT IS BEST TO USE A SMALL DRILL BIT FIRST AND
THEN THE FULL SIZE BIT. THE BOLTS SHOULD BE A SNUG FIT
IN THE HOLES.

Fig. 6

To check that the level is accurate the following procedure is carried out.

Make a pencil line in the centre of the upright. Aim the level at the stickman with the plumbline over the mark.

Signal to the stickman to move the sightstick up or dawn until the equal height mark is level with the sight bar.
Mark both positions and change ends.

Repeat the process, the equal height mark should be in line with the sight bar, if it is not the level needs adjusting. The adjustment is simple.

Aim the sight bar until it is in line with

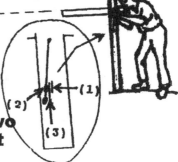

the equal height mark.
Now make a second pencil mark (2) under the bobline. Next draw a third pencil line exactly between the two lines (3). This is the correct line.

Finally repeat the process using the centre line, and adjust if needed. Now make a permanent line over correct pencil line and the level is accurate.

To check and if needed, adjust The Good News Level.

Fig. 7

<u>Marking out level contour lines.</u> The stickman stands facing the level man

1. with his hand and forearm level with the equal height mark. The level man aims the level at the stick while a helper not shown checks that the plumbline is over the plumb mark. The level man lowers his eyes until the top of the sight bar looks like a single line __. He then signals the stickman to move the stick up or down the slope until the top of the stick mans hand is level with the sighting bar. In the illustration the stick is too low.

2. ## <u>Using the Good News Level for measuring the slope of a hillside.</u>

1. Hold the level upside down with the further end of the sight bar touching the slope to be measured, and the plumbline over the plumb mark.

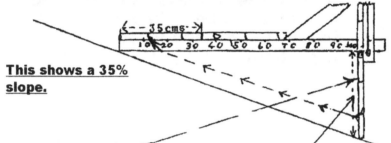

<u>This shows a 35% slope.</u>

2. Hang a measuring thing such as a peice of grass, string, or a stick down from the bottom of the sight bar to the ground. Next put the measuring thing on the scale to see how many centimetres it is from the sighting bar to the ground. The number of centimetres is equal to the percentage slope.

3. Equal height mark

<u>For marking out or measuring a 1% slope.</u>

1 metre below the equal height mark

|——————————————100 metres ——————————————|

The average slope between the to positions is 1%   For a 2% slope use 50 metres or reverse the positions and use a higher sighing stick.
If a 2% slope is required for a pipeline the level can be aimed at a mark 2 metres higher than the sighting bar at 100 metres distance, then a 2% mark ⌐ is drawn on the upright below the plumbline and that mark used.

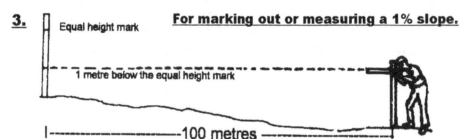

4. For steeper slopes an extra height pole is placed over and above the normal sighting stick.so that the difference in height to the equal height mark can be noted. In this case it is three metres. The lower metre of the extra height pole is marked off in centimetres so that parts of a metre difference can also be noted.

In this case the level may be used to measure the difference in height between two positions such as in a planned drinking water pipeline.

Fig. 8

5/.

<u>Using the Good News Level for setting the difference in the height of contour lines on a slope.</u>

1. First you need to find the percentage of slope.
   If the slope is 50% or more you need to set the vertical fall between the terraces as three times the depth of the good soil. Otherwise the terraces will be too narrow and the steps into the hard material insufficient to hold against downward slip; if 30-40% the vertical fall should be two and half times the depth of the good soil. Below 30% - twice the depth of the good soil.

2. You need the normal sighting stick plus an extra tall sighting stick. You also need something to dig with and a piece of string or similar material.

3. Dig some holes in the field in order to find the average depth of the good or reasonable soil and how far to the hard strong soil, shale or rock. If there is no strong material within 2 feet of the surface, dig deeper to see what lies beyond. If still no strong material do not make normal terraces as they will probably slide down the hill when the soil becomes saturated with water. Mini terraces 1.3 m (4ft) wide or less should be made or plant bamboo, trees, or grass and/or fodder legumes, unless the good soil is deep. If there is hard rocky material just under the good soil, it is again not good to make normal terraces. The work in digging out the rock will not be worth the small surface area of terrace produced. Mini terraces should be made or plant bamboo, trees or grasses and/or fodder legumes.

4. If the land is suitable for terracing we take our piece of string and use it to measure the depth of the good soil. Suppose the angle is below 30% we should have twice the depth of the reasonable soil between this terrace and the next terrace. We use the string measure to mark off twice the depth of the reasonable soil from the normal equal height mark. It may be necessary to use the extra tall sighting stick.

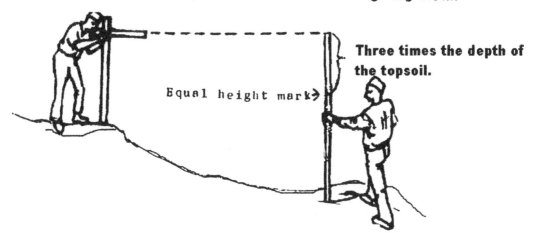

Equal height mark→

Three times the depth of the topsoil.

The stick man goes down the hill until the top of the extra sighting stick is level with the sight bar. A marker stick is put to mark the new contour line. And then the new contour line can be marked out.

Fig. 9

DESCRIPTION OF WHY THE DISTANCE BETWEEN TERRACES SHOULD BE SET ACCORDING TO THE DEPTH OF THE TOPSOIL.

As with most recommendations these are a guide. If rocks are close to the surface, or the topsoil is very deep (more than 7" (18 cm's)) the vertical distance should be less. If the subsoil is of good quality or the soil can easily be moved for example by a bulldozer or bullock dozer, the vertical distance may be greater. It should be remembered that if the terrace is made too large the wall will be higher and so will need more care in building.

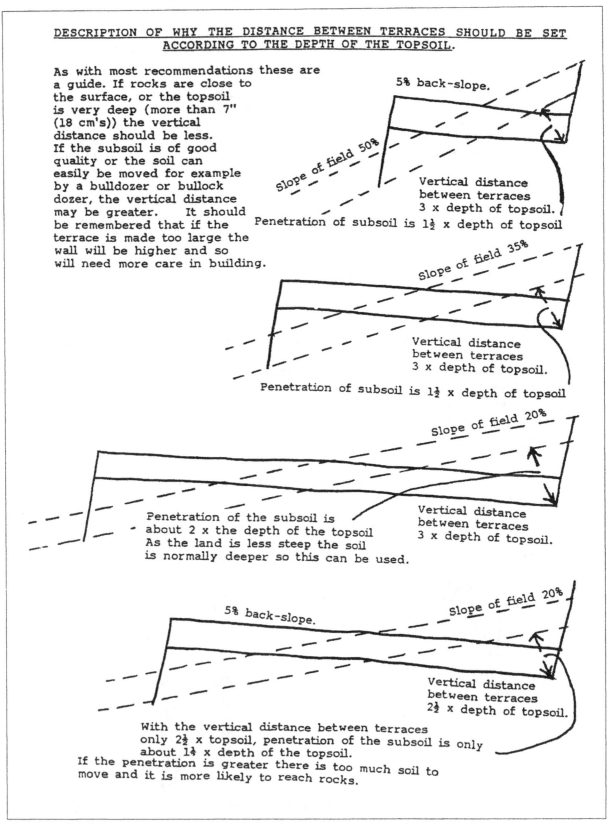

5% back-slope.

Slope of field 50%

Vertical distance between terraces 3 x depth of topsoil.

Penetration of subsoil is 1½ x depth of topsoil

Slope of field 35%

Vertical distance between terraces 3 x depth of topsoil.

Penetration of subsoil is 1½ x depth of topsoil

Slope of field 20%

Penetration of the subsoil is about 2 x the depth of the topsoil As the land is less steep the soil is normally deeper so this can be used.

Vertical distance between terraces 3 x depth of topsoil.

5% back-slope.   Slope of field 20%

Vertical distance between terraces 2½ x depth of topsoil.

With the vertical distance between terraces only 2½ x topsoil, penetration of the subsoil is only about 1¼ x depth of the topsoil. If the penetration is greater there is too much soil to move and it is more likely to reach rocks.

Fig. 10. Too many terraces are made without considering the depth of soil, and often it results in much harder work, as they come to rocky layers.

6/. <u>To Use the Good News Level to set the angle for building terrace walls.</u>

The level is used plus some bamboo or similar poles.

The level has marks on the sight bar, they are for making rock walls, and clod walls, strong soil walls and sandy soil walls. Angles or slopes can be set for either of them or you may find other angles suitable for your own conditions and you may mark them on the bar.

1. Place the bottom of the leg of the Good News Level at the place where the wall will start, with the sight bar facing towards the slope and the bob line over the plumb mark.

2. Take one of the poles and put one end of it at the bottom of the leg of the level.

3. Lean the pole against the appropriate mark on the sight bar and then fix the pole at that angle. Repeat with other poles in other positions along the contour and then link up the poles with string lines. Now you have the guides for building the walls.

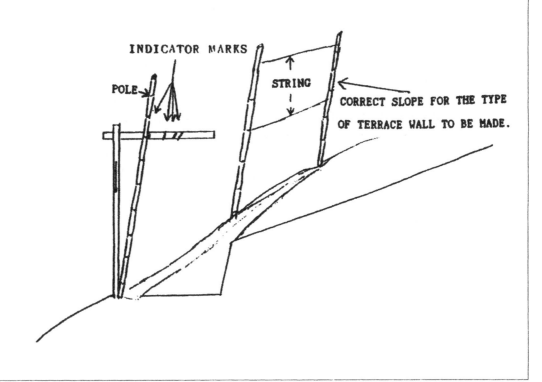

Fig. 11

7/.   COMPLETING A TERRACE - THE BACK SLOPE.

When levelling a terrace we need to remember that there should be a back slope, from the outside edge to the inside corner of the terrace.

There are two reasons for this.
1/ Any excess water should flow to the inside of the terrace rather than the outer edge, if it flowed to the outer edge and over the wall it would cause erosion and spoil the terrace wall and the bank below it. The water flowing down from above could also start erosion of the land below.

2/ As the soil towards the outer part of the terrace is all loose soil, it will settle lower. This settling must be allowed for or later the terrace will have an undesirable outer slope.

The percentage of slope required varies with the condition of the soil. If the soil is lumpy the lumps will collapse when they are wetted in the wet season, they will settle more than sandy soil, and so more slope should be allowed for this. The range of slope will be between 3% and 5%.

The sight bar of the level can be used to measure the width of the terrace in metres. Next multiply the width by the percentage of slope which is required. For example, if the terrace is three metres wide and the appropriate slope is 5% the result will be 15. This 15 is the number of centimetres of height difference between the inside of the terrace and the outside.

Make a pencil mark 15 centimetres above the equal height mark on the sighting stick. Now the stick man takes the sighting stick to the inside edge of the terrace and holds his hand level with the pencil mark. The level man sights along the level from the outer edge at the top of the stick mans hand. The slope of the soil is then adjusted till the pencil mark matches the level of the sight bar.

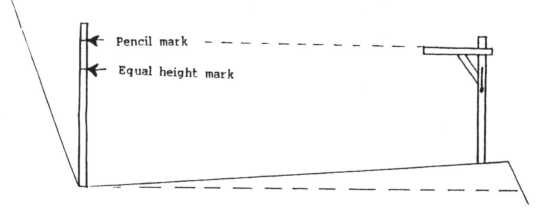

Fig. 12

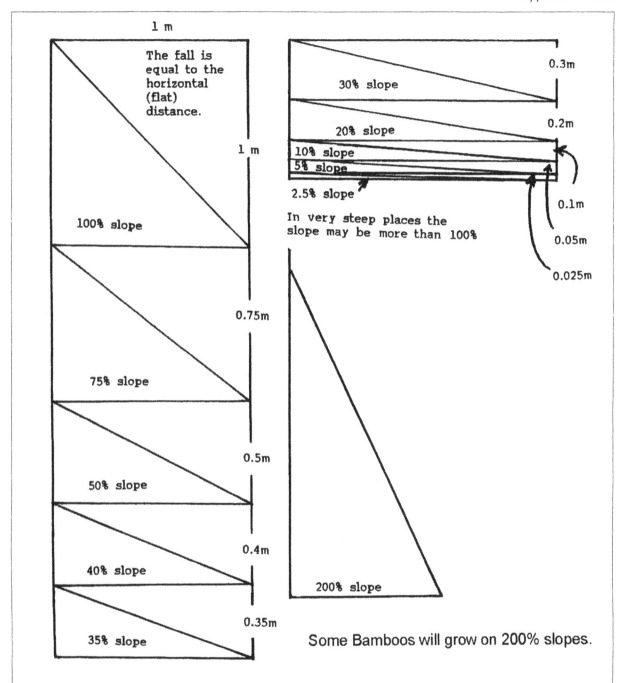

1 m

The fall is equal to the horizontal (flat) distance.

1 m

100% slope

75% slope

0.75m

50% slope

0.5m

40% slope

0.4m

35% slope

0.35m

30% slope

0.3m

20% slope

0.2m

10% slope

5% slope

2.5% slope

0.1m

0.05m

0.025m

In very steep places the slope may be more than 100%

200% slope

Some Bamboos will grow on 200% slopes.

Fig. 13 For people who have not used percentage slope here is a diagram to show how it works.
We can see that the distance in centimetres from a level bar to the ground is the same as the percentage slope.

Conversion Chart

For people accustomed to the measurement of slope being expressed in proportion, or in degrees, I have made the two conversion tables shown here.

Table of proportion compared with the system used in this book.

It is easy to realise that if the difference in height is 50 and the level distance is 100 which in our system it is 50% the slope is 100 divided by 50 = 2 so the slope is 1 in 2
To calculate the other way round we divide 100 by the proportion needed, so if we want to know what a slope of 1 in 2 was in percentage slope we divide 100 by 2 and find the answer is 50% slope.

TO AVOID THE NEED TO CALCULATE HERE IS A TABLE.

1 IN 1 SLOPE= 100%SLOPE			
1-1.1 = 90.9	1-2.75 = 36.36	1-6.5 = 15.38	1-17 = 5.88
1-1.2 = 83.3	1-3 = 33.33	1-7 = 14.28	1-18 = 5.55
1-1.3 = 76.92	1-3.25 = 30.76	1-7.5 = 13.33	1-19 = 5.26
1-1.4 = 71.42	1-3.5 = 28.57	1-8 = 12.5	1-20 = 5
1-1.5 = 66	1-3.75 = 26.66	1-8.5 = 11.76	1-25 = 4
1-1.6 = 62.5	1-4 = 25	1-9 = 11.11	1-30 = 3.33
1-1.7 = 58.82	1-4.25 = 23.52	1-10 = 10	1-40 = 2.5
1-1.8 = 55.55	1-4.5 = 22.22	1-11 = 9.09	1-50 = 2
1-1.9 = 52.65	1-4.75 = 21.05	1-12 = 8.33	
1-2 = 50	1-5 = 20	1-13 = 7.69	
1-2.1 = 47.6	1-5.25 = 19.04	1-14 = 7.14	
1-2.2 = 45.45	1-5.5 = 18.18	1-15 = 6.66	
1-2.25 = 44.44	1-5.75 = 17.39	1-16 = 6.25	
1-2.5 = 40	1-6 = 16.66		

For people accustomed to using the degrees of slope based on 360 degrees in a circle I have prepared the following table:-

% slope		number of degrees	% slope		number of degrees
150 %	=	135 degrees	40 %	=	36
125	=	112.5	35	=	31.5
110	=	99	30	=	27
90	=	81	25	=	22.5
85	=	76.5	20	=	18
80	=	72	15	=	13.5
75	=	67.5	10	=	9
70	=	63	7	=	6.75
65	=	58.5	5	=	4.5
60	=	54	4	=	3.6
55	=	49.5	3	=	2.7
50	=	45	2	=	1.8
45	=	40.5	1	=	0.9

Fig. 14

SOILS

BRIEF DEFINITIONS OF SOIL UNITS* And soil terms from FAO/UNESCO. SOIL MAP OF THE WORLD. Note:- Headings prefixed by * With additional comments and descriptions. Including some from a chart of the USDA Ref 14.p61, some from[16.]

*Acrisols (from Latin acris, very acid). Soils having an argillic B horizon and of low base saturation.

Albic bleached, (from Latin albus adj, white).

Alfisols Soils of medium, age "formed under a hardwood forest become alfisols"[16.]

With grey to brown surface horizon, medium to high base supply and subsurface horizons of clay accumulation, usually moist but may be dry during the warm season.

Andosols (from Japanese an, dark and do, soil; connotative of soils formed from materials rich in volcanic glass and commonly having a dark surface horizon). Weakly developed soils, rich in allophane ("comprises amorphous hydrated aluminium silicates of rather variable composition as the first weathering product of some kinds of fresh volcanic material."[18]) and having a low bulk density.

Aquults (from Latin aqua, water) seasonally saturated with water.

Arenosols (from Latin arena, sand). Strongly weathered sandy soils of tropical and subtropical areas.

Argilic (from Latin argilla clay) soils with a horizon of clay accumulation.

Aridosols (from Latin aridus, dry parched). Soils of medium age with separate horizons, low in organic matter, and dry for more than six months of the year in all horizons.

Calcic (from Latin calx-lime) Soil containing lime/chalk calcium Ca CO_3

Cambisols (from Latin camiare, change; connotative of soils in which changes in colour, structure and consistence have taken place as a result of weathering). Soils formed by a weak alteration of the parent material.

*Chernozems (from Russian chern, black and zemlja, earth land). Soils of the grassland steppes showing strong accumulation of organic matter in the surface horizons and an accumulation of calcium carbonate at shallow depth.

Entisols (from Greek entelekh the becoming or being actual of what was potential) Young soils which have not yet differentiated clear horizons. As they mature they develop into inceptisols. Their agricultural value and use depends on the parent material, if alluvial or volcanic origin they will be potentially fertile soils.

*Ferralsols (from Latin Ferrum and aluminium; connotative of a high content of sesquioxides. Strongly weathered soils consisting mainly of kaolinite, quartz and hydrated oxides.

*Fluvisols (from Latin Fluvius, river, connotative of flood plains and alluvial deposits). Weakly developed soils from alluvial deposits in active flood plains.

*Glevsols (from Russian local name gley' meaning mucky soil mass; connotative of reduced or mottled layers resulting from an excess of water). Soils in which the hydromorphic processes are dominant.

*Grevzems (from grey and Russian zemlja, earth, land). Soils of the forest steppe transition, rich in organic matter and having a grey colour caused by white silica powder or structure faces.

*Histasols. (from Greek Histos, tissue). Organic soils.

"If the parent material is highly organic, such as a filled-in bog or swamp, the soil that develops, regardless of climatic type and age, is called a histosol. Because of their high organic matter content, histasols can be highly productive, although they tend to be difficult to manage"[16]

Humults (from Latin humus earth) with high or very high organic matter content.

Inceptisols (from Latin inceptio, beginning) young soils. Soils that are usually moist with soil horizons showing alteration-of the parent materials, not by accumulation.

Their value for agriculture depends on the parent material.

Kasstanozems (from Latin castaneo, chestnut and from Russian zemlja, earth, land). Soils of the semi arid steppes showing an accumulation of inorganic matter in the surface horizons, often calcareous throughout.

Latosols. see at the foot of this table.

Lithosols. (from Greek Lithos, stone). Shallow soils over hard rock.

Luvisols (from Latin luvi, from Greek luo, to wash, connotative of illuvial accumulation of clay). Soils having an argillic (claylike) B horizon and of low base saturation.

Mollisols (from Latin mollis, soft supple) soils of medium age with nearly black, organic rich surface horizons and high base supply. They develop under conditions of prairie, steppe or savanna.

Natric (from Arabic Naturn-Native sesquicarbonate of soda) Containing sodium.

Nitosols (from Latin nitidus, shiny, bright, lustrous; connotative of shiny ped faces). Soils having deeply developed argillic B horizon and showing features of strong weathering.

Oxisols (from Greek Oxus sharp or acid, oxidisation. To cause to combine with oxygen. This comes from the understanding at first that oxygen was the essential principle in the formation of acids.)

"Oxisols as the order name suggests, are highly weathered, nutrient poor soils with an oxic (showing oxidation) layer. They occur on some of the oldest landscapes on earth such as the Brazilian and Guyannan Shields in the Amazon Basin and the African shield of the Congo Basin High Annual rainfall leaches the bases from the soil, yielding relatively high levels of iron and aluminium oxides (sesquioxides.) As a result, oxisols are acid in reaction. Extreme weathering removes even the silica component producing Kaolinitic clays with a cation exchange capacity of only 2-4 meq/100 g soil. Base saturation levels of oxisols in the Amazon Basin typically fall below 15%.

Oxisols range in colour from yellows to reds to dark browns. Unlike with temperate zone soils, however, colour has little to do with organic-matter content and fertility. Rapid decomposition and uptake of organic matter nutrients leaves little chance for clay-humus complex formation, and excessive weathering results in weatherable minerals occurring at levels below 1%. Deep weathering of some oxisols places the hard rock parent material 10-40 metres below the surface."[16]

Phaeozems (from Greek phaios, dusky, and Russian zemlja, earth. Soils of the forest steepes showing a strong accumulation of organic matter in the surface but a deep leaching of calcium carbonate.

Planosols (from Latin planus, flat, level). Connotative of soils generally developed in level or depressed topography with poor drainage.

Podzols (from Russian local name derived from pod, under and zola, ash). Soils with a strongly bleached horizon having B horizons with iron or humus accumulation, or both.

Podzoluvisols (combined Podzol and Luvisol). Soils having an argillic B horizon but also showing features of Podsols.

Rankers (from Austrian rank, steep slope; connotative of shallow soils). Soil which develop a surface horizon enriched in organic matter over siliceous materials.

Regasols (from Greek rhegos, blanket; connotative of mantle of loose material.)

Rendzinas (from Polish rzedzic, noise; connotative of plough noise in shallow soils).

Solic (from Latin Sal, salt) Salty containing Sodium Chloride, Na Cl.

Solonchaks (from Russian sol, salt). Soils showing strong salinity.

Solonetz (from Russian sol, salt). Soils developed under the influence of high sodium saturation.

Spodosols (from Greek, spodos, wood ashes) "middle aged soils which develop under coniferous forest"[16]. With accumulations of amorphous (without definite shape or structure) materials in sub surface horizons. They are not normally not very good for agriculture.

Udults (from Latin udus, humid, damp) low organic matter content, temperate or warm and moist.

Ultisols (from Latin ultimare, the last, come to an end). Older more leached soils that are usually moist with horizon of clay accumulation and a low base supply. As they are not able to store plant foods they are normally very infertile unless managed well.

"Two common features of ultisols are plinthite and fragipans. Plinthite is a subsoil phenomenon dependent on alternating wet and dry periods, like a fluctuating water table. The drying out of a wetter iron rich layer can harden irreversibly into plinthite. (A kind of brick red clay). Fragipans occur in poorly drained ultisols, where they restrict water movement and are sometimes responsible for perched water tables."[16]

Vertisols (from Latin verto, turn; connotative of turn over of a surface soil), self mulching soils, with a high content of clay which swells and becomes very sticky and impervious (water proof) when moist, and forms deep, large, cracks when dry. Surface material fills the cracks in the dry season, this causes an inverting process and is the reason for the name. This kind of soil is not good for crops other than paddy crops or pasture, or perhaps some species of Bamboo.

Xerosols (from Greek Xeros, dry). Soils of semi arid environments. Dry with winter rains.

Yermosols (from Spanish yermo. desert, derived from Latin eremus, solitary, desolate). Soils of desert environments.

--- --- --- --- --- --- ---

Latosols though not mentioned in the above should I think be included though separately, PJS. (from Latin later, a brick, because blocks of it dug out set so hard as to be able to be used as building material). Found in the Humid tropics (also in areas with monsoon rain but otherwise dry, PJS). Such soils have a high content of iron and aluminium, in these soils leaching tends to be considerable. They contain a high percentage of clay but are not as heavy working as would be soils with the same clay percentage in temperate latitudes. They are acid and have a clay fraction dominated by kaolinite or iron and aluminium oxides. Their red colour is due to the iron hydroxides formed under tropical conditions of weathering; this yellow colour is from aluminium hydroxides.

Anthony Young pointed out in a paper "The nature and management of the poorer tropical latosols" that:-

> Within the latosols there are some 10-12 distinctive soil types. The classification employed in the FAO-Unesco Soil Map of the World (1974) contains eleven classes which fall within the group; the six classes of ferralsols and three of nitrosols, plus the ferric luvisols and ferric acrisols---. In the US Soil taxonomy (USDA 1975) the whole order of oxisols together with Ustic and Udic suborders of both alfisols and untisols are included. In a recent review by the present author (Young 1976) six main types of latosols with eleven subdivisions are recognized.

> By no means all of the latosols are agriculturally poor. Those derived from basic rocks, the ferrisols (FAO nitosols) can be highly productive, with a capacity to sustain continuous cultivation. The ferruginous soils (FAO ferric luvisols) have a potential for fairly continuous arable use, given a reasonable level of inputs. Many of the humic latosols of higher altitudes, with higher organic matter contents associated with the cooler temperatures, are not problem soils except where steep slopes give a high erosion hazard.

In monsoon regions the leaching of the soil is slower as it is interrupted by the dry season when some of the bases washed down during the rains move back up as the soil moisture moves up and is evaporated or used by plants. The way the land is managed will also effect the leaching and whether there are deep rooted plants which also bring up minerals which have been leached. Generally leaching continues and in the dry season the oxides accumulate near and on the surface and hard crusts and concretions or even pebble like structures develop. This crust can be almost like concrete, certainly as hard as brick. When the land is cultivated this is not so obvious but in the dry season the soil is more like broken brick than soil. Even in the wet season it is not as hospitable to fine root systems as soil should be.

Someone who I have not been able to identify gave what I feel is a good description of fully developed laterite, when he stated :-

We may have

1. A few inches of soil retaining some organic constituents, or a surface crust of iron oxide, often forming extensive pavements of pebbly gravel or even a continuous carapace (Shell or crust, like the shell of a tortoise.)

2. Red and yellow, more or less crumbly and sticky, (when wet) subsoil, with iron concretions (hard stony like lumps) around old roots.

3. Deeper less weathered red and yellow mottled clays; this is Buchanan's original laterite, and often has a reticulate (netlike pattern) structure with vermiform (worm shaped) iron nodules enclosing clay.

There are many other names for soils, I do not think this is the place for more than those listed. However I think it is useful to have a list of the sources of the names of soils and some of the terms used, as this knowledge is helpful in understanding the terminology of the subject. It is not easy for the field worker to discover these meanings and so much of what he sees written will be more difficult to understand than it need be. Eg Acr in a description means at the end, very much weathered. Morph comes from the Greek Morphe and means form, body. These descriptions are found in the glossary of terms appendix 6.

Most authorities point out that there are very many systems of soil classification, and much argument regarding the pros and cons of each.

Another table of classification by Kellogg 1950 found in "Tropical Agriculture" by Gordon Wrigley shows more similarity to the above than many of the systems described.

CLASSIFICATION OF TROPICAL SOILS:

(A) *Zonal*   *Soil group*

    (1) Latosols
- (1) Red Latosol.
- (2) Earthy Red Latosol-more open structure, crumbly.
- (3) Reddish-Yellow Latosol.
- (4) Reddish brown Latosol.
- (5) Yellow Latosol.
- (6) Black and Red Latosol.

    (2) Chenozemic soils   (7) Reddish Prairie.
    (3) Podzolic soils
- (8) Reddish-Yellow Podzol.
- (9) Latosolic Brown Podzol.
- (10) Grey Podzolic.

(B) *Intrazonal*

    (1) Halomorphic
- (1) Saline (Solonchak), contains an excess of salts, pH 7.0-8.5
- (2) Alkali (Solonetz), containing alkali salts pH over 8.5
- (3) Solodise-Solonetz

    (2) Hydromorphic
- (4) Latosolic Wiesenboden (Humic Gley)
- (5) Bog and half bog, peaty soils of humid and sub humid areas
- (6) Latosolic Planosol
- (7) Ground Water Podzol
- (8) Ground Water Laterite-Buchanan's Profile
- (9) Grey Hydromorphic (Low Humic Gley)

    (3) Calcimorphic
- (10) Latosolic Rendzina-not so black
- (11) Ando-fresh volcanic ash or yellow earth
- (12) Brown forest soil

(C) *Azonal*
    Young, scarcely changed
- (1) Lithosols-freshly and incompletely weathered rock mass
- (2) Alluvial soils-formed by recent deposit
- (3) Regosols (unconsolidated parent material eg Loess, dry sands, Fresh Volcanic Ash).

Zonal soils. Have well developed characteristics resulting from the influence of the active soil-forming factors, climate and vegetation.

Intrazonal soils. The characteristics are developed under the dominating influence of some local factor of relief or parent rock over climate and vegetation.

Azonal soils. which have no soil profile, are usually of recent origin.

With the high temperature and heavy rain in the tropics the soil formation process is quicker. The resultant soils are frequently deep but the nitrites and bases (substances which acted upon by an acid produce a salt and water only,) have often been leached out and the resultant soils are generally poorer than in temperate zones.

The inorganic or mineral portion of the soil, which results from the weathering of the parent rock, or alluvium deposit, has particles varying in size from inert stones and rocks to the fine colloidal clays. In 1927 international agreement classified the particle sizes, for the purpose of mechanical analysis, as follows:

Gravel	over 2 mm diameter
Coarse sand	2-0.2 mm diameter
Fine sand	0.2-0,02 mm diameter
Silt	0.02-0.002 mm diameter
Clay	less than 0.002 mm diameter

The particles less than 0.002 mm in diameter form the very important colloidal fraction. The smaller the particle size the more active they are both chemically and physically.

Essential elements obtained from air and water	Macronutrients	Micronutrients	Elements essential to animals
Hydrogen	Nitrogen	Chlorine	Sodium
Oxygen	Phosphorus	Manganese	Iodine
Carbon	Potash	Copper	Cobalt
	Sulphur	Iron	
	Calcium	Zinc	
	Magnesium	Boron	
		Molybdenum	

<Some plants such as rice also need Silicon PJS>

Note: Tables vary with most authorities placing Sulphur Magnesium and sometimes Calcium as micronutrients, as in chapters 3 and 4 of this book.

REFERENCES

1. Vetiver Grass. 3rd Edition. 1990. The World Bank. Washington DC. USA.
2. Tinau Watershed Management Plan. Vol 1. June 1980. Pages 69-60. Swiss Association for Technical Assistance. Nepal.
3. Officers of the Dep. of Agriculture and Stock, Brisbane Queensland The Queensland Agricultural and Pastoral Handbook. Vol 1. 1962. The Dep of Agriculture and Stock, Brisbane Queensland Australia.
4. Masefield. G.B., A Handbook of Tropical Agriculture. 1962. Oxford University Press. UK.
5. Hurni. Hans. (Prepared by) Soil Conservation in Ethiopia. 1986. Published by Community Forests and Soil Conservation. Dept. Ministry of Agric., Ethiopia.
6. FAO. Soil Erosion by Water. Some Measures for its Control on Cultivated Lands. 1965. FAO. Rome.
7. Jacks G.V. and Whyte R.O., 1939. The Rape of the Earth. A World Survey of Soil Erosion, Faber. London.
8. Schwab. Frebert. Barnes. Edminster. 1957. Elementary Soil and Water Engineering. John Wiley. USA.
9. Moody. J.E. 1948. Plant Nutrients Erode Too. Virginia Agricultural Experimental Station quoted in Ref 8.
10. Hudson. N.W. 1975. Field Engineering for Agricultural Development. Clarendon Press, Oxford. UK.
11. St Barbe Bake. Richard. 1956. Land of Tane-the threat of Erosion-Lutterworth Press. London UK.
12. Abstracts from the British Society of Soil Scientists Easter Meeting. 1992 The subject - Sustainable Land Management in the Tropics, what role for Soil Science.
13. Sanchez. Pedro. A. 1976. Properties and Management of Soils in the Tropics. North Carolina State University. USA. John Wiley & Sons
14. Olson. G.W. 1981 Soil and the Environment, Chapman and Hall. New York. USA.
15. Sidway F.H. and Barnett. A.P. The Article on Wind and Water erosion; Control aspects of multiple cropping in Multiple Cropping, Pub. by American Society of Agronomists. University of California (Berkeley) and John Vande Meer Michigan University. USA.
16. Rice. Robert. A. Climate and the Geography of Agriculture in the book Agroecology, McGraw Hill. USA.
17. Tivy, Joy. Agricultural Ecology. Longmans, UK.
18. Webster. C.C. AND Wilson P.N. 1980. Agriculture in the tropics, Longmans, London.
19. Hall. Sir Daniel. and Robinson G.W. 1945. The Soil, an introduction to the scientific study of the growth of crops. 5th Edition. John Murray. London. UK.
20. United States Department of Agriculture. Soil: 1957 Yearbook of the United States Department of Agriculture.
21. Blake. Michael. 1970. Down to earth. Crosby Lockwood and Sons Ltd. London UK.
22. Subba Rao. N.S. 1982. Biofertilizers in Agriculture. Oxford and I B H Co India.
23. Jeavons. John. 1979 How to grow more vegetables. Ten Speed Press California. USA.
24. Storey P.J. 1990 Basic Vegetable Growing In Nepal. United Mission to Nepal. PO 126 Kathmandu. Nepal.
25. Russell. 1961. Soil Conditions and Plant Growth. 1961.
26. Arakeri. H.R. Chalam. G.V. Satyanaraya. P. and Donahue. R.L. 1962. Soil management in India, Asia Press. The Strand. London. U.K.
27. Bautiste. O.F. and R.C. Mabesa. 1977. Vegetable production. University of the Phillipines College of Agriculture.
28. Storey. P.J. 1990. Bamboo a valuable crop for the hills. Helvetas P.O. Box 113. Kathmandu.
29. Storey P.J. 1992. Bamboo a valuable soil conserving/improving and income generating crop.
30. Varmah. J.C. and Bahadur K.N. 1980. Country Report and Status of Research on Bamboos in India. - India Forester Record (new Series) (Bot) 6. Manager of Publication, Delhi. India.
31. Gaur. J.C. Bamboo Research in India. Systematic Botany Branch. Forest Research Institute and College. Dehra Dun 248006. U.P. India.
32. Raizada M.B. and Chatterjee, R.N. 1956 World Distribution of Bamboos, with special reference to the Indian Species and their more important uses. Indian Forester 82.215.
33. McClure. F. A. 1966 The Bamboos. Harvard University Press, Cambridge. Massachusetts.
34. Reganold. J.P. (Dep't fo Agronomy and Soil Washington State University) Elliot. Lyoyd. F. (Agricultural Research Service, U.S. Dep't of Agriculture. Pullman, Washington.) Unger. Yvonne. L. (Dept. of Plant and Soil Sciences (University of Maine, Orono, Maine) USA. 1987. Long Term effects of Organic and Conventional Farming on Soil erosion. Pub. Nature. Vol. 330 26 November 1987.
35. Reganold. John. June 1989. Farmings Organic Future. New Scientist. 19 June 1989.
36. Fukuoka Masanobu. 1985. The Natural Way of Farming. The theory and practice of Green Philosophy. Japan Publications Inc; Tokyo and New York.
37. Hue. N.V. and Amien. I. 1989. Aluminium Detoxification with Green Manures. Commun. in Soil Sci. Plant Anal.20 (15 & 16) 1499-1511. For more information N.V.Hue, Department of Agronomy and Soil science, College of Tropical Agriculture and Human resources. University of Hawaii. Hawaii 96822 USA. Quoted in International Ag'Sieve Vol IV (4).
38. Indian Journal of Agronomy, vol 1, number 2, 1956.
39. Memoirs of the Department of Agriculture, Madras, no 36 1954.
40. Handbook of Agriculture. Indian Council of Agricultural Research. New Delhi. 1984
41. Knott. J.E. 1966 Handbook for Vegetable Growers. John Wiley and Sons, New York.
42. Vickery, Deborah & James. 1981. Intensive gardening for profit and self sufficiency. Training Manual Number 25. Peace Corps. USA.
43. Morrison. F.B. 1957 Feeds and Feeding. 22nd edition.
44. Mollison. Bill. Permaculture. A Designers Manual. Tagari Press. Australia.
45. See 18.
46. Perry. Mac. 1978. The Gro-box Method. Pub. by The Farm. Foreign Agricultural Relief Mission. Florida 33541.

47. Pratt, Robert. 1958. Florida Guide To Citrus Insects, Diseases and Nutritional Disorders In Color. Agricultural Experiment station. Gainesville. Floride. USA.

48. Doth, H.D. Fundamentals of Soil Science. 7th Edition. John Wiley and Sons.

49. Wilcox, Larry. 1972. The Answers. Wilcox Enterprises Inc. Okla USA.

50. New Farmer and Grower. Winter 1991/92 p23.

51. Hansen, Israelsen & Stringham. 1980. Irrigation Principles and Practices. John Wiley and Sons.

52. Water: The Yearbook of Agriculture 1955. The United States Dept. of Agriculture.

53. Phipps, McColly, Scranton & Cook. 1955. Farm Mechanics Text and Handbook. Interstate publishers, Danville. Illinois, USA.

54. El-Sharkawy, Mabrook. CIAT International Centre for Tropical Agriculture Report AA 67 13 Quoted in International Agricultural Development Sep/Oct 1992.

55. Stoesz, Edgar. 1972. Beyond Good Intentions. United Printing Inc; Newton Kansas. Px11 USA.

56. Raymond Dick. Down to Earth Vegetable Gardening Know-how. Garden Way Publishing, USA.

57. Joshi R.M. & Khatiwada M.K. 1986. Agri Publication Series. P.O Box 3143 Kathmandu. Nepal.

58. Cox, Jeff. 1988. Rodale Press. Emmaus. Pennsylvania.

59. Kasasian, L.1971. Weed Control in The Tropics. Leonard Hill. London.

60. List of Approved Products and their use for farmers and growers 1984. Agricultural Chemicals Approval Scheme. H.M. Stationary Office. UK.

61. Tropical Pasture Manual. 1967. Terranova Tropical Pastures. Anderson Seeds Ltd. Brisbane. Queensland. Australia.

62. Pasture Legumes and Grasses. 1965. Bank of New South Wales. Australia.

63. Humphreys L.R., 1969. A Guide to Better Pastures for the Tropics and Sub Tropics. Wright Stephenson and Co Ltd. Melbourne. Australia.

64. Grass. Yearbook of Agriculture. 1948. US Department of Agriculture. Washington. USA.

65. Gerold. P O.S.B. 1989 Sunnhemp. Benedictine Publications. Ndanda. Perimiho. Tanzania.

66. Wrigley. G. 1961 Tropical Agriculture. B.T. Batsford Ltd. London.

67. International Agricultural Development; (an excellent journal.) Nov/Dec 1992. From a summary of a video "Building on traditions- Conserving Land and Alleviating Poverty" By The International Fund for Agricultural Development.

68. Dairy Pastures in Queensland Agricultural Journal Sept' 1970.

69. Elders Pastures Seeds Information Service. No date. Elder Smith Goldsborough Mort LTD Western Australia.

70. Queensland Agricultural Journal. Feb 1970.

71. The Agricultural Gazette of New South Wales. March 1970. New South Wales Government Printer. Australia.

72. The Teach Yourself Concise Encyclopedia of General Knowledge. 1956. English Universities Press Ltd London.

73. Pedro Sanchez. Tropical soils and land use.

74. Greenland and Dart. 1972. Quoted in ref 74.

75. Hans, Ruthenberg. 1971. Farming systems in the tropics. Oxford University Press.

76. International Agricultural Development (Journal) Jan/Feb 1993.

77. Shen. T.H., Agricultural Development on Taiwan since World War II. Second edition. 1971. Meiya Publications Inc. Taipei Taiwan.

78. Hills. L. 1976. Comfrey Past Present and Future. Faber.

79. Hills. L. 1975. Fertility without fertilizers. Henry Doubleday Research Association. Ryton on Dunsmore. Conventry CV8 3LG.

80. Marianne Sarrantonio 1991. How to choose a soil building legume. The New Farm (Journal) USA Jul/Aug'91.

81. International Agricultural Development; November/December 1991.

82. Papers presented to the British Society of Soil Scientists Easter 1992 at Newcastle University.

83. International Ag-Sieve Vol II, Number 4, 1989.

84. Food Matters Worldwife No 8 October 1990.

85. Echo Development notes. 17430 Durrance Road. North Ft. Myers Fl 33917 USA. Issue 36 April 1992.

86. " " " " " Issue 38 September 1992.

87. Agroforestry Systems. Vol 4. No 2/ 1986.

88. International Agricultural Development Jan/Feb 1991.

89. Jean-Louis Chleq and Hugues Dupriez. Vanishing Land and Water 1988. Macmillan, London, and CTA, Vageningen, Netherlands.

90. Graeme, Thomas, The Access route for peasants still some roadblocks. In "CERES" FAO review on Agricultural Development, Vol 16. No 8. Nov/Dec 1983.

91. Vijay Shah. Making the most of little rain. ILELA Newsletter March 1992. Kastanjelaan. 5 PO Box 64 3830 AB. Leusden. Netherlands.

92. Michael Redclift. Raised Bed Agriculture in pre Columbian Central and South America. Quoted in International Ag Sieve Vol V(3) 1993. Rodale Institute. Kutztown PA: USA.

93. Daniel C. Mountjoy and Stephen R. Gliessman. A traditional solution to the problem of Sustainable Farming Systems. Quoted in International Ag Sieve Vol V (3) 1993.

94. N.F. Oebker, R.W. Peebles and C. Brentcluff. University of Arizona, Tucson, Arizona. A Mulch Water Harvest Technique for Growing Vegetables in Arid Lands.

95. International Agricultural Development Jul/Aug 92. p7

96. The future is what we make it. B.A.S.F. Agriculture publication.

97. Dr B.R. Gupta and B.K. Dwivedi. Mycorrhiza - An Alternative to inorganic Phosphatic Fertilisers. Farmer and Parliament (Indian Journal) May 1989.

98. International Ag-sieve, Vol II 8 November 1989. Rodale Institute 222 Main Street, Emmaus PA, 18098 USA.

99. N.R. Hulagalle, IITA/SAFGRAD Project, Ouagadougou Field Crops Research. 17, 219-228 No 314 Dec 1987 Quoted in International Ag-Sieve Vol 1 (2) Aug/Sept 1988. Rodale International.

100. Lester R. Brown. 1985 A false sense of security. New World Outlook. June 1985. Adapted from State of the World-1985 a World Watch Institute report on progress towards a sustainable society, published in the US by WW Norton and Company.

101. Lester R. Brown and Edqard C. Wolf. 1984. Soil Erosion; Quiet Crisis in the World Economy. Wordwatch Paper 6. Sept' 1984.

102. World Development Movement leaflet "It's a matter of life and death'. 1993.

103. John, F.A. Russell. Developing Farmer-Extension-Research Linkages to address the needs of Resource-poor Farmers in Rainfed Environments.
A paper prepared for the International Conference on 'Extension Strategy for Minimising Risk in Rainfed Agriculture' sponsored by the International Fund for Agricultural Development. New Delhi, April 1991.

104. International Agricultural Development Nov/Dec 1993 page 23.

105. " " " pages 7-10. "
106. From a leaflet accompanying the video "Looking after our land" produced by IIED 3 Endsleigh Street. London. WC lH 0DD, UK.
107. Sandra Postel. "The Last Oasis." (Earthscan publications) From a summary of the book in the Tropical Agriculture Association Newsletter. September 1993.
108. Conservation Policies for Sustainable Hill Side Farming" published by the Soil and Water Conservation Society. Amkeny. Iowa. USA.
109. A Sand County Almanac with essays on conservation, from Round River. Aldo Leopold. 1970 Ballantine Books. New York.
110. David Pierce Smyder. Inevitable forces for change. Insight magazine (4) 7 4-15 (part 1); (4)8: 2-10 (Part 11) 1987
111. "W.C Molderhuaer, N.W. Hudson. T.C. Sheng and San-Wei Lee editors Development of conservation farming on hill-slopes. published by Soil and Water conservation Society. 1991. Ankeny. Iowa. USA.
112. Rao Radhkrishna. International Agricultural Development January/February 1998 p14.
113. Michael Pickstock. Director of WREN media in an article in "International Agricultural Development. Jan/Feb 1998.
114. International Agricultural Development. JAN/FED 1995 page 19.
115. Ghose S K "Women and crime, crime against women and crime for women" Ashish Publishing House. New Delhi. India.
116. Reginold. J. Associate Professor in the Department of Agronomy and Soils at Washington State University, Pullman, Washington. USA. The article appearead in the 'New Scientist'. June 1989. p52.
117. Stocking, M. Reader in Natural Resource Development And Dean of the School of Development Studies at the UK's University of East Anglia. Norwich. "SOIL CONSERVATION." from a paper presented to the British Society of Soil Science. Easter 1992.
118. Chalk. S. Interveiwing Claire Short for Christianity Magazine. June 1999. ISSN 1365-3695.
119. Madeley, John. Broad Brush. Appropriate Technology. Vol 21. No.1. June 1994.
120. Swift. M.J. Sanginga. N & Mulongay. K. of the International Institute of Tropical agriculture, Ibadan, Nigeria. From an abstract of a paper presented at The British Society of Soil Scientists Easter Meeting. 1992.
121. Lucas. R.E. Proc, Soil SC. Amer; 1946. 10.269
122. Pretty, Jules N. 1995 Regenerating Agriculture. Published by Earthscan.
123. Liao, Mien-Shun; Hu, Su-Cherng; and Wu, Huei-Long. 'Soil conservation research and development for hillside farming in Taiwan'.
124. Anderson, J,M; 'Integrated Nutrient Management in Multispeies Farming Systems at the Forest/Agriculture interface.' Tropical Agriculture Association Newsletter, June 1999, ISSN 0954-6790.
125. Dent, Barry, Professor; Welcome address at the South West Regional Seminar 'Improving Soil Management' reported in Tropical Agriculture Association Newsletter, June 1999. ISSN 0954-6790.
126. Jackson, David. 'The role of mulch and mulching material in combating soil degradation in the Sahel, a symposium held in Burkino Faso 14-16 July 1998. In the Tropical Agriculture Newsletter (Journal). June 1999.
127. Hien F.G., Slingerland MA and Hien V. Poster Communication for the Regional Workshop on Soil Fertility Management in West Africa Land Use Systems, NIAMEY, 4-8 MARCH 1997.
128. Mando, A. 1997 'The role of termites and mulch in the rehabilitation of crusted Sahelian soils' Tropical Resource Management Paper No.16.
129. Slingerland, M. and Masdewel, M. 1996. Mulching on the Central Plateau of Burkino Faso. Chapter 10, Sustaining the Soil; Indigenous Soil and Water Conservation in Africa. Edited by Chris Reij, Ian Scoones and Camilla Toulmin.
130. Coulibaly, I.S. (1998) Promotion of Mulch in Niger. Experience of the PASP/GTZ. Paper presented at the Mulch Symposium, Burkino Faso, 14-16 July 1998.
131. Research News in 'International Agricultural Development' ISSN No 0261-4413. May/June 1999.
132. Cherrett, Ian and Hellin, Jon. 'Soil conservation by stealth' an article in 'International Agricultural Development' ISSN No 0261-4413 May/June 1999.
133. New Scientist, 5/12/88, Vol. 118, #1612.
134. Hatfield, J,L, 'Cover crops summary'. Papers from the 1997 Conference "Cover crops, soil quality, and Ecosystems" of the Soil and Water Conservation society. From a report in the Journal of the fourth quarter 1998. The author in Laboratory Director of the National Soil Tilth Laboratory, Ames, Iowa. USA.
135. Mausbach, M, J; "The Soil Resource" The Journal of Soil and Water Conservation. Volume 54. Number 2. Second quarter 1999.
136. Lowdermilk, W,C; (1953) "Conquest of the Land through Seven Thousand Years." Agriculture Information Bulletin No.99 USDA, Soil Conservation Service, Washington, D.C., USA. 30 pp.
137. Cooke, G, W, Journal of Agric. Sci; 50 253; 1959, 53, 46.
138. Fried, M, Proc. Soil Sci. Soc, Amer; 1953, 17, 357.
139. July/Aug 1999. International Agricultural Development. ISSN: 0261-4413 Research Information Ltd 222 Maylands Ave Hemel Hempstead, HP2 7TD. Herts. UK.
139. An article 'Cover crops critical to sustainable agriculture' in the Journal 'International Agricultural Development' July/August 1999.
140. in 'Agroforestry today' January/February 1995 - the Journal of ICRAF the International Centre for Research in Agroforestry.
141. Underexploited Tropical Plants with Promising Economic Value. 1975 Prepared and published by The National Academy of Sciences. Washington DC. Library of congress catalog number 78-71595.
142. Farmer Innovators in Soil and Water Conservation. Newsletter of the programme on Indigenous Soil and Water Conservation in Africa, Phase II Issue 1, June 1997. Published by CDCS/Vrije Univeriseit. De Boelelaan 1115. 1081 HV Amsterdam. The Netherlands.
143. Altieri, Miguel A. The Science of Sustainable agriculture, in Agroecology. Published by Intermediate Technology Publications. London UK. and by McGraw Hill. USA.
144. Water and Soil Protection. (Shwei Tu Bau Chr) (in Chinese Script) April 14th 1968 Taiwan Provincial Government Department of Forestry and Soil Conservation. Taichung. Taiwan.
145. Watershed Conditions, problems and Research needs in Taiwan. 1964. By Robert E Dills. Chinese American Joint Commission on Reconstruction. Taipei. Taiwan.
146. Thammincha, Songkram. 1980. The Role of Bamboos in Rural Development and Socio economies. A Case Study in Thailand. A paper published in Bamboo research in Asia. Published by the International Department Research centre. Box 8500 Ottawa. Canada.

147. Aina. P,O. et al. Quoted in Lal, R, "Effective Conservation Farming Systems for the Humid Tropics," in American Society of Agronomy, *Soil Erosion and Conservation in the Tropics,* ASA Special Publication No 43 (Madison, Wisc: 1982.)

148. Brown, Lester and Wolf, Edward C. 1984 'Soil Erosion: Quiet Crisis in The World economy.' World Watch Institute. 1776 Massachusetts Avenue. N. W. Washington, DC. 20036 USA.

149. Pretty, J and Shaxson, F 'The Potential of Sustainable Agriculture September 1998. Enable. The Journal of the Association for Better Land Husbandry. C/O Greensbridge, Sackville Street, Winterborne Kingstone. Dorset DT11 9BJ

150. Enable No 10 August 1999. The Newsletter of the Association for Better Land Husbandry.

151. Landers, J.N. 1999: Policy and Organisational Dimensions of the Process of Transition Towards Sustainable Intensification in Brazilian Agriculture. 24pp.

152. 'Erosion control in the tropics' an article in Agrodok II published by Agromisa. P.O. Box 41. 6700AA Wageningen. The Netherlands.

153. Bennet, A, J. Chief Natural Resources Advisor, The British Overseas Development Administration. 1991. Aid to Natural Resources–A Forward Look. The Ralph Melville Memorial Lecture. The Tropical Agricultural Association.

154. Farmer Innovators in Land Husbandry. The Joint Newsletter of Indigenous Soil and Water Conservation in Africa, Phase II and Promoting Farmer Innovation. Issues 8. April 2000.

155. Douglas, M. The Machobane Farming System in Lesotho. In Enable. The Newsletter of the Association for Better Land Husbandry. No. 10. August 1999.

156. Farmer Innovators in Land Husbandry. The joint newsletter of the programme 'Indigenous Soil and Water Conservation in Africa, Phase II (ISWCII)' and 'Promoting Farmer Innovators (PFI)' Issue 4 & 5, August 1998.

157. The Shape of Agriculture a summary of a document "Land resources: on the edge of a Malthusian precipice?" prepared for the British Royal Society meeting December 1996 in the 'International Agricultural Development' journal. September/October 1997. The full original paper is available from Science Promotion Section, The Royal society, 6 Carlton House Terrace, London SWIY 5AG, UK.

158. Tropical Agricultural Association Newsletter. March 2000.

159. International Agricultural Development Journal September/October 1997.

160. Annual Report 1993/94. The Lumle Agricultural Research Centre. PO Box 1. Pokhara, Kaski. Nepal.

161. Knott, J.E. 1996 Handbook for VEGETABLE GROWERS, John Wiley and Sons Inc. USA.

162. Pettygrove, G.S. How to perform an agricultural experiment. Published by VITA 3706 Rhobe Island Ave. Mt. Rainier, Maryland, USA. 20822.

163. ST. Barbe Barker, Richard 'I planted trees' 1994. Lutterworth Press. London.

164. Jackson Bob 'Higher than the hills', ISBN 1 897913 48 6 Published by Highland Books. Two High Pines. .Knoll Road. Godalming. Surrey. GU7 2EP.

165. Brown Lester, Flavin Christopher, and French Hilary. The State of the World 2000. Published by the Worldwatch Institute 1776 Massachusetts Avenue, NW, Washington<DC 20036-1904, USA. ISBN 1 85383 680 X. REFSOURCE. ES 9/6/00

166. Rainwater Harvesting. A summary in Appropriate Technology. Vol 27 No1 ISSN: 0305-0920 Research Information Ltd 222 Maylands Ave.Hemel Hempstead, Herts HP2 7TD, UK.

167. Flowerdew, Bob. Bob Flowerdew's organic bible.

168. Tropical Legumes: Resources for the Future. National Academy of Sciences. Washington. D. C. 1979.

169. Food Legumes. Tropical Products Institute. 56-62 Gray's Inn Road. London WCIX 8LU. UK.

170. Gupta, B.R. and Dwivedi, B.K. Chandra Shekar Azad University of Agriculture and Technology, Kanpur-208002. Mycorrhiza-An Alternative to Inorganic Fertilizers. Farmer and Parliament May 1989. India.

171. Joy, D. and Wibberley, E.J.A. Tropical Hand Book. Cassell. London. UK.

172. Gerold, P. OSB Sunnhemp "Marejea" ISBN 9976 67 037 0 Published by Benedictine Publications Ndanda–Perimiho Tanzania.

173. Sanders, D., Huszar, P.C., Sombatpanit, S. and Enters, T. Incentives in Soil Conservation. 1999. Science Publishers, Inc; USA. ISBN 1-57808-061-4.

174. Smith, A. 1994 'Incentives in Community Forestry Projects: A Help or Hindrance?' Network Paper no. 17c. Rural Development Forestry Network, Overseas Development Institute, London.

175. Storey, P.J. 'Quality Rural Development' hopefully to be published in 2002.

176. Pierce, Smyder, David (1987). "Inevitable Forces for Change". Insight Magazine (4) 7 4-15 (part1); (4) 8: 2-10 (Part 11) 1987.

177. FAO. 1990. 'The Conservation and Reclamation of African Lands—An International Scheme. The FAO, Rome.

178. Sanders, D. President's Report, in the October 2000 edition of the World Association of Soil and Water Conservation NEWSLETTER. Available from 7515 NE Ankeny Road Ankeny, Iowa 50021-9764 USA.

179. Newsletter of the Association of Better Land Husbandry. C/0 F. Shaxson, Greensbridge. Winterborne Kindston, Dorset DT11 9BJ U.K.

180. Addressing the Water Crisis—Healthier and More Productive Lives for Poor People. March 2000 Department For International Development. UK Government Whitehall,. London.

181. Derpsch, R. 'Frontiers in Conservation Tillage and Advances in Conservation Practice', a paper presented at the ISCO Conference held at Purdue University, West Lafayette, Indiana, USA, May 23-28, 1999. Reviewed by W C Moldenhauer and Francis Shaxson. In The World Association of Soil and Water Conservation Newsletter. January 2001.

182. Critchley, W. Farmer innovators in Land Husbandry. Experience from Ease Africa. In The World Association of Soil and Water Conservation. Newsletter. January 2001.

183. New Scientist, 5/12/88, Vol. 118,# 1612; pp. 50-51.

184. Liu, S.Y. and Li Y.Q. Rainwater Harvesting and Courtyard Production: An Emerging Land Husbandry System in the Chinese Loess Plateau. An article in Number 11, July 2000 of 'Enable' the newsletter of the Association for better land husbandry. Editor Francis Shaxson. E mail f.shaxson@cwcom.net

185. Meitzner, L.S. and Price M.L. 'Amaranth to Zai holes' Ideas for Growing Food Under Difficult Conditions. ISBN 0-9653360-O-X Published by ECHO 17430 Durrance Road. Noth Fort Myers. FL 33917-2239, USA.

186. Chambers, Robert. 1994 "Challenging the Professions" ISBN 185339 194 81T Publications. London.

187. Bunch, Roland. "Two Ears of Corn", A Guide to People Centred Agricultural Improvement. 1985. ISBN 0-942716-

03-05. World Neighbours 5116 North Portland, Oklahoma City, Oklahoma 73112 USA.

188. (Paul and Clark 1989 in Soil Microbiology and Biochemistry. California Academic Press, Inc.

189. Rouschel, C. and Strugger, S. Naturwie:1943 31 300; Canad. J. Res; 1948, 26 C, 188.

190. Appropriate Technology. July/Sept. 01

191. Shaxson, F. Rethinking effects and causes of land degradation At the Soil and Water Conservation Workshop of the Tropical Agricultural Association at Duham University in September 2001. It is summarised on the website of the association *www.taa.org.uk* and will be published in the December 2001 issue of the TAA Newsletter.

192. Soils and Men. Yearbook of Agriculture 1938. U. S. Department of Agriculture.

193. Young Anthony. The Nature and Management of the Poorer Tropical Latosols. In Outlook on Agriculture Volume 10. 1979 3/12/01

194. Sanders David, Presidents report. In the Newsletter of the World Association of Soil and Water Conservation Volume 16. Number 4 (October 2000).

195. Shaxson, Francis. Land Husbandry and the Social Science/Natural Science Divide. Editorial in 'Enable' the Newsletter of The Association for Better Land Husbandry. Number 12. July 2001.

196. Paul and Clark, 1989 in Soil Microbiology and Biochemistry. California Academic Press Inc.

197. Kilduff Jim Reducing Forage Wastage and Improving Manure Quality in Stall-fed Ruminant Livestock. In the Newsletter of the Tropical Agriculture Association. March 1993.

198. Pretty, J. Thompson, J. and Hinchcliffe, F. The Productivity myth. Sustainable Agriculture: Impacts on Food Production and Challenges for Food Security. Published by the International Institute for Environment and Development. 3 Endsleigh Street London WCIH 0DD U.K.

199. Gajurel Deepak. A freelance journalist and a teacher with Tribhuwan University, Kathmandu. 'Hungry Days Ahead?' writing in Orbit the Journal of the VSO (Voluntary Service Overseas) fourth quarter 1996.

200. Morgan RPC. 1986 Soil Erosion and Conservation. Scientific and Technical, England.

201. Kusumandari Ambar, and Mitchell Bruce in 'Soil erosion and sediment yield in forest and agroforestry areas in West Java, Indonesia'. In the Journal of Soil and Water Conservation, September-October 1997.

202. Hudson, N.W. 1975 Field Engineering for Agriculture Development. ISBN 0 19 859442 9 Clarendon Press, Oxford.

203. Ong Chin, Where alleys work–and do not work. International Agricultural Development January/February 1995.

204. Fathoming the potential of low input agriculture. International Agricultural Development. March/April 1998.

205. Pereira Charles, 'Can Research on Tropical Agriculture Reduce Poverty'. Tropical Agriculture Association Newsletter September 1999.

206. Dixon David. 'Farmers embrace a creeper' FAO Ceres 158 March/April 1996.

207. Yaholnitsky, Ivan. 'Regenerating the soils of Lesotho.' Permaculture magazine. Issue No 25. ISSN 0967-5663 Available by referring to www.permaculture.co.uk.

208. Kaley.Vinoo Venu Bharati Publisher Aproop Nirman. B2 Pushpagandha Dharampeth. Nagpur 440010. Maharashtra. India.

209. Hellin. J. Land Degradation and Development. 13: 233-250 (2002) Published on line 26 April 2002 in Wiley Interscience (www.interscience.wiley.co). DOI: 10.1002/ldr.501.

210. Appropriate Technology. Vol. 29. No. 1, March 2002.

211. Appropriate Technology. Vol. 29. No. 3, September 2002.

212. Lipangile T.N. 'Bamboo and Woodstave Technology' can be obtained from Wood Bamboo Project, P.O. Box 570, Iringa. Tanzania. Telex : 52235 (Umemetz).

213. Overseas Development Administration of 'Manual of Environmental Appraisal' UK Government Publication 1989.

PLATES

Plate 1

T intro1 The 3,000-year-old Ifugao Rice Terraces, Banaue, may be rightfully counted among the wonders of the ancient world. Carved out of mountain sides and employing an ingenious irrigation system, this particular group covers an area of 400 sq. klms. If placed end to end the terrace walls would cover an estimated distance of 14,000 miles, or more than half-way round the world.

Plate 2

These south facing slopes were once very fertile, without conservation they are now only able to grow occasional poor crops. However terracing can still make a huge and lasting difference. This land can grow good crops again.

T intro 2.

Plate 3

T intro 3. From the Philippines Ref 67 p133.

Plate 4

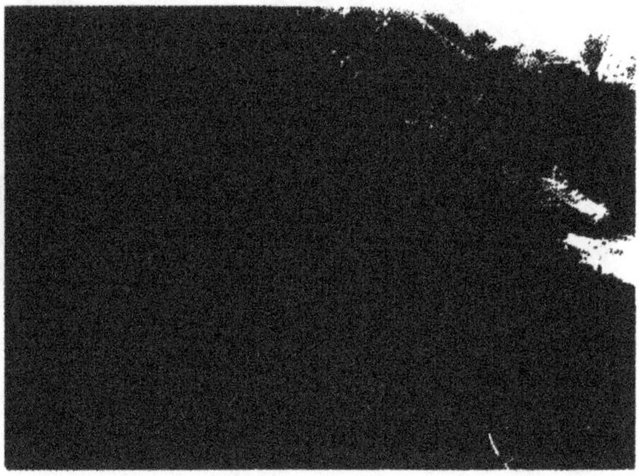

As the top of the strips are infertile manure is added to compensable, but monsoon rains will wash out the plant foods. T intro 4.

Plate 5

This hillside was contour stripped, the end result is it's fertility has been stripped and it is eroded. **T intro 5.**

Plate 6

This is what happens with contour strips in time, now the area hardly grows grass. Even the land in the foreground was cropland. **T intro 6**

Plate 7

This shows part of a soil conservation trial from 1965-1968. It showed that only terracing was a durable method.

This stony land was terraced in 1967, a visitor in 2002 found the land was still as fertile.

Plate 8

B1. The bulldozer is raised as the rotovator is lowered to break up the soil ready for the bulldozer to push it to where it is needed.

Plate 9

B2. The Good News Bulldozer Conversion. After the rotovator has dug up the soil it is raised and the bulldozer lowered to push the loose soil to where needed.

Plate 10

B3. The topsoil from this strip was pushed to the terrace below, then the subsoil levelled. Next the topsoil from the strip above is pushed down and spread over this terrace.

Plate 11

B4. While the first terrace are growing crops, during a slack time more slope is being converted to a lasting more fertile future.

Plate 12

This terrace admittedly easier than most, has all the topsoil on the top, and yet was built averaging 10 meters per hour for two men, so for a six hour day 60 metres. Two farmers working for one week of six days 360 metres, the following week another 360 metres on the other farm would stop the loss of soil and nutrients on the areas worked. The crops would be so much better on the terraces, and encourage the two farms to do more terracing the following year till all was terraced. T40/b.

Plate 13

Good news terracing the first step. Soil and water moving down the slope is stopped by the terrace below it, so there is no loss. The terraces produce better crops which encourage more terracing until the first step is finished, then the slope land is also terraced. T40/c.

Plate 14

Fig.8. Preparing to plant bamboo cuttings on a landslide. Such landslides are now stabilised and producing fodder, fuel and income.

Plate 15

Fig.9 Bamboo will grow on infertile landslides like this one where not even weeds will grow. This was one year from a culm (stem) cutting being planted. In such conditions growth is slow for the first 3 years, then it is quite amazing.

Plate 16

Fig.10 This acid soil has become crusted and hard, yet as we have proved it could be planted with bamboo and become an income generating resource, in addition to the bamboo the leaves used for fodder and so adding to manure for the local fields.

Plate 17

Fig.11 Bamboo has an amazing ability to hold soil.

Plate 18

INDEX